THERAPEUTIC APPLICATIONS OF ADENOVIRUSES

GENE AND CELL THERAPY SERIES

Series Editors

Anthony Atala & Nancy Templeton

PUBLISHED TITLES

Regenerative Medicine Technology: On-a-Chip Applications for Disease Modeling,
Drug Discovery and Personalized Medicine
Sean V. Murphy and Anthony Atala

Therapeutic Applications of Adenoviruses
Philip Ng and Nicola Brunetti-Pierri

Cellular Therapy for Neurological Injury
Charles S. Jr., Cox

Placenta: The Tree of Life
Ornella Parolini

GENE AND CELL THERAPY

THERAPEUTIC APPLICATIONS OF ADENOVIRUSES

Edited by
Philip Ng
Nicola Brunetti-Pierri

CRC Press
Taylor & Francis Group
Boca Raton London New York

CRC Press is an imprint of the
Taylor & Francis Group, an **informa** business

Cover illustration:

Adenovirus vector particles undergo multiple interactions with non-cellular and cellular host components. The upper left part shows the natural situation of wildtype Ad5 particles, which bind blood coagulation factor X (red) that protects to some degree from natural IgMs (grey) and complement proteins (blue), but mediates hepatocyte tropism. After genetic modification of the capsid to ablate factor X binding and the particle's hepatocyte tropism (yellow point mutations, upper right part), FX cannot shield the vector particles and complement and natural IgM efficiently inactivate the vector particles. A combination of genetic ablation of FX binding and chemical shielding by a synthetic polymer (green) can generate vector particles resistant to complement and detargeted from liver (lower part). See Chapter 4 for more details. This figure has been designed by Moritz and Lea Krutzke.

CRC Press
Taylor & Francis Group
6000 Broken Sound Parkway NW, Suite 300
Boca Raton, FL 33487-2742

First issued in paperback 2020

© 2017 by Taylor & Francis Group, LLC
CRC Press is an imprint of Taylor & Francis Group, an Informa business

No claim to original U.S. Government works

ISBN-13: 978-1-4987-4548-2 (hbk)
ISBN-13: 978-0-367-65818-2 (pbk)

Visit the Taylor & Francis Web site at
http://www.taylorandfrancis.com

and the CRC Press Web site at
http://www.crcpress.com

Contents

Series Preface

Gene and cell therapies have evolved in the past several decades from a conceptual promise to a new paradigm of therapeutics, able to provide effective treatments for a broad range of diseases and disorders that previously had no possibility of cure.

The fast pace of advances in the cutting edge science of gene and cell therapy, and supporting disciplines ranging from basic research discoveries to clinical applications, requires an in-depth coverage of information in a timely fashion. Each book in this series is designed to provide the reader with the latest scientific developments in the specialized fields of gene and cell therapy, delivered directly from experts who are pushing forward the boundaries of science.

In this volume of the cell and gene therapy book series *Therapeutic Advances in Adenoviruses*, a remarkable group of authors comprehensively cover the latest developments in adenovirus biology, immune responses, and adenoviruses as platforms for gene-editing and gene delivery. The authors also cover crucial areas of adenovirus therapeutic applications, such as vaccines and cancer. In addition, this volume also addresses current hurdles in the field, the development of adenovirus hybrid vectors, and improvements to overcome barriers for *in vivo* vector delivery.

We would like to thank the volume editors, Philip Ng and Nicola Brunetti-Pierri, and the authors, all remarkable experts, for their valuable contributions. We also would like to thank our senior acquisitions editor, C.R. Crumly, and members of the CRC Press staff, for all their efforts and dedication to the cell and gene therapy book series.

Preface

The adenoviral vector remains the most frequently used vector in clinical gene therapy. The chapters in this book focus on the most up-to-date developments in the therapeutic applications of adenoviruses. The intended audience is individuals in the life sciences interested in therapeutic applications of adenoviruses. These include undergraduate, graduate, and medical students, basic and applied scientists at all levels, physicians, and physician scientists. Each chapter in this book emphasizes the latest developments in the field.

Chapter 1, written by Leslie A. Nash and Robin J. Parks, serves as a foundation for those less familiar with the adenovirus, and provides a comprehensive review of the life cycle, virion structure, and biology of the adenovirus, as well as a historical perspective of its development and evolution as a gene transfer vector. The tragic death of a patient in 1999 as a result of an innate inflammatory response to the injected adenoviral vector focused great attention on the safety and efficacy of adenoviral vectors. Chapter 2, written by Svetlana Atasheva and Dmitry M. Shayakhmetov, reviews the critical topic of the interaction of the adenoviral vector with the innate and adaptive immune systems. Helper-dependent adenoviral vectors are the most advanced class of adenoviral vectors, with improved safety and efficacy compared to the earlier generation adenoviral vectors. Chapter 3, written by Nicola Brunetti-Pierri and Philip Ng, reviews helper-dependent adenoviral vectors, including methods for their production, and their use in a variety of gene and cell therapy applications. Not only is high efficiency transduction of the desired target cells important for clinical gene therapy, but the avoidance of unwanted or unintended transduction of non-target cells can also be equally important from a safety perspective. Chapter 4, written by Florian Kreppel, identifies the barriers that limit successful transduction of the desired target cells *in vivo*, and modifications that have been made to the viral capsid to overcome these barriers, including avoidance of unwanted vector uptake by non-target cells. The adenoviral vector is a non-integrating vector with its genome existing episomally in the nucleus of transduced cells. While this avoids the potential danger of insertional mutagenesis, it also limits its efficacy in applications targeting dividing cells, where the vector genome would be lost, resulting in transient transgene expression. Chapter 5, written by Anja Ehrhardt and Wenli Zhang, reviews the use of adenoviruses as a platform for the development of novel hybrid gene transfer vectors that permit vector genome persistence in dividing target cells, either through episomal maintenance or genomic integration. The ability to introduce specific, predetermined changes into the genome of target cells offers a powerful tool for gene and cell therapy. Chapter 6, written by Kamola Saydaminova, Maximilian Richter, Philip Ng, Anja Ehrhardt, and André Lieber reviews the use of adenovirus to deliver engineered endonucleases for efficient genome editing. With the constant threat of bioterrorism, the ongoing battle against cancer, and emerging life-threatening infectious diseases, the development of effective vaccines remains a high priority, and adenoviral vectors have proven to be very effective in this regard. Chapter 7, written by Michael A. Barry, reviews the use of adenoviral vectors for vaccination.

Importantly, Dr. Barry puts into perspective the recent high profile "failure" of an adenoviral vaccine against HIV, separating fact from fiction. Adenoviral vectors remain the most utilized vector type for clinical cancer gene therapy. In Chapter 8, Cristian Capasso, Manlio Fusciello, Erkko Ylösmäki, and Vincenzo Cerullo review the use of oncolytic adenovirus for cancer gene therapy.

The therapeutic application of adenoviral vectors is a growing and fast moving field, and it is impossible to predict what the future might bring. Nevertheless, the contributors of this work are all acknowledged experts in their field and their chapters convey the current state-of-the-art in their indicated topics.

Editors

Philip Ng earned his BSc (1990) from the University of Toronto (Ontario, Canada), and his MSc (1994) and PhD (1999) from the University of Guelph (Ontario, Canada). He completed his post-doctoral fellowship at McMaster University (Ontario, Canada). Dr. Ng is currently an associate professor in the Department of Molecular and Human Genetics at Baylor College of Medicine (Houston, Texas).

Nicola Brunetti-Pierri earned his MD in 1997 from the University of Naples Federico II, Italy. From 1998 to 2002, he was a resident in Pediatrics at the Federico II University Hospital in Naples, Italy. From 2002 to 2005, he was a postdoctoral research fellow in the Department of Molecular and Human Genetics at Baylor College of Medicine (Houston, Texas). He completed clinical genetics and biochemical genetics training in the Department of Molecular and Human Genetics at Baylor College of Medicine (Houston, Texas). Currently, he is an associate investigator at the Telethon Institute of Genetics and Medicine, Pozzuoli (Naples) and associate professor at Federico II University of Naples, Italy.

Contributors

Svetlana Atasheva
Lowance Center for Human
　Immunology
Emory University School of Medicine
Atlanta, Georgia

Michael A. Barry
Department of Medicine, Immunology,
　and Molecular Medicine
Mayo Clinic
Rochester, Minnesota

Nicola Brunetti-Pierri
Telethon Institute of Genetics and
　Medicine
Pozzuoli, Italy

and

Department of Translational Medicine
Federico II University
Naples, Italy

Cristian Capasso
ImmunoViroTherapy Laboratory
Division of Pharmaceutical Biosciences
Faculty of Pharmacy
University of Helsinki
Helsinki, Finland

Vincenzo Cerullo
ImmunoViroTherapy Laboratory
Division of Pharmaceutical Biosciences
Faculty of Pharmacy
University of Helsinki
Helsinki, Finland

Anja Ehrhardt
Institute for Virology and Microbiology
Center for Biomedical Education and
　Research (ZBAF)
Department for Human Medicine
Faculty of Health
University Witten/Herdecke
Witten, Germany

Manlio Fusciello
ImmunoViroTherapy Laboratory
Division of Pharmaceutical Biosciences
Faculty of Pharmacy
University of Helsinki
Helsinki, Finland

Florian Kreppel
Department of Gene Therapy
Ulm University
Ulm, Germany

André Lieber
Division of Medical Genetics
Department of Medicine
University of Washington
Seattle, Washington

Leslie A. Nash
Regenerative Medicine Program
Ottawa Hospital Research Institute
Ottawa, Ontario, Canada

Philip Ng
Department of Molecular and Human
　Genetics
Baylor College of Medicine
Houston, Texas

Robin J. Parks
Regenerative Medicine Program
Ottawa Hospital Research Institute
Ottawa, Ontario, Canada

Maximilian Richter
Division of Medical Genetics
Department of Medicine
University of Washington
Seattle, Washington

Kamola Saydaminova
Division of Medical Genetics
Department of Medicine
University of Washington
Seattle, Washington

Dmitry M. Shayakhmetov
Lowance Center for Human
    Immunology
Emory University School of Medicine
Atlanta, Georgia

Erkko Ylösmäki
ImmunoViroTherapy Laboratory
Division of Pharmaceutical Biosciences
Faculty of Pharmacy
University of Helsinki
Helsinki, Finland

Wenli Zhang
Institute for Virology and
    Microbiology
Center for Biomedical Education and
    Research (ZBAF)
Department for Human Medicine
University Witten/Herdecke
Witten, Germany

1 Adenovirus Biology and Development as a Gene Delivery Vector

Leslie A. Nash and Robin J. Parks

CONTENTS

ABSTRACT

Adenovirus (Ad)-based vectors are one of the most commonly used gene delivery vehicles in molecular biology and gene therapy applications. For over 60 years, researchers have studied the basic biology of Ad, unraveling the subtle, yet profound, interactions between the virus and the host. With our increased understanding of the biology of the virus, we have refined the function of Ad as a gene delivery tool, and tailored it to disease-specific applications. For example, removal of all viral genes from the vector, generating helper-dependent Ad, can result in lifelong correction of genetic defects in mouse models of human disease, whereas oncolytic Ads contain more modest modifications of the viral genome that allows them to specifically replicate in, and kill, cancer cells. In this chapter, we will discuss the development and evolution of Ad as a gene therapy vector.

1.1 INTRODUCTION

Adenovirus (Ad) vectors are used extensively in molecular biology applications to achieve high-level gene expression of a desired transgene in mammalian cells. Ads are also the most commonly used delivery vehicles in gene therapy applications. As of July 2015, 22% (n = 506) of all human gene therapy clinical trials used Ad vectors to deliver a therapeutic gene of interest (GOI), for a variety of indications (JGM 2015). The first commercially available gene therapeutic, an Ad-based vector expressing p53 for the treatment of head and neck cancer, became approved for use in China in October 2003. This same vector is now in advanced stage clinical trials in the United States and other countries around the world for the treatment of a variety of cancers, including head and neck, lung, and ovarian cancer. Ad has progressed as a favored delivery vehicle, despite the significant notoriety Ad achieved in the popular press due to the death of a patient entered in a clinical trial for treatment of ornithine transcarbamylase (OTC) deficiency in 1999. This death was directly attributed to a systemic inflammatory response due to the Ad vector that was administered. This event raised questions regarding whether Ad should be used in gene therapy applications and, more broadly, whether we should reconsider the rationale and validity of gene therapy as an experimental therapeutic. Regardless of this perceived major setback to the field of gene therapy, research on Ad and its use in clinical trials continues.

The goal of this chapter is to provide more of a historical perspective of the development of Ad as a gene delivery tool. We will briefly highlight a number of significant milestones (Figure 1.1), and describe how the identification of Ad vector limitations spurred their further development and refinement.

1.2 Ad BIOLOGY

Ads were first discovered in the early 1950s as novel viral agents associated with respiratory ailments in human patients (Rowe et al. 1953; Hilleman and Werner 1954). Their name derives from the original source of tissue from which the prototype member was isolated, the adenoids. Since that time, over 100 family members have been identified and characterized, from a wide variety of mammalian and avian species, in addition to reptiles and amphibians. The discovery that some Ads are tumorigenic in rodents (Trentin et al. 1962; Yabe et al. 1962) stimulated intensive research into the physiology, genetics, and molecular biology of Ads which has continued over the past 60 years. These studies have given us a great deal of information about DNA replication and control of gene expression of the virus, and have provided significant insight into these processes in the host cell itself. For example, alternative splicing, an ubiquitous process in mammalian cells, was first identified in

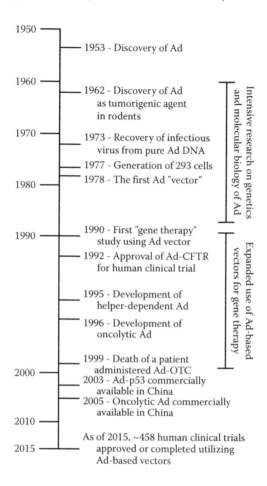

FIGURE 1.1 Timeline of Ad development as a gene delivery vehicle. Shown are the major milestones in the history of Ad, from its discovery to the present day.

Ad (Berget et al. 1977; Chow et al. 1977). It is this broad knowledge of Ad biology that laid the foundations for the later development of Ads as gene transfer vectors.

Of the human Ads, serotypes 2 (Ad2) and 5 (Ad5), both of subclass C, are the most extensively characterized (reviewed in Berk 2007). Their genomes have been sequenced, and are ~95% identical at the nucleotide level, with a similar arrangement of transcriptional units. Unless otherwise indicated, the remainder of this chapter will focus on Ad5 (with information gleaned for Ad2 assumed to also be valid for Ad5).

1.2.1 Ad Virion Structure

All Ads have the same general structural characteristics: an icosahedral, non-enveloped capsid (~70–100 nm in diameter) surrounding a nucleoprotein core containing a linear double-stranded genome (~30–40 kbp) (Figure 1.2) (Greber 1998).

Therapeutic Applications of Adenoviruses

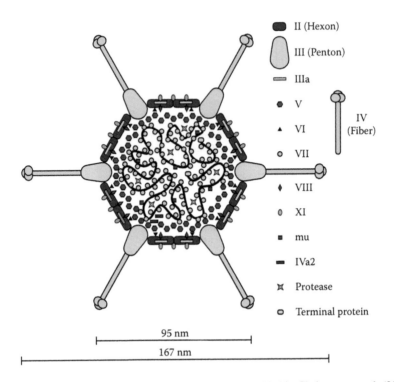

II (Hexon)

III (Penton)

IIIa

V

VI

VII

VIII

XI

mu

IVa2

Protease

Terminal protein

IV (Fiber)

95 nm

167 nm

FIGURE 1.2 Model of the Ad5 virion, based on data provided in Christensen et al. (2008), Russell (2009), Liu et al. (2010), Reddy et al. (2010), Reddy and Nemerow (2014), as adapted by Giberson, Davidson, and Parks (2012), Saha, Wong, and Parks (2014). (Reproduced with permission under a Creative Commons Attribution License from Saha, B., C. M. Wong, and R. J. Parks, *Viruses*, 6 (9), 3563–83, 2014. doi: 10.3390/v6093563 v6093563 [pii].)

The atomic structure of the outer Ad capsid has been refined to a significant degree; however, the inner nucleoprotein core does not appear to have an ordered structure (Christensen et al. 2008; Russell 2009; Liu et al. 2010; Reddy et al. 2010; Perez-Berna et al. 2015). The Ad capsid is composed of eight polypeptides, named in order of decreasing size. Hexon (a trimer of protein II) assembles into a sheet-like structure called the "group-of-nine," which forms the 20 facets of the icosahedron. Protein III clusters into groups of five (known as pentons) at the vertices of the icosahedron, from which extend trimers of protein IV, known as fiber. These proteins which make up the major capsid are supported by five minor capsid proteins (IIIa, IVa2, VI, VIII, and IX). Within the viral capsid, the viral DNA associates with three basic proteins, V, VII, and Mu (μ), which function to neutralize the charge on DNA, permitting its tight packing within the virion. Protein V is postulated to create a shell and coat the protein VII–DNA complex (Everitt et al. 1973; Brown et al. 1975). Protein VII functions similarly to that of cellular protamines, in that it is responsible for wrapping and condensing the viral DNA (Mirza and Weber 1982) with the help of pre-Mu (Anderson et al. 1989). Ad-encoded proteases can reverse this process and relax the viral DNA nucleoprotein structure by cleavage of pre-Mu prior to entering the nucleus (Perez-Berna et al. 2009). Once inside the nucleus, remodeling of the nucleoprotein structure containing the Ad DNA must occur, as the tightly packed DNA structure is inhibitory to viral transcription (Matsumoto et al. 1993; Okuwaki and Nagata 1998), and its relaxation allows for efficient gene expression and DNA replication (Giberson et al. 2012).

1.2.2 THE Ad GENOME

The length of the wild-type Ad5 genome is approximately 36 kb, yet the capsid can accommodate DNA up to 105% of the wild-type genome length (Bett et al. 1993). However, increasing the genome size to this upper limit tends to lead to instability and can result in spontaneous rearrangement of the genome to reduce the size of the DNA closer to wild-type length. This observation has important implications to the design of Ad vectors with large transgenes (Kennedy and Parks 2009). No lower limit for DNA packaging has been identified, and genomes as small as 9 kb have been found packaged into virions (Lieber et al. 1996). Nonetheless, similar to large genomes, small genomes tend to rearrange or multimerize to bring the overall size of the genome closer to the wild-type Ad (Parks and Graham 1997; Morsy et al. 1998).

Genes encoded by Ad are classified as early or late, depending on whether they are expressed before or after DNA replication (Figure 1.3) (Davison et al. 2003). The early regions E1A, E1B, E2, E3, and E4 are the first regions transcribed and encode proteins involved in activating transcription of other viral regions and altering the cellular environment to promote viral production. The E1A proteins induce mitogenic activity in the host cell and stimulate the expression of other viral genes (Berk 2007). The E2 proteins mediate viral DNA replication, while E3 and E4 proteins alter host immune responses and cell signaling, respectively (Horwitz 2004; Weitzman 2005). Activation of the major late promoter (MLP) following the start of virus DNA synthesis, allows expression of the late genes which encode primarily virion structural proteins. These late regions (L1–L5) are transcribed from an alternatively spliced transcript. Recently, it was shown that the regions encoding the L4-22K and -33K proteins are initially expressed at low levels from a novel promoter located within the L4 region (Morris et al. 2010), and these proteins are responsible for fully activating the MLP (Morris and Leppard 2009). There are also four small products produced at intermediate/late times of infection, including the structural protein IX (pIX) and the IVa2 protein that helps package viral DNA into immature virions (Christensen et al. 2008). The late products, VA RNA I and II, inhibit activation of the interferon response, impede cellular micro-RNA processing, and may influence expression of host genes (O'Malley et al. 1986; Aparicio et al. 2006). The viral DNA also contains the origins of replication (the inverted terminal repeats [ITR], ~100 bp located at both the left and right end of the genome) and the packaging sequence (~150 bp, located immediately adjacent to the left ITR).

Perhaps somewhat surprisingly, our knowledge of genes encoded by the virus is still expanding. Tollefson et al. (2007) identified a new open reading frame (ORF) that they termed U exon protein (UXP), which was located between the fiber-coding sequence and E3. UXP is expressed from a unique promoter during late stages of

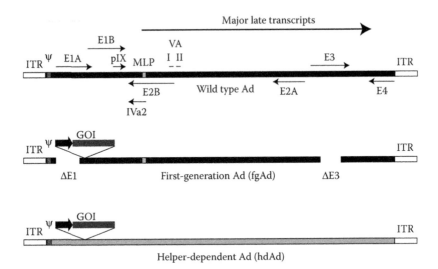

FIGURE 1.3 Schematic representation of the Ad5 genome and vectors. Top panel: The Ad5 genome is divided into four early transcription units, E1–E4, which are expressed before DNA replication and five late transcription units, L1–L5 (not shown), which are expressed after DNA replication and are alternative splice products of a common major late transcript. Four smaller transcription units are also produced: pIX, IVa2, and VA RNA's I and II. Also shown are the viral ITR, approximately 100 bp at each end of the viral genome, and the viral packaging signal (Ψ), located from nucleotides 190–380 at the left end of the genome, which are cis-acting elements involved in Ad replication and packaging, respectively. The location of the MLP is also shown. Middle panel: A typical early region (E1)-deleted Ad vector is shown. The GOI replaces the E1 deletion and its expression is driven by a heterologous promoter (dark arrow). Most E1-deleted vectors also have the E3 region removed as it is not essential for viral replication and allows for insertion of an ~8 kb foreign expression cassette. Bottom panel: HDAd schematic. All viral protein-coding regions are removed, and only the ITRs and packaging sequence are required. HDAd vectors usually have non-coding stuffer DNA (gray) to ensure the genome is stable. (Reproduced with permission under a Creative Commons Attribution License from Saha, B., C. M. Wong, and R. J. Parks, *Viruses*, 6 (9), 3563–83, 2014. doi: 10.3390/v6093563 v6093563 [pii].)

infection, and may play a role in virus DNA replication or RNA transcription (Tollefson et al. 2007; Ying et al. 2010). A recent study using deep cDNA sequencing identified many new alternatively spliced transcripts originating from the Ad genome (Zhao et al. 2014), suggesting that there may be many undiscovered, new, or altered polypeptides produced by Ad in the infected cell.

1.2.3 Ad INFECTION

In most cells, Ad5 infection initiates with the fiber knob domain binding to the primary receptor for Ad5 on the cell surface, termed the Coxsackie-Adenovirus receptor (CAR) (Bergelson et al. 1997; Tomko et al. 1997). However, it should be noted that different serotypes have different cell surface receptors and these can be used to "naturally" alter Ad vector infection to enhance uptake by specific cell types (Havenga et al. 2002). After initial binding, the Ad5 penton interacts with a secondary receptor comprised of $\alpha_v\beta_3$ or $\alpha_v\beta_5$ integrins (Wickham et al. 1993). The efficiency with which Ad binds to and enters cells is directly related to the level of the primary and secondary receptors found on the cell surface (Goldman et al. 1996; Wickham et al. 1996). More recent studies have suggested that some Ad serotypes, including Ad5, can enter cells using heparin sulfate proteoglycans as an alternative receptor, either through direct binding to sequences in the Ad fiber shaft (Smith et al. 2003), or bridged through interaction of Ad with various blood factors or the complement component C4-binding protein (Shayakhmetov et al. 2005; Kalyuzhniy et al. 2008; Waddington et al. 2008).

Upon systemic delivery in most species, human Ad5 preferentially accumulates in the liver (Guo et al. 1996; Worgall et al. 1997; Nicol et al. 2004). This preferential uptake is due in part to the physical architecture of this tissue, as Ad becomes trapped in the liver sinusoids and fenestrations (Bernt et al. 2003; Ross and Parks 2003). However, in the liver, Ad shows a nonlinear uptake by hepatocytes—at low doses, little Ad is taken up by hepatocytes whereas at higher doses disproportionately more is taken up (Tao et al. 2001). Initially, Ad tends to be preferentially taken up by Kupffer cells within the liver, and it is only after the receptors on these cells are saturated that the virus is available to infect hepatocytes. Uptake of Ad by Kupffer cells leads to death of the cells (Smith et al. 2008). Consequently, enhanced hepatocyte transduction can be achieved *in vivo* using two sequential vector injections (the first with an irrelevant Ad vector that saturates and kills the Kupffer cells, allowing the second therapeutic vector to effectively transduce hepatocytes; Tao et al. 2001) or through pretreatment with Kupffer cell-killing compounds such as clodronate liposomes (Tao et al. 2001; Ziegler et al. 2002).

Ad5 is taken up by hepatocytes using a rather unique mechanism. The Ad5 hexon protein interacts with coagulation factor X which provides a bridging interaction for binding to heparin sulfate proteoglycans expressed on the surface of hepatocytes, allowing for internalization (Kalyuzhniy et al. 2008; Waddington et al. 2008; Bradshaw et al. 2010; Alba et al. 2011). However, Ad vectors containing mutations in the hexon protein that prevent its interaction with factor X still localize to the liver, although uptake is dramatically reduced (Kalyuzhniy et al. 2008, Alba et al. 2010). These results suggest that physical constraints, other than receptor binding, contribute significantly to vector biodistribution.

Internalization of Ad occurs through endocytosis, triggered by the penton–integrin interaction (Wickham et al. 1993). Acidification of the endosome alters the Ad capsid structure, allowing for the release of protein VI from the inner capsid. Protein VI possesses membrane lytic activity (Wiethoff et al. 2005), and mediates the rupture of the endosome and release of the virus (Mellman 1992; Wickham et al. 1994; Leopold et al. 1998; Wiethoff et al. 2005; Wiethoff and Nemerow 2015). The virion translocates to the nucleus along the microtubule network (Leopold et al. 1998), during which time there is a sequential disassembly of the Ad virion and, as a final step, the Ad hexon remains outside the nuclear membrane while the DNA bound to protein VII passes into the nucleus (Chatterjee et al. 1986; Greber et al. 1993; Strunze et al. 2011). Viral DNA replication and assembly of progeny virions occur within the nucleus of infected cells, and the entire life cycle takes approximately 24–36 h, providing an output of approximately 10^4 virions per cell for the wild-type virus. In humans, Ads are not associated with neoplastic disease, and only cause relatively mild, self-limiting illness in immunocompetent individuals; primarily respiratory illnesses, keratoconjunctivitis, or gastroenteritis (depending on the serotype) (Wold and Horwitz 2007). For a more comprehensive discussion of Ad biology, the reader is referred to other excellent reviews (McConnell and Imperiale 2004; Berk 2007; Russell 2009).

1.3   FIRST-GENERATION Ad VECTORS

1.3.1   THE FIRST Ad VECTORS

The first suggestion that Ad could be used to express foreign genes in mammalian cells came with the identification of spontaneous recombinants between Ad and simian virus 40 (SV40) (Hassell et al. 1978; Tjian 1978), in which large regions of the Ad genome, mainly the E3 and E4 transcription units, were replaced with SV40 sequences. One of these recombinants, Ad2 + D2, produced large quantities of T antigen (fused to an Ad structural protein) from the MLP, which allowed for the purification of large quantities of a T antigen-related protein that retained biological activity. Due to the loss of essential E4 functions, Ad2 + D4 could only be propagated in the presence of the wild-type helper virus, usually present in at least a 10-fold excess. To ensure that Ad2 + D2 was maintained in the mixed virus culture, the viruses were propagated in COS cells, which are only permissive for Ad replication in the presence of T antigen, thus providing a strong selection for maintenance of Ad2 + D2. Due to the difficulty in maintaining stable vector stocks in the absence of a helper virus, the use of this type of vector was limited.

1.3.2   E1-COMPLEMENTING CELL LINES

Graham and van der Eb (1973) developed the calcium phosphate technique for introducing DNA into mammalian cells. This technique allowed for the recovery of an infectious virus from purified DNA of Ads and other viruses, in addition to providing a relatively efficient method to transform any DNA into almost any cell type *in vitro*. Using this technique, they were also able to map the transforming genes of Ad5 (i.e., the E1 region) (Graham et al. 1974, 1975). In 1977, Graham et al. (1977) reported the isolation and characterization of a human embryonic cell line transformed by Ad5 after transfection with sheared viral DNA. These cells were originally designated 293-31 cells, because Graham sequentially numbered all his experiments, this was his 293rd experiment, and the transformed colony appeared in experimental dish 31. Now simply known as 293 cells, they have nucleotides 1-4344 of the Ad5 genome inserted in chromosome 19 (Louis et al. 1997), and express both the E1A and E1B transcription units. The pIX transcription unit is also present in these cells; however, they do not express this protein at a detectable level (Sargent et al. 2004). Although 293 cells were derived after transfection of human embryonic kidney cells (giving them the common "HEK" designation), the cell line actually expresses many protein markers consistent with a neuronal lineage (Shaw et al. 2002). This and other data have led to the speculation that 293 cells may not be derived from kidney cells, but rather from rare neuronal lineage cells present in human kidney cultures (Shaw et al. 2002; Toth and Wold 2002). A separate study suggested that the 293 cell line transcription profile was more consistent with being derived from adrenal lineage (Lin et al. 2014).

Regardless of their cell type of origin, because 293 cells complement Ads with mutations or deletions in E1, these cells have proven to be extremely useful for researchers interested in production of Ad vectors expressing foreign genes. The E1 region of Ad could now be replaced with a GOI, and the resulting vector could be easily propagated in the E1-complementing 293 cell line. The significance of this was twofold: first, Ad vectors could now be generated that were not dependent on the presence of a helper virus, so very pure preparations of the vector could be produced and, second, these E1-deleted vectors can infect many different cell lines with the efficiency of wild-type Ad and express high levels of the GOI. However, since the vectors do not replicate in the absence of E1 functions, cells not expressing E1 (i.e., virtually all mammalian cells) can be transduced with the Ad vector and maintained for a period of time without cell death due to virus replication. Not only are 293 cells used for generation of Ad vectors, but they are also frequently used as the base cell line for production of adeno-associated virus (AAV) and retrovirus/lentivirus vectors, and because of a number of other useful properties, 293 cells have become one of the most commonly utilized mammalian cell lines in molecular biology.

1.3.3 THE PROBLEM OF REPLICATION-COMPETENT Ad

Although the 293 cell line is extremely useful for production of Ad vectors, it does have one limitation: E1-deleted Ad vectors propagated on 293 cells can recombine with the E1 sequences contained within the cell line, thus transferring E1 to the vector and yielding replication-competent Ad (RCA) (Lochmuller et al. 1994; Hehir et al. 1996). The presence of RCA, essentially a wild-type virus, in vector stocks is undesirable, and was a serious concern and obstacle in early gene therapy clinical trials with Ad. As a result, new E1-complementing cell lines have been generated which contain little or no sequence overlap with the E1-deleted vector (Fallaux et al. 1998; Gao et al. 2000; Schiedner et al. 2000). Although these new cell lines do prevent RCA formation, some reports have shown that other E1+, but replication-defective, recombinant viruses can be generated in these cell lines (Murakami et al. 2002, 2004). These recombinants contain and express E1 but lack other essential viral sequences, and thus cannot replicate by themselves. However, co-infection of the recombinant E1+ "virus" and the E1-deleted vector results in cross-complementation, permitting virus replication. The E1+ viruses likely arise through recombination between very short stretches of homology between the vector and the cell line, or through nonhomologous recombination events, and clearly illustrate how adept viruses are at circumventing replication blocks. Although the formation of RCA (and other less well-defined recombinant viruses) appears to be an inevitable consequence when using Ad vectors, this has not prevented Ads rise to the forefront of gene delivery vectors.

1.3.4 Methods for Generation of E1-Deleted Ad Vectors

One of the by-products of Ad DNA replication is the formation of covalently closed circles of full-length Ad genomic DNA (Ruben et al. 1983). This observation was followed by a study showing that if an antibiotic resistance gene and bacterial origin of replication were inserted in the otherwise wild-type Ad genome, these circles could be recovered and propagated in bacteria, and are capable of producing an infectious virus when re-introduced into appropriate mammalian cells (Graham 1984). Other methods for cloning infectious viral DNA in bacterial plasmids involved insertions of linear Ad genomes into plasmid DNA, from which the viral DNA must be released by restriction enzyme digestion in order to produce an infectious virus following transfection (Hanahan and Gluzman 1984). These advances facilitated wide-scale manipulation of the Ad genome using common cloning techniques. However, the large size of these plasmids (typically 30–40 kbp) makes standard cloning procedures difficult. As a result, several methods have been developed in order to simplify the process of viral construction, typically taking advantage of the recombination pathways present in prokaryotic and eukaryotic cells (reviewed in Danthinne and Imperiale 2000). One of the earliest systems for creating first-generation Ad employed the natural recombination pathway in mammalian cells following the transfection of two DNAs: (1) a plasmid containing the left-hand portion of the virus (including the packaging signal and 5′ ITR) with the desired transgene expression cassette replacing the E1 sequences and, (2) the right-hand portion of the virus genome (either purified virus DNA, digested with appropriate restriction endonucleases to remove the packaging signal, or a plasmid containing a circularized genome) (reviewed in Graham and Prevec 1995). Recombination between homologous Ad sequences on the two DNAs leads to the formation of recombinant viral genomes. The efficiency of this system has been improved 100-fold by using a site-specific recombinase (Cre or FLP) to mediate recombination between the two plasmids instead of the natural homologous recombination pathways in mammalian cells (Ng et al. 1999; Ng, Cummings et al. 2000; Ng, Parks et al. 2000).

Alternatively, recombinant Ad DNA molecules can be generated using the highly efficient homologous recombination machinery present in RecA+ bacteria (Chartier et al. 1996; He et al. 1998). In this system, bacteria are transformed with two plasmids: (1) a shuttle plasmid, similar to the one used for rescue in mammalian cells and, (2) a circularized Ad genome with a deletion of the entire left-end region of the virus, including the 5′ ITR, packaging signal, and E1 region. Recombination between homologous sequences in these two plasmids occurs through the *Escherichia coli* RecA recombination pathway. Appropriate recombinant genomes are then screened by identifying bacterial clones with the correct plasmid. Once the correct plasmid is identified and amplified, the viral DNA must be released from the plasmid by cleavage with an appropriate restriction enzyme and transfected into an E1-complementing cell line, where it will generate the desired recombinant Ad. The simplicity of these systems, and the fact that all of the necessary reagents are commercially available, has made the generation and use of Ad vectors a common technique in many laboratories that do not necessarily specialize in virological techniques.

1.3.5 E1-Deleted Ads in Gene Therapy

Which gene therapy study was the first to use an Ad-based vector to deliver a therapeutic gene *in vivo* is somewhat debatable, and may largely depend on the definition of "gene therapy." In 1990, a study was published in one of the first issues of the journal *Human Gene Therapy* that described the use of an Ad vector to deliver the gene-encoding OTC to the Spf-ash strain of mice, which express reduced levels of the OTC protein (Stratford-Perricaudet et al. 1990). Normal levels of hepatic OTC were reached in approximately 25% of these animals, and expression persisted for 2 months. One animal continued to show protein expression and a phenotypic effect for over 1 year. These results suggested that long-term correction of a genetic defect could be achieved using Ad-mediated delivery of a therapeutic gene. This study was followed over the next few years by a flurry of high-profile studies describing the use of Ad vectors in animal models to deliver a variety of genes to a range of tissues (for a review of these early studies, see Kozarsky and Wilson 1993).

Since Ad2 and Ad5 are respiratory viruses, several of these early gene therapy studies explored Ad-mediated delivery of a therapeutic gene to the lungs of mice, rats, and nonhuman primates (Rosenfeld et al. 1992; Engelhardt, Simon et al. 1993; Engelhardt, Yang et al. 1993; Simon et al. 1993; Zabner et al. 1994). The resulting data provided the rationale for proposing the first Ad-mediated human gene therapy clinical trials using the virus to deliver the cystic fibrosis transmembrane conductance regulator (CFTR) gene to patients with cystic fibrosis (Welsh et al. 1994; Wilson et al. 1994; Crystal et al. 1995), with three protocols receiving formal approval in late 1992. In general, these studies showed that low-dose (5×10^7 IU and 2×10^9 pfu) administration of an Ad vector was well tolerated and that some evidence of phenotypic correction was noted (Zabner et al. 1993; Crystal et al. 1994). However, therapeutic gene expression from the Ad vector was only transient. Ad has now been used in over 450 clinical trials worldwide to explore therapies for a variety of genetic and acquired diseases (Table 1.1).

1.3.6 Limitations of E1-Deleted Ad Vectors

Throughout the 1990s, Ad continued to be used prominently in many gene therapy-based studies. However, much of the data suggested that E1-deleted Ad had a significant limitation: gene delivery was accompanied by strong anti-Ad immune response (Ahi et al. 2011; Thaci et al. 2011; Hendrickx et al. 2014). Systemic delivery of Ad vectors to mice results in immediate activation of innate immunity, and this response seems to be solely dependent on the Ad capsid and/or the infection process, since viral gene expression or replication is not required (Zhang et al. 2001). It appears that almost every aspect of the Ad entry process is under scrutiny by the innate immune system. This includes binding of Ad to CAR (Tamanini et al. 2006) or integrins (Di Paolo et al. 2009) within the endosome, detection of capsid-bound blood factors by TLR4 (Doronin et al. 2012) or the viral DNA by TLR9 (Zhu et al. 2007; Appledorn et al. 2008); and rupture of the endosome by Ad (Barlan et al. 2011). Once in the cytoplasm, the Ad DNA can be detected by DNA-dependent

TABLE 1.1
Gene Therapy Clinical Trials Using Ad Vectors (1989–2015)[a]

	Clinical Phase						
Disease	**I**	**I/II**	**II**	**II/III**	**III**	**IV**	**Total**
Cancer							
Immune modulator	75	14	19				108
Tumor suppressor	33	10	24	5	6	2	80
Cell suicide gene	44	9	6		2		61
Oncolytic	29	12	5	1	1		48
Vaccine	17	4	4				25
Other	2	15	4		2		23
Vascular	20	7	8	4	3		42
Infectious Disease	30	7	3				40
Monogenic	15	2					17
Other	11	1	2				14
Total	276	81	75	10	14	2	458

[a] Data in this table were compiled from the Gene Therapy Clinical Trial database maintained by Wiley and Sons (http://www.abedia.com/wiley/). An effort was made to exclude trials which had incomplete data on the current status of their approval, or which were abandoned. This table provides a rough outline of human clinical trials utilizing Ad-based vectors, but should not be considered definitive or complete.

activator of interferon (IFN)-regulatory factors (DAI) (Schulte et al. 2013) or the nucleotide oligomerization domain (NOD)-like receptors (Muruve et al. 2008). These events activate several signal transduction pathways that culminate in the production and release of a plethora of cytokines and chemokines, such as IP-10, MIP-2, MCP-1, RANTES, IL-8, IL-6, IL-12, IFN-γ, and tumor necrosis factor (TNF)-α (Muruve et al. 1999, 2008; Zhang et al. 2001). Many of these same responses have been noted after Ad injections into nonhuman primates and humans (Schnell et al. 2001). Within hours of induction, many of the chemokine/cytokine molecules act to recruit neutrophils, macrophages, and T-lymphocytes (including cytotoxic T-lymphocytes, CTL) (Muruve et al. 1999; Nayak and Herzog 2010). The presence of these cells correlates with the onset of acute toxicities, such as rises in hepatocyte derived transaminases (AST and ALT) and pan-hepatitis within the serum, leading to an immune-mediated loss of the Ad transduced cells (Gao et al. 1996; Dedieu et al. 1997; Christ et al. 2000). Delivery of large doses of Ad vector to primates results in significant transduction of the liver, but this is also accompanied by hepatic inflammation, thrombocytopenia, and hematological indications of disseminated intravascular coagulation (Sullivan et al. 1997; Raper et al. 1998; Morral et al. 2002). Primates can become moribund after a high-dose vector administration (1.2×10^{13} virus particles/kg), while administration of vector at a 10-fold lower dose results in no symptoms (Morral et al. 2002). Taken together, these data clearly show that activation of the innate immune response by Ad vectors can have serious deleterious consequences that limit vector efficacy and can significantly affect the health and safety of the host.

In addition to innate immunity, E1-deleted Ad induce adaptive immunity, classically shown by the formation of anti-Ad CTL and antibodies in treated animals (Yang, Ertl, and Wilson 1994; Yang et al. 1994). Formation of adaptive immunity coincides with a continual production of some of the inflammatory cytokines and chemokines noted during the early innate response (Muruve et al. 1999; Muruve 2004). It is believed that the long-term inflammatory response is due, at least in part, to low-level expression of viral proteins from the vector backbone (Muruve et al. 2004); however, anti-Ad CTL can also be generated in the absence of viral gene expression (Kafri et al. 1998). Although transgene expression can be extended by several months using T-cell deficient animals, implicating acquired immunity as a major limiting factor in long-term gene expression (Yang et al. 1994; Barr et al. 1995; Michou et al. 1997), other studies showed that Ad can persist long term, even in the presence of a robust cellular response to the virus (Wadsworth et al. 1997). In addition to the CTL response, high-dose intravenous injection of Ad vectors results in activation of the humoral immune response and the generation of high-titer, neutralizing anti-Ad capsid antibodies that prevent transgene expression following re-infection with the same serotype of Ad vectors (Kass-Eisler et al. 1996; Mastrangeli et al. 1996; Mack et al. 1997). Anti-Ad antibodies have also been elicited in human studies (Zuckerman et al. 1999), although not in all cases (Crystal et al. 1994). More recently, Ads have been shown to activate the complement pathway (Jiang et al. 2004; Shayakhmetov et al. 2005), which can also lead to a virus induced inflammatory response. Ad activation of the various components of the host immune response ultimately limits their ability to provide long-term therapeutic gene expression, which is required for correction of most genetic diseases. However, E1-deleted Ads have proven valuable in applications directed toward cancer, where, in many cases, only short-term expression is desired since once the tumor is eradicated, the vector is no longer necessary.

1.4   MULTIPLY-ATTENUATED Ads

Although the early inflammation and innate immune responses to Ad vector are a consequence of interaction of the Ad capsid proteins with the infected cell, long-term inflammation and toxicity is likely a result of low-level viral gene expression. To further reduce expression of viral genes, many researchers have developed Ad vectors with deletions of other essential genes (reviewed in Amalfitano and Parks 2002). For example, viruses have been generated with additional mutations or deletions in the E2 (Fang et al. 1996; Gorziglia et al. 1996; Zhou et al. 1996; Amalfitano and Chamberlain 1997) or E4 (Armentano et al. 1995; Krougliak and Graham 1995; Wang et al. 1995; Brough et al. 1996; Gao et al. 1996; Yeh et al. 1996; Lusky et al. 1998) regions, both of which are required for normal viral replication. These vectors, termed "second-generation" Ad vectors, must be propagated in cell lines which complement both E1 and the second missing function, and are generated using similar methods as described for first-generation Ads. Results using second-generation

Ads have been somewhat mixed, ranging from no improved function (Fang et al. 1996; Lusky et al. 1998) to significantly improved, long-term transgene expression and reduced immunogenicity and toxicity (Dedieu et al. 1997; Wang et al. 1997; Hu et al. 1999; Everett et al. 2003) compared to vectors with only E1-deleted. One interesting observation arising from the analysis of transgene expression from E4-deleted vectors is the influence that this region can have on the persistence of expression of transgenes controlled by viral promoters contained in these vectors (e.g., cytomegalovirus immediate-early enhancer/promoter). Vectors deleted of most or all of E4 showed reduced transgene expression over time, which was not accompanied by a loss of vector DNA (Armentano et al. 1997; Brough et al. 1997). Subsequently, it was determined that the Ad E4 open reading frame 3 (E4ORF3) was able to prevent viral promoter down-regulation, which can occur over time in transduced cells (Armentano et al. 1999), and this appears to be a generalized phenomenon for viral, but not cellular, promoters contained in Ad vectors. Inclusion of an E4ORF3 expression cassette in plasmid constructs also resulted in an improved duration of transgene expression *in vivo* (Yew et al. 1999).

Perhaps the best example of the utility of a second-generation Ad is from Amalfitano and co-workers (Amalfitano et al. 1999; McVie-Wylie et al. 2003; Xu et al. 2005), who showed that administration of an E1/E2B (Ad DNA polymerase)-deleted vector-encoding human acid-α-glucosidase (GAA) to GAA-knockout mice or in a quail model of GAA-deficiency could result in systemic correction of muscle glycogen storage disease. In these experiments, a single injection of the vector resulted in efficient uptake of the virus by hepatic cells, and GAA proenzyme produced from the hepatic "protein factories" was secreted into the serum and, subsequently, taken up by skeletal and cardiac muscle. Although second-generation Ads are easier to generate than fully deleted Ad vectors (see below), they have not gained widespread use because of their inconsistent performance and the only marginal increase in cloning capacity over first-generation Ads.

1.5 HELPER-DEPENDENT Ads

An alternative approach to eliminating the complications of the immune response brought upon by the use of the Ad vectors would be to completely remove all viral protein-coding sequences from the vector backbone, giving rise to fully deleted or helper-dependent Ad (HDAd) vectors. Such vectors must be propagated in the presence of a helper virus which provides all of the required functions for HDAd vector replication and packaging. Early studies suggested that HDAd vectors were capable of delivering a therapeutic transgene to cells; however, problems with vector production meant that these stocks were contaminated with large quantities of the helper virus (Mitani et al. 1995; Fisher et al. 1996; Haecker et al. 1996; Kochanek et al. 1996; Kumar-Singh and Chamberlain 1996). In 1996, two independent research groups developed the Cre/loxP system for propagating HDAd (Parks et al. 1996; Hardy et al. 1997). This system utilized an E1-deleted helper virus with the viral DNA packaging sequence flanked by loxP sites such that, upon infection of a 293-derived cell line that stably expressed the Cre recombinase, the packaging signal was excised thus rendering the helper virus DNA unpackagable. Removal of the packaging element does not interfere with replication or viral gene expression from this virus, or its ability to co-replicate with and support the packaging of the HDAd genome. The problem of RCA formation was circumvented by designing the helper virus to contain a relatively large E3 insert that prevented packaging of E1+ derivatives (Parks et al. 1996). Utilizing this system, it was possible to produce large quantities of relatively pure HDAd. Similar systems have been developed utilizing the FLP recombinase (Ng et al. 2001; Umana et al. 2001), and the Cre/lox system has now been adapted to non-adherent spinner culture for very large-scale HDAd vector production (Palmer and Ng 2003). Most HDAd vectors in current use are generated using this latter system.

HdAd have consistently demonstrated distinct advantages over E1-deleted Ad vectors (reviewed in Palmer and Ng 2005, Brunetti-Pierri and Ng 2008, 2011; Cots et al. 2013). Numerous studies have shown very long-term gene expression in both mice and nonhuman primates with HDAd, typically complemented with reduced toxicity and immune reaction compared to E1-deleted Ad. Perhaps the best early example of the potential of HDAd was provided by Kim et al. (2001) who showed that a single injection of a HDAd-encoding apolipoprotein E (HDAd-ApoE) led to a significant reduction in blood cholesterol and prevented formation of atherosclerotic lesions in ApoE-deficient mice for the life of the animal (2.5 years). Studies in non-human primates have shown that HDAd can persist and express a transgene for more than 7 years (Brunetti-Pierri et al. 2013), the longest that any gene transfer vector has been followed.

Since HDAd have a cloning capacity of approximately 36 kbp, they can accommodate additional "options" to further improve vector function. These include large upstream regulatory regions to achieve tissue-specific expression (Schiedner et al. 1998; Pastore et al. 1999; Shi et al. 2002, 2004), or gene regulatory systems to achieve on/off transgene expression (Burcin et al. 1999; Aurisicchio et al. 2001; Xiong et al. 2006). "Hybrid" HDAd vector systems have also been produced that combine the high transduction efficiency of Ad vectors with elements from other vector systems that permit stable persistence of vector DNA within the transduced cell. Examples of these hybrid systems include utilizing the integrative machinery of AAV (Recchia et al. 1999; Goncalves et al. 2002) or retrovirus/lentivirus (Soifer et al. 2002; Kubo and Mitani 2003), transposition (Yant et al. 2002), or episomal maintenance (Tan et al. 1999). Also, since the HDAd genome is packaged into capsid proteins provided by the helper virus, genetically identical HDAd have been generated using helper viruses based on different serotypes (e.g., Ad2, Ad5, or Ad6). The sequential use of HDAd based on different serotypes allows for the evasion of neutralizing antibodies in previously immunized animals, thereby allowing HDAd vector readministration (Morral et al. 1999; Parks, Evelegh, and Graham 1999). Alternatively, helper viruses with altered tropism have been used to enhance HDAd transduction of target tissues (Biermann et al. 2001; Bramson et al. 2004). Taken together, these studies indicate that HDAd are a very versatile platform to achieve long-term gene expression *in vivo*.

The HDAd genome should be constructed between approximately 27 and 36 kb to accommodate upper and lower size constraints of the DNA for efficient packaging into the Ad capsid (Parks and Graham 1997). Since many therapeutic transgenes are smaller than this optimal size, non-coding "stuffer" DNA must be included in the vector, preferably derived from eukaryotic sources (Parks et al. 1999; Sandig et al. 2000; Schiedner et al. 2002). The question addressed in several studies was whether the inclusion of large contiguous regions of chromosomal DNA would alter the frequency of vector insertion, which is normally considered very low for the wild-type or E1-deleted Ad. In one study (Hillgenberg et al. 2001), inclusion of a 27.4 kb fragment of DNA derived from the X chromosome into an HDAd did not lead to integration of the vector by homologous recombination. Rather, low-frequency integration occurred by nonhomologous events, resulting in almost perfect insertion of the entire HDAd (i.e., in many cases the vector ends remained intact). Random integration did occur at a slightly higher frequency for the HDAd compared to a control E1-deleted Ad (166 vs. 26 neo-resistant colonies per 10^6 cells for HDAd-neo and E1-deleted Ad-neo, respectively). However, integration of an entire genome of an E1-deleted vector could be slightly deleterious or toxic to the host cells, possibly explaining the enhanced frequency of recovery of clones transformed by HDAd-neo compared to a first-generation vector. In a second study, HDAd were used to target gene correction of the hypoxanthine-guanine phosphoribosyl transferase (HPRT) locus in mouse embryonic stem cells (Ohbayashi et al. 2005). The frequency of homologous recombination between the vector and the locus was relatively high (0.2% with an 18.6 kb stretch of homologous sequence), however the frequency of nonhomologous integration was over 10-fold higher. The nonhomologous insertion sites were distributed in both coding and noncoding DNA, which is in contrast to retrovirus, lentivirus, and AAV vectors which tend to integrate within active genes (Nakai et al. 2003; Mitchell et al. 2004). It should be noted that Ad and HDAd vectors seem to persist in cycling cells much better than would have been predicted based on their normally episomal location (Ehrhardt et al. 2003), suggesting that the virus has evolved a mechanism to remain associated with the nucleus even during cell division.

At least part of the reason that HDAd perform better than E1-deleted Ad is due to their improved immunological profile. In contrast to E1-deleted Ad, which stimulates both an early and late inflammatory response, HDAd do not induce late after delivery *in vivo* (Muruve et al. 2004), likely due to the lack of viral gene expression from the vector backbone. However, the early inflammatory response against HDAd can be substantial and lethal (Brunetti-Pierri et al. 2004). Systemic delivery of 5.6×10^{12} vector particles per kg to nonhuman primates resulted in 50% hepatocyte transduction, accompanied by relatively mild, acute toxicity which resolved within 24 hr (Brunetti-Pierri et al. 2004). At a twofold higher dose (1.1×10^{13} vector particles per kg), 100% of the liver was transduced; however, there was severe and lethal acute toxicity. This is perhaps not surprising, since this early inflammatory response is driven solely by the infecting capsid, and does not require viral gene expression. Fortunately, early inflammation from HDAd can be reduced somewhat through coating the virus with masking agents, such as polyethylene glycol (Croyle et al. 2005; Mok et al. 2005), suggesting that even this limitation to HDAd can be overcome.

1.6   ONCOLYTIC Ad FOR CANCER-DIRECTED THERAPY

Although one arm of Ad vectorology has focused on trying to create stealth-type vectors that provide minimal insult during gene delivery, other researchers have focused upon creating Ad that actively replicate in certain tissues of the host. Oncolytic or conditionally replicating Ad (CRAd) is designed to replicate in cancer cells, but not normal tissue. Perhaps the best characterized CRAd is Addl1520, also known as ONYX-015, which is deficient for the E1B-encoded 55 kDa protein (Barker and Berk 1987; Bischoff et al. 1996). During infection with wild-type Ad, one of the functions of proteins encoded within the E1B region is to sequester p53, and promote its degradation, thus preventing p53 activation at the G2 checkpoint, which might lead to premature cell death (thereby causing premature termination of the virus lifecycle) (Steegenga et al. 1998). E1B also aids in viral replication by forming a complex with E4 open reading frame 6 (ORF6) and allowing for the accumulation and transportation of viral mRNA, while inhibiting host mRNA transportation and translation (Pilder et al. 1986; Ornelles and Shenk 1991). In the absence of E1B 55 kDa, Ad cannot replicate in normal cells; however, since p53 is mutated in almost half of all tumors (Lane 1992), the virus will replicate in, and kill, tumor cells, but not spread to adjacent normal tissue (Harada and Berk 1999; O'Shea et al. 2004; Cheng et al. 2015). CRAd is becoming very popular for cancer-directed therapy, and is evolving and maturing in its design (reviewed in Choi et al. 2012; Larson et al. 2015; Uusi-Kerttula et al. 2015). For example, the E1A-coding region in these vectors can be placed under regulation by a tumor-specific promoter, thereby increasing their specificity for replication in a particular tumor type (Glasgow et al. 2004). Alternatively, "armed" CRAd can be generated through inclusion of a therapeutic transgene in the vector, such as a cytokine, to enhance immune reaction to the tumor cell, or a cell suicide gene such as the herpes simplex virus thymidine kinase gene (HSV TK, used in combination with ganciclovir administration) or p53 (Choi et al. 2012). The promising results observed for ONYX-015 in preclinical studies spurred its movement into phase I and II trials, either as a lone therapeutic or in combination with radiation or chemotherapy (McCormick 2003; Larson et al. 2015). Unfortunately, one of the main biotechnology companies exploring this mode of cancer therapy, Onyx Pharmaceuticals, closed its therapeutic virus program in 2003 and, with it, cancelled many of their ongoing clinical trials, including a Phase III trial (Kirn 2006; Larson et al. 2015). However, in November 2005, Shanghai Sunway Biotech (Shanghai, China) announced that it had received regulatory approval from the Chinese government for use of an oncolytic Ad deleted of E1B-55kDa, in combination with cisplatin, for treatment of nasopharyngeal carcinomas (Kirn 2006), making this the second Ad-based gene therapeutic that is commercially available (see below for a discussion of the first Ad-based gene therapy commercial product).

1.7 HUMAN CLINICAL TRIALS UTILIZING Ad-BASED VECTORS

Currently, Ads are being used in about 22% of all gene therapy clinical trials, with the vast majority of these trials directed toward the treatment of cancer (JGM 2015). Some of the shortcomings of E1-deleted Ad vectors, mainly short-term transgene expression and induction of an immune response, are not of concern in most cancer trials, since the therapeutic is usually only required short term (i.e., once the tumor is gone, the virus is no longer needed). As shown in Table 1.1, clinical trials with Ad vectors follow a trend found throughout the field of gene therapy: many Phase I trials are conducted, few of which progress to Phase II or III. Failure to proceed to a Phase II or III trial is not usually a result of safety or toxicity issues but, rather, a lack of evidence for efficacy in the Phase I study (although technically Phase I trials are not designed to examine efficacy, almost all clinical researchers look for some measure of efficacy to provide support for Phase II trials). This is a common problem in the gene therapy community, where promising results are obtained in inbred mouse models of the disease, but similar results are not observed in the more diverse human population.

1.7.1   Ad-Mediated Clinical Trials for Cancer

Novel gene therapeutics utilizing Ad-based vectors have advanced to 24 Phase II/III or III clinical trials and 2 trials for post-marketing surveillance in phase IV (both Ad-p53). Of these 24 clinical trials utilizing Ad, 17 are directed toward cancer. Eleven of these 17 trials explore the use of Ad encoding p53 to treat a variety of cancers, including ovarian, nasopharyngeal, non-small cell lung carcinoma, head and neck, hepatocellular, glioblastoma multiforme, prostate, and cervical cancer. The two Phase IV studies will assess Ad-p53 for thyroid cancer. Involved in detecting DNA damage, p53 controls cell fate by deciding on whether the damage is repairable, or whether the cell should undergo apoptosis (Lane 1992). Understandably, approximately half the number of tumors contain a mutated p53 which allows the cells to accumulate DNA damage while circumventing activation of the p53 checkpoint during G2 of the cell cycle. Numerous studies have shown that reintroducing a functional copy of p53 into p53-defective cancer cells leads to reestablishment of appropriate damage surveillance and, frequently, death of the cell (Lane, Cheok, and Lain 2010). When examining all human clinical gene therapy trials, including all vector systems, approximately 80 out of the 2210 total trials involve the delivery of p53 for treatment of a variety of cancer types (reviewed in Chen et al. 2014), underscoring the perceived importance of this gene. Nine of the 11 phase II/III or III clinical trials for Ad-p53 involve evaluation in combination with chemotherapy and/or radiation therapy. The available data from early clinical trials indicates that the virus is well tolerated, although some relatively minor adverse events were reported, including transient fever, pain at the injection site, and fatigue. The phase I and II clinical trials with Ad-p53 have used a variety of doses and administration routes, resulting in variable rates of tumor regression or stabilization of disease, ultimately prolonging survival in some patients (Li et al. 2015). Taken together, the data from these early trials were sufficiently compelling to warrant progression to Phase III study.

In October 2003, the Shenzhen SiBono GeneTech biotech company received government approval to produce and distribute their Ad-p53 virus (trademarked as Gendicine) for treatment of head and neck squamous cell carcinoma (Peng 2005). Gendicine was the world's first commercially available gene therapy product. In the 16 months following its official launch in April 2004, 2600 patients were treated with Gendicine, and this number was expected to increase to 50,000 by 2006 (Wilson 2005). The exact number of patients that have been treated with Gendicine is not readily available. Although currently approved only for head and neck squamous cell carcinoma, the product is in late-stage testing for a variety of other cancer indications (Peng 2005). Given the large number of patients that will receive Gendicine, the gene therapy community should look forward to very solid, statistically significant data regarding efficacy and safety of Ad and the Ad-p53 product. As of 2009, data on 2500 Gendicine treated patients had been published (Shi and Zheng 2009).

Of the remaining six cancer-directed trials, three utilize traditional E1-deleted vectors and three involve conditionally replicating vectors. Two of the E1-deleted vectors are designed to specifically target the tumor vasculature, either to inhibit angiogenesis or kill endothelial cells. Endostatin is a 20 kDa C-terminal fragment of collagen XVIII, and has been shown to have potent anti-angiogenic activity (O'Reilly et al. 1997). Ad-mediated delivery of endostatin has shown efficacy in preclinical and early clinical trials in a variety of cancers (Lin et al. 2007; Ye et al. 2014). The second vector that targets vasculature expresses a chimeric Fas and human TNF receptor 1 protein (Fas signaling is activated only upon TNF-α binding to the receptor component of the chimeric protein), under regulation in the vector by the pre-proendothelin-1 promoter (Triozzi and Borden 2011). This vector is being evaluated in combination with bevacizumab for treatment of glioblastoma. Two additional trials deliver classic cell suicide genes, either the herpes simplex virus thymidine kinase gene (HSV TK) in a replication-defective vector or an RCA vector expressing a fusion protein of yeast cytosine deaminase (yCD) and HSV TK. Both these trials are for prostate cancer, and both are in combination with conventional radiation therapy. HSV TK acts through converting the prodrug ganciclovir into a nucleotide analog that causes DNA synthesis chain termination and replication arrest. yCD converts 5-fluorocytidine into 5-fluorouracil, which is an irreversible inhibitor of thymidylate synthase, thus impairing production of dTMP and ultimately significantly impacting the levels of dTTP within the cell. Both vectors showed sufficient efficacy in Phase II study to warrant advancement to Phase III (Teh et al. 2004; Fujita et al. 2006; Freytag et al. 2014).

The final two cancer-directed phase III trials utilize oncolytic vectors. The first is rendered tumor-selective due to placement of E1A expression under regulation by E2F-1 promoter. E2F-1 expression is misregulated in cancer cells that are defective in the retinoblastoma pathway (Sherr and McCormick 2002), thus making this virus selective for replication in these specific cell types. The virus, designated CG0070, also expresses granulocyte–macrophage colony-stimulating factor (GM-CSF), which can aid in establishing anti-tumor immunity (Bristol et al. 2003). The other oncolytic vector in Phase III trial is deficient in E1B, similar to ONYX-015 (Larson et al. 2015).

1.7.2 Ad-MEDIATED CLINICAL TRIALS FOR VASCULAR DISEASE

The remaining seven Ad-mediated clinical trials were directed toward vascular disease, six trials using Ad to deliver fibroblast growth factor-4 (FGF-4) and one exploring the efficacy of vascular endothelial growth factor (VEGF). FGF-4 is involved in the growth and migration of many cell types in the developing vessel wall, and Ad-mediated delivery of FGF-4 showed efficacy in various animal models of ischemia and early human clinical trials to treat myocardial and critical limb ischemia

(Grines et al. 2002, 2003; Matyas et al. 2005). Most patients in these trials experienced only relatively mild adverse side-effects, including minor fever and asymptomatic elevation in liver enzymes in a few patients (Grines et al. 2003; Matyas et al. 2005). However, 14 severe adverse events occurred, including accelerated toe pain, myalgia, and peripheral edema, and several patients required toe or limb amputation as a result of advancement of disease. Interestingly, two Phase IIb/III trials utilizing the Ad-FGF-4 vector uncovered a gender-specific bias in efficacy. Patients (n = 532) were treated using two doses of Ad5-FGF-4, 1×10^9 or 1×10^{10} viral particles per patient, but the trial was halted when a preliminary analysis of the data indicated that the primary end point, an increase in exercise treadmill time (ETT) from baseline at 12 weeks, would not reach significance (Henry et al. 2007). Subsequent additional analysis showed that although in male treated patients the placebo effect was large and not different from Ad5-FGF-4 treated patients, the placebo effect in women was negligible and the treatment effect was significantly greater than placebo. At least one additional Phase III trial has been proposed to test Ad5-FGF-4 as a therapeutic for angina pectoris, which will include examining potential gender-specific effects (Kaski and Consuegra-Sanchez 2013).

In summary, a number of Ad-based vectors have advanced to clinical trial and shown efficacy in human patients. It is hoped that these vectors will provide sufficient efficacy in the Phase III studies to promote their advancement to widespread availability to provide new treatment options for some of these devastating conditions.

1.8   Ad-OTC

With the wide-scale testing and availability of Ad-based vectors for vascular disease or cancer, the future of Ad in gene therapy appears very bright, at least for some applications. However, several years ago, Ad clinical and basic researchers faced serious concerns regarding the use of Ad in gene therapy, and even broader criticisms regarding the validity of the "gene therapy" approach to treating disease. These legitimate questions arose as a result of the first death of a gene therapy clinical trial participant that was directly attributed to delivery of the gene therapy vector. The Phase I trial involved escalating dose delivery of an E1/E4-deleted Ad vector-encoding OTC for treatment of OTC deficiency (OTCD) (Batshaw et al. 1999). Patients with OTCD develop severe hyperammonemia and excessive elevation of ammonia in the brain, which can cause encephalopathy, coma, and brain damage. The OTC gene is located on the X chromosome and males born with no OTC activity usually die within a few weeks of birth. Patients with partial OTC activity are susceptible to hyperammonemic crisis throughout their life, and dietary restriction is only partially effective at managing OTCD. Patients entered into the trial received between 1.4×10^{11} and 3.8×10^{13} total virus particles of the Ad-OTC vector. The second last patient entered in the trial, an asymptomatic female, received a dose of 3.6×10^{13} of Ad-OTC, which was accompanied by only a transient rise in fever and liver enzymes (Raper et al. 2002). The final patient received a similar dose of vector, 3.8×10^{13} virus particles. However, shortly after vector administration, the patient experienced a severe adverse event, with the following report logged with the Recombinant DNA Advisory Committee (RAC) and the National Institutes of Health (NIH 1999):

> Patient death due to adult respiratory distress syndrome, multiple organ failure, and disseminated intravascular coagulation. Within 12 hours of receiving the intrahepatic adenoviral vector administration, patient experienced fever, nausea, and back pain. The following morning after vector administration, patient experienced elevated ammonia levels and jaundice. During days 2–4 after vector injection, patient experienced disseminated intravascular coagulation, adult respiratory distress syndrome, and kidney and liver failure. Patient died four days after vector administration.

After an intensive review of clinical and postmortem findings, it was determined that the patient death was most likely a result of a systemic, Ad-vector induced shock syndrome, culminating in the events described above (Raper et al. 2003). There has also been significant discussion regarding possible clinical procedural issues and conflict of interest (Wilson 2009; Yarborough and Sharp 2009). Immediately following this death, the NIH ordered the suspension of several technically similar gene therapy clinical trials (several more were suspended voluntarily), and the establishment of an RAC Working Group on Ad vector Safety and Toxicity (AdSAT). The report from AdSAT has been released (NIH 2002), and included several recommendations such as improved methods for standardizing vector dose and potency, improved collection and dissemination of data on vector safety and toxicity, and increased clarity of the risks and benefits on the informed consent documents used in gene therapy trials. With respect to Ad vector use in gene therapy, AdSAT concluded that "human gene transfer experiments using Ad-based vectors should continue—with caution (NIH 2002)." Since the release of this report, Ad-mediated gene transfer has continued, and we now have two commercially available gene therapy products based on an Ad platform.

1.9 CONCLUDING REMARKS

The death of a participant in a gene therapy trial was, perhaps, not unexpected. In 1995, an NIH-sponsored panel was asked to evaluate the rationale and potential of gene therapy, and their findings were published as the Orkin-Motulsky Report (NIH 1995). They concluded that, "Somatic gene therapy is a logical and natural progression in the application of fundamental biomedical science to medicine and offers extraordinary potential, in the long-term, for the management and correction of human disease." However, it was clearly recognized that all gene therapy transfer vectors have shortcomings, and that we have an inadequate understanding of the biological interactions between the host and the vector. Ad serves as a prime example of a vector system that has evolved over time in an attempt to reduce or eliminate identified deficiencies in order to improve vector function and safety. Since Ad was first proposed as a gene therapy transfer vehicle, there has been a significant amount of research aimed at characterizing the many interactions between the virus, the infected cell, and the host organism. Our knowledge is far from complete; however, it is hoped that the continued efforts of virologists and gene therapy researchers will improve the safety and efficacy of this gene transfer vehicle for gene therapy applications, and provide new hope for patients afflicted with these devastating, currently incurable diseases.

ACKNOWLEDGMENTS

R.J.P. wishes to extend his gratitude to Dr. Frank L. Graham, a great mentor and friend. Over his entire career, Dr. Graham has provided key contributions that have greatly expanded our knowledge of Ad biology and vectorology, only some of which are highlighted in this chapter. Research in the Parks laboratory is supported by grants from the Canadian Institutes of Health Research (CIHR), the National Sciences and Engineering Research Council (NSERC), and the Cancer Research Society. L.A.N was supported by an Ontario Graduate Scholarship from the Ontario government.

REFERENCES

Ahi, Y. S., D. S. Bangari, and S. K. Mittal. 2011. Adenoviral vector immunity: Its implications and circumvention strategies. *Curr Gene Ther* 11 (4):307–20.

Alba, R., A. C. Bradshaw, L. Coughlan et al. 2010. Biodistribution and retargeting of FX-binding ablated adenovirus serotype 5 vectors. *Blood* 116 (15):2656–64. doi: 10.1182/blood-2009-12-260026.

Alba, R., A. C. Bradshaw, N. Mestre-Frances et al. 2011. Coagulation factor X mediates adenovirus type 5 liver gene transfer in non-human primates (*Microcebus murinus*). *Gene Ther* 19 (1):109–13. doi: 10.1038/gt.2011.87.

Amalfitano, A., and J. S. Chamberlain. 1997. Isolation and characterization of packaging cell lines that coexpress the adenovirus E1, DNA polymerase, and preterminal proteins: Implications for gene therapy. *Gene Ther* 4 (3):258–63.

Amalfitano, A., A. J. McVie-Wylie, H. Hu et al. 1999. Systemic correction of the muscle disorder glycogen storage disease type II after hepatic targeting of a modified adenovirus vector encoding human acid-alpha-glucosidase. *Proc Natl Acad Sci U S A* 96 (16):8861–6.

Amalfitano, A., and R. J. Parks. 2002. Separating fact from fiction: Assessing the potential of modified adenovirus vectors for use in human gene therapy. *Curr Gene Ther* 2 (2):111–33.

Anderson, C. W., M. E. Young, and S. J. Flint. 1989. Characterization of the adenovirus 2 virion protein, mu. *Virology* 172 (2):506–12. doi: 10.1016/0042-6822(89)90193-1.

Aparicio, O., N. Razquin, M. Zaratiegui, I. Narvaiza, and P. Fortes. 2006. Adenovirus virus-associated RNA is processed to functional interfering RNAs involved in virus production. *J Virol* 80 (3):1376–84. doi: 80/3/1376 [pii] 10.1128/JVI.80.3.1376-1384.2006.

Appledorn, D. M., S. Patial, A. McBride et al. 2008. Adenovirus vector-induced innate inflammatory mediators, MAPK signaling, as well as adaptive immune responses are dependent upon both TLR2 and TLR9 *in vivo*. *J Immunol* 181 (3):2134–44. doi: 181/3/2134 [pii].

Armentano, D., M. P. Smith, C. C. Sookdeo et al. 1999. E4ORF3 requirement for achieving long-term transgene expression from the cytomegalovirus promoter in adenovirus vectors. *J Virol* 73 (8):7031–4.

Armentano, D., C. C. Sookdeo, K. M. Hehir et al. 1995. Characterization of an adenovirus gene transfer vector containing an E4 deletion. *Hum Gene Ther* 6 (10):1343–53.

Armentano, D., J. Zabner, C. Sacks et al. 1997. Effect of the E4 region on the persistence of transgene expression from adenovirus vectors. *J Virol* 71 (3):2408–16.

Aurisicchio, L., H. Bujard, W. Hillen et al. 2001. Regulated and prolonged expression of mIFN(alpha) in immunocompetent mice mediated by a helper-dependent adenovirus vector. *Gene Ther* 8 (24):1817–25.

Barker, D. D., and A. J. Berk. 1987. Adenovirus proteins from both E1B reading frames are required for transformation of rodent cells by viral infection and DNA transfection. *Virology* 156 (1):107–21.

Barlan, A. U., T. M. Griffin, K. A. McGuire, and C. M. Wiethoff. 2011. Adenovirus membrane penetration activates the NLRP3 inflammasome. *J Virol* 85 (1):146–55. doi: 10.1128/JVI.01265-10 JVI.01265-10 [pii].

Barr, D., J. Tubb, D. Ferguson et al. 1995. Strain related variations in adenovirally mediated transgene expression from mouse hepatocytes *in vivo*: Comparisons between immunocompetent and immunodeficient inbred strains. *Gene Ther* 2 (2):151–5.

Batshaw, M. L., J. M. Wilson, S. Raper, M. Yudkoff, and M. B. Robinson. 1999. Recombinant adenovirus gene transfer in adults with partial ornithine transcarbamylase deficiency (OTCD). *Hum Gene Ther* 10 (14):2419–37.

Bergelson, J. M., J. A. Cunningham, G. Droguett et al. 1997. Isolation of a common receptor for Coxsackie B viruses and adenoviruses 2 and 5. *Science* 275 (5304):1320–3.

Berget, S. M., C. Moore, and P. A. Sharp. 1977. Spliced segments at the 5′ terminus of adenovirus 2 late mRNA. *Proc Natl Acad Sci U S A* 74 (8):3171–5.

Berk, A. J. 2007. Adenoviridae: The viruses and their replication. In *Fields Virology*, edited by D.M. Knipe and P. M. Howley, 2355–2394. Philadelphia, PA: Lippincott Williams & Wilkins.

Bernt, K. M., S. Ni, Z. Y. Li, D. M. Shayakhmetov, and A. Lieber. 2003. The effect of sequestration by nontarget tissues on anti-tumor efficacy of systemically applied, conditionally replicating adenovirus vectors. *Mol Ther* 8 (5):746–55.

Bett, A. J., L. Prevec, and F. L. Graham. 1993. Packaging capacity and stability of human adenovirus type 5 vectors. *J Virol* 67 (10):5911–21.

Biermann, V., C. Volpers, S. Hussmann et al. 2001. Targeting of high-capacity adenoviral vectors. *Hum Gene Ther* 12 (14):1757–69.

Bischoff, J. R., D. H. Kirn, A. Williams et al. 1996. An adenovirus mutant that replicates selectively in p53-deficient human tumor cells. *Science* 274 (5286):373–6.

Bradshaw, A. C., A. L. Parker, M. R. Duffy et al. 2010. Requirements for receptor engagement during infection by adenovirus complexed with blood coagulation factor X. *PLoS Pathogens* 6 (10):e1001142. doi: 10.1371/journal.ppat.1001142.

Bramson, J.L., N. Grinshtein, R.A. Meulenbroek et al. 2004. Helper-dependent adenoviral vectors containing modified fibre for improved transduction of developing and mature muscle cells. *Hum Gene Ther* 15:179–188.

Bristol, J. A., M. Zhu, H. Ji et al. 2003. *In vitro* and *in vivo* activities of an oncolytic adenoviral vector designed to express GM-CSF. *Mol Ther* 7 (6):755–64. doi: S1525001603001035 [pii].

Brough, D. E., C. Hsu, V. A. Kulesa et al. 1997. Activation of transgene expression by early region 4 is responsible for a high level of persistent transgene expression from adenovirus vectors *in vivo*. *J Virol* 71 (12):9206–13.

Brough, D. E., A. Lizonova, C. Hsu, V. A. Kulesa, and I. Kovesdi. 1996. A gene transfer vector-cell line system for complete functional complementation of adenovirus early regions E1 and E4. *J Virol* 70 (9):6497–501.

Brown, D. T., M. Westphal, B. T. Burlingham, U. Winterhoff, and W. Doerfler. 1975. Structure and composition of the adenovirus type 2 core. *J Virol* 16 (2):366–87.

Brunetti-Pierri, N., and P. Ng. 2008. Progress and prospects: Gene therapy for genetic diseases with helper-dependent adenoviral vectors. *Gene Ther* 15 (8):553–60. doi: 10.1038/gt.2008.14.

Brunetti-Pierri, N., and P. Ng. 2011. Helper-dependent adenoviral vectors for liver-directed gene therapy. *Hum Mol Genet* 20 (R1):R7–13. doi: 10.1093/hmg/ddr143.

Brunetti-Pierri, N., T. Ng, D. Iannitti et al. 2013. Transgene expression up to 7 years in nonhuman primates following hepatic transduction with helper-dependent adenoviral vectors. *Hum Gene Ther* 24 (8):761–5. doi: 10.1089/hum.2013.071.

Brunetti-Pierri, N., D. J. Palmer, A. L. Beaudet et al. 2004. Acute toxicity after high-dose systemic injection of helper-dependent adenoviral vectors into nonhuman primates. *Hum Gene Ther* 15 (1):35–46.

Burcin, M. M., G. Schiedner, S. Kochanek, S. Y. Tsai, and B. W. O'Malley. 1999. Adenovirus-mediated regulable target gene expression *in vivo. Proc Natl Acad Sci U S A* 96 (2):355–60.

Chartier, C., E. Degryse, M. Gantzer et al. 1996. Efficient generation of recombinant adenovirus vectors by homologous recombination in *Escherichia coli. J Virol* 70 (7):4805–10.

Chatterjee, P. K., M. E. Vayda, and S. J. Flint. 1986. Identification of proteins and protein domains that contact DNA within adenovirus nucleoprotein cores by ultraviolet light crosslinking of oligonucleotides 32P-labelled *in vivo. J Mol Biol* 188 (1):23–37.

Chen, G. X., S. Zhang, X. H. He et al. 2014. Clinical utility of recombinant adenoviral human p53 gene therapy: Current perspectives. *Onco Targets Ther* 7:1901–9. doi: 10.2147/OTT.S50483 ott-7-1901 [pii].

Cheng, P. H., S. L. Wechman, K. M. McMasters, and H. S. Zhou. 2015. Oncolytic replication of E1b-deleted adenoviruses. *Viruses* 7 (11):5767–79. doi: 10.3390/v7112905 v7112905 [pii].

Choi, J. W., J. S. Lee, S. W. Kim, and C. O. Yun. 2012. Evolution of oncolytic adenovirus for cancer treatment. *Adv Drug Deliv Rev* 64 (8):720–9. doi: 10.1016/j.addr.2011.12.011 S0169-409X(11)00305-X [pii].

Chow, L. T., R. E. Gelinas, T. R. Broker, and R. J. Roberts. 1977. An amazing sequence arrangement at the 5′ ends of adenovirus 2 messenger RNA. *Cell* 12 (1):1–8.

Christ, M., B. Louis, F. Stoeckel et al. 2000. Modulation of the inflammatory properties and hepatotoxicity of recombinant adenovirus vectors by the viral E4 gene products. *Hum Gene Ther* 11 (3):415–27.

Christensen, J. B., S. A. Byrd, A. K. Walker et al. 2008. Presence of the adenovirus IVa2 protein at a single vertex of the mature virion. *J Virol* 82 (18):9086–93.

Cots, D., A. Bosch, and M. Chillon. 2013. Helper dependent adenovirus vectors: Progress and future prospects. *Curr Gene Ther* 13 (5):370–81. doi: CGT-58030 [pii].

Croyle, M. A., H. T. Le, K. D. Linse et al. 2005. PEGylated helper-dependent adenoviral vectors: Highly efficient vectors with an enhanced safety profile. *Gene Ther* 12 (7):579–87.

Crystal, R. G., A. Jaffe, S. Brody et al. 1995. A phase 1 study, in cystic fibrosis patients, of the safety, toxicity, and biological efficacy of a single administration of a replication deficient, recombinant adenovirus carrying the cDNA of the normal cystic fibrosis transmembrane conductance regulator gene in the lung. *Hum Gene Ther* 6 (5):643–66.

Crystal, R. G., N. G. McElvaney, M. A. Rosenfeld et al. 1994. Administration of an adenovirus containing the human CFTR cDNA to the respiratory tract of individuals with cystic fibrosis. *Nat Genet* 8 (1):42–51.

Danthinne, X., and M. J. Imperiale. 2000. Production of first generation adenovirus vectors: A review. *Gene Ther* 7 (20):1707–14.

Davison, A. J., M. Benko, and B. Harrach. 2003. Genetic content and evolution of adenoviruses. *J Gen Virol* 84 (Pt 11):2895–908.

Dedieu, J. F., E. Vigne, C. Torrent et al. 1997. Long-term gene delivery into the livers of immunocompetent mice with E1/E4-defective adenoviruses. *J Virol* 71 (6):4626–37.

Di Paolo, N. C., E. A. Miao, Y. Iwakura et al. 2009. Virus binding to a plasma membrane receptor triggers interleukin-1 alpha-mediated proinflammatory macrophage response *in vivo. Immunity* 31 (1):110–21. doi: 10.1016/j.immuni.2009.04.015.

Doronin, K., J. W. Flatt, N. C. Di Paolo et al. 2012. Coagulation factor X activates innate immunity to human species C adenovirus. *Science* 338 (6108):795–8. doi: 10.1126/science.1226625.

Ehrhardt, A., H. Xu, and M. A. Kay. 2003. Episomal persistence of recombinant adenoviral vector genomes during the cell cycle *in vivo. J Virol* 77 (13):7689–95.

Engelhardt, J. F., R. H. Simon, Y. Yang et al. 1993. Adenovirus-mediated transfer of the CFTR gene to lung of nonhuman primates: Biological efficacy study. *Hum Gene Ther* 4 (6):759–69.

Engelhardt, J. F., Y. Yang, L. D. Stratford-Perricaudet et al. 1993. Direct gene transfer of human CFTR into human bronchial epithelia of xenografts with E1-deleted adenoviruses. *Nat Genet* 4 (1):27–34.

Everett, R. S., B. L. Hodges, E. Y. Ding et al. 2003. Liver toxicities typically induced by first-generation adenoviral vectors can be reduced by use of E1, E2b-deleted adenoviral vectors. *Hum Gene Ther* 14 (18):1715–26.

Everitt, E., B. Sundquist, U. Pettersson, and L. Philipson. 1973. Structural proteins of adenoviruses. X. Isolation and topography of low molecular weight antigens from the virion of adenovirus type 2. *Virology* 52 (1):130–47.

Fallaux, F. J., A. Bout, V. van d et al. 1998. New helper cells and matched early region 1-deleted adenovirus vectors prevent generation of replication-competent adenoviruses. *Hum Gene Ther* 9 (13):1909–17.

Fang, B., H. Wang, G. Gordon et al. 1996. Lack of persistence of E1-recombinant adenoviral vectors containing a temperature-sensitive E2A mutation in immunocompetent mice and hemophilia B dogs. *Gene Ther* 3 (3):217–22.

Fisher, K. J., H. Choi, J. Burda, S. J. Chen, and J. M. Wilson. 1996. Recombinant adenovirus deleted of all viral genes for gene therapy of cystic fibrosis. *Virology* 217 (1):11–22.

Freytag, S. O., H. Stricker, M. Lu et al. 2014. Prospective randomized phase 2 trial of intensity modulated radiation therapy with or without oncolytic adenovirus-mediated cytotoxic gene therapy in intermediate-risk prostate cancer. *Int J Radiat Oncol Biol Phys* 89 (2):268–76. doi: 10.1016/j.ijrobp.2014.02.034 S0360-3016(14)00278-8 [pii].

Fujita, T., B. S. Teh, T. L. Timme et al. 2006. Sustained long-term immune responses after *in situ* gene therapy combined with radiotherapy and hormonal therapy in prostate cancer patients. *Int J Radiat Oncol Biol Phys* 65 (1):84–90. doi: 10.1016/j.ijrobp.2005.11.009 S0360-3016(05)02867-1 [pii].

Gao, G. P., R. K. Engdahl, and J. M. Wilson. 2000. A cell line for high-yield production of E1-deleted adenovirus vectors without the emergence of replication-competent virus. *Hum Gene Ther* 11 (1):213–9.

Gao, G. P., Y. Yang, and J. M. Wilson. 1996. Biology of adenovirus vectors with E1 and E4 deletions for liver- directed gene therapy. *J Virol* 70 (12):8934–43.

Giberson, A. N., A. R. Davidson, and R. J. Parks. 2012. Chromatin structure of adenovirus DNA throughout infection. *Nucl Acids Res* 40 (6):2369–76. doi: 10.1093/nar/gkr1076.

Glasgow, J. N., G. J. Bauerschmitz, D. T. Curiel, and A. Hemminki. 2004. Transductional and transcriptional targeting of adenovirus for clinical applications. *Curr Gene Ther* 4 (1):1–14.

Goldman, M., Q. Su, and J. M. Wilson. 1996. Gradient of RGD-dependent entry of adenoviral vector in nasal and intrapulmonary epithelia: Implications for gene therapy of cystic fibrosis. *Gene Ther* 3 (9):811–8.

Goncalves, M. A., I. van der Velde, J. M. Janssen et al. 2002. Efficient generation and amplification of high-capacity adeno-associated virus/adenovirus hybrid vectors. *J Virol* 76 (21):10734–44.

Gorziglia, M. I., M. J. Kadan, S. Yei et al. 1996. Elimination of both E1 and E2 from adenovirus vectors further improves prospects for *in vivo* human gene therapy. *J Virol* 70 (6):4173–8.

Graham, F. L. 1984. Covalently closed circles of human adenovirus DNA are infectious. *Embo J* 3 (12):2917–22.

Graham, F. L., P. J. Abrahams, C. Mulder et al. 1975. Studies on *in vitro* transformation by DNA and DNA fragments of human adenoviruses and simian virus 40. *Cold Spring Harb Symp Quant Biol* 39 Pt 1:637–50.

Graham, F. L., and L. Prevec. 1995. Methods for construction of adenovirus vectors. *Mol Biotechnol* 3 (3):207–20.

Graham, F. L., J. Smiley, W. C. Russell, and R. Nairn. 1977. Characteristics of a human cell line transformed by DNA from human adenovirus type 5. *J Gen Virol* 36 (1):59–74.

Graham, F. L., and A. J. van der Eb. 1973. A new technique for the assay of infectivity of human adenovirus 5 DNA. *Virology* 52 (2):456–67.

Graham, F. L., A. J. van der Eb, and H. L. Heijneker. 1974. Size and location of the transforming region in human adenovirus type 5 DNA. *Nature* 251 (5477):687–91.

Greber, U. F. 1998. Virus assembly and disassembly: The adenovirus cysteine protease as a trigger factor. *Rev Med Virol* 8 (4):213–22.

Greber, U. F., M. Willetts, P. Webster, and A. Helenius. 1993. Stepwise dismantling of adenovirus 2 during entry into cells. *Cell* 75 (3):477–86.

Grines, C. L., M. W. Watkins, G. Helmer et al. 2002. Angiogenic Gene Therapy (AGENT) trial in patients with stable angina pectoris. *Circulation* 105 (11):1291–7.

Grines, C. L., M. W. Watkins, J. J. Mahmarian et al. 2003. A randomized, double-blind, placebo-controlled trial of Ad5FGF-4 gene therapy and its effect on myocardial perfusion in patients with stable angina. *J Am Coll Cardiol* 42 (8):1339–47.

Guo, Z. S., L. H. Wang, R. C. Eisensmith, and S. L. Woo. 1996. Evaluation of promoter strength for hepatic gene expression *in vivo* following adenovirus-mediated gene transfer. *Gene Ther* 3 (9):802–10.

Haecker, S. E., H. H. Stedman, R. J. Balice-Gordon et al. 1996. In vivo expression of full-length human dystrophin from adenoviral vectors deleted of all viral genes. *Hum Gene Ther* 7 (15):1907–14.

Hanahan, D., and Y. Gluzman. 1984. Rescue of functional replication origins from embedded configurations in a plasmid carrying the adenovirus genome. *Mol Cell Biol* 4 (2):302–9.

Harada, J. N., and A. J. Berk. 1999. p53-Independent and -dependent requirements for E1B-55K in adenovirus type 5 replication. *J Virol* 73 (7):5333–44.

Hardy, S., M. Kitamura, T. Harris-Stansil, Y. Dai, and M. L. Phipps. 1997. Construction of adenovirus vectors through Cre-lox recombination. *J Virol* 71 (3):1842–9.

Hassell, J. A., E. Lukanidin, G. Fey, and J. Sambrook. 1978. The structure and expression of two defective adenovirus 2/simian virus 40 hybrids. *J Mol Biol* 120 (2):209–47.

Havenga, M. J., A. A. Lemckert, O. J. Ophorst et al. 2002. Exploiting the natural diversity in adenovirus tropism for therapy and prevention of disease. *J Virol* 76 (9):4612–20.

He, T. C., S. Zhou, L. T. da Costa et al. 1998. A simplified system for generating recombinant adenoviruses. *Proc Natl Acad Sci U S A* 95 (5):2509–14.

Hehir, K. M., D. Armentano, L. M. Cardoza et al. 1996. Molecular characterization of replication-competent variants of adenovirus vectors and genome modifications to prevent their occurrence. *J Virol* 70 (12):8459–67.

Hendrickx, R., N. Stichling, J. Koelen et al. 2014. Innate immunity to adenovirus. *Hum Gene Ther* 25 (4):265–84. doi: 10.1089/hum.2014.001.

Henry, T. D., C. L. Grines, M. W. Watkins et al. 2007. Effects of Ad5FGF-4 in patients with angina: An analysis of pooled data from the AGENT-3 and AGENT-4 trials. *J Am Coll Cardiol* 50 (11):1038–46. doi: 10.1016/j.jacc.2007.06.010 S0735-1097(07)01999-7 [pii].

Hilleman, M. R., and J. H. Werner. 1954. Recovery of new agents from patients with acute respiratory illness. *Proc Soc Exp Biol Med* 85:183–8.

Hillgenberg, M., H. Tonnies, and M. Strauss. 2001. Chromosomal integration pattern of a helper-dependent minimal adenovirus vector with a selectable marker inserted into a 27.4- kilobase genomic stuffer. *J Virol* 75 (20):9896–908.

Horwitz, M. S. 2004. Function of adenovirus E3 proteins and their interactions with immunoregulatory cell proteins. *J Gene Med* 6 (Suppl 1):S172–83. doi: 10.1002/jgm.495.

Hu, H., D. Serra, and A. Amalfitano. 1999. Persistence of an [E1-, polymerase-] adenovirus vector despite transduction of a neoantigen into immune-competent mice. *Hum Gene Ther* 10 (3):355–64.

JGM. 2015. Gene therapy clinical trials worldwide. Accessed January 4. http://www.abedia. com/wiley/vectors.php.

Jiang, H., Z. Wang, D. Serra, M. M. Frank, and A. Amalfitano. 2004. Recombinant adenovirus vectors activate the alternative complement pathway, leading to the binding of human complement protein C3 independent of anti-ad antibodies. *Mol Ther* 10 (6):1140–2.

Kafri, T., D. Morgan, T. Krahl et al. 1998. Cellular immune response to adenoviral vector infected cells does not require *de novo* viral gene expression: Implications for gene therapy. *Proc Natl Acad Sci U S A* 95 (19):11377–82.

Kalyuzhniy, O., N. C. Di Paolo, M. Silvestry et al. 2008. Adenovirus serotype 5 hexon is critical for virus infection of hepatocytes *in vivo. Proc Natl Acad Sci U S A* 105 (14):5483–8.

Kaski, J. C., and L. Consuegra-Sanchez. 2013. Evaluation of ASPIRE trial: A Phase III pivotal registration trial, using intracoronary administration of Generx (Ad5FGF4) to treat patients with recurrent angina pectoris. *Expert Opin Biol Ther* 13 (12):1749–53. doi: 10.1517/14712598.2013.827656.

Kass-Eisler, A., L. Leinwand, J. Gall, B. Bloom, and E. Falck-Pedersen. 1996. Circumventing the immune response to adenovirus-mediated gene therapy. *Gene Ther* 3 (2):154–62.

Kennedy, M. A., and R. J. Parks. 2009. Adenovirus virion stability and the viral genome: Size matters. *Mol Ther* 17 (10):1664–6.

Kim, I. H., A. Jozkowicz, P. A. Piedra, K. Oka, and L. Chan. 2001. Lifetime correction of genetic deficiency in mice with a single injection of helper-dependent adenoviral vector. *Proc Natl Acad Sci U S A* 98 (23):13282–7.

Kirn, D. H. 2006. The end of the beginning: Oncolytic virotherapy achieves clinical proof-of-concept. *Mol Ther* 13:237–8.

Kochanek, S., P. R. Clemens, K. Mitani et al. 1996. A new adenoviral vector: Replacement of all viral coding sequences with 28 kb of DNA independently expressing both full-length dystrophin and beta-galactosidase. *Proc Natl Acad Sci U S A* 93 (12):5731–6.

Kozarsky, K. F., and J. M. Wilson. 1993. Gene therapy: Adenovirus vectors. *Curr Opin Genet Dev* 3 (3):499–503.

Krougliak, V., and F. L. Graham. 1995. Development of cell lines capable of complementing E1, E4, and protein IX defective adenovirus type 5 mutants. *Hum Gene Ther* 6 (12):1575–86.

Kubo, S., and K. Mitani. 2003. A new hybrid system capable of efficient lentiviral vector production and stable gene transfer mediated by a single helper-dependent adenoviral vector. *J Virol* 77 (5):2964–71.

Kumar-Singh, R., and J. S. Chamberlain. 1996. Encapsidated adenovirus minichromosomes allow delivery and expression of a 14 kb dystrophin cDNA to muscle cells. *Hum Mol Genet* 5 (7):913–21.

Lane, D. P. 1992. Cancer. p53, guardian of the genome. *Nature* 358 (6381):15–6.

Lane, D. P., C. F. Cheok, and S. Lain. 2010. p53-based cancer therapy. *Cold Spring Harb Perspect Biol* 2 (9):a001222. doi: 10.1101/cshperspect.a001222 cshperspect.a001222 [pii].

Larson, C., B. Oronsky, J. Scicinski et al. 2015. Going viral: A review of replication-selective oncolytic adenoviruses. *Oncotarget* 6 (24):19976–89. doi: 10.18632/oncotarget.5116 5116 [pii].

Leopold, P. L., B. Ferris, I. Grinberg et al. 1998. Fluorescent virions: Dynamic tracking of the pathway of adenoviral gene transfer vectors in living cells. *Hum Gene Ther* 9 (3):367–78.

Li, Y., B. Li, C. J. Li, and L. J. Li. 2015. Key points of basic theories and clinical practice in rAd-p53 (Gendicine) gene therapy for solid malignant tumors. *Expert Opin Biol Ther* 15 (3):437–54. doi: 10.1517/14712598.2015.990882.

Lieber, A., C. Y. He, I. Kirillova, and M. A. Kay. 1996. Recombinant adenoviruses with large deletions generated by Cre-mediated excision exhibit different biological properties compared with first- generation vectors *in vitro* and *in vivo*. *J Virol* 70 (12):8944–60.

Lin, X., H. Huang, S. Li et al. 2007. A phase I clinical trial of an adenovirus-mediated endostatin gene (E10A) in patients with solid tumors. *Cancer Biol Ther* 6 (5):648–53. doi: 4004 [pii].

Lin, Y. C., M. Boone, L. Meuris et al. 2014. Genome dynamics of the human embryonic kidney 293 lineage in response to cell biology manipulations. *Nat Commun* 5:4767. doi: 10.1038/ncomms5767 ncomms5767 [pii].

Liu, H., L. Jin, S. B. Koh et al. 2010. Atomic structure of human adenovirus by cryo-EM reveals interactions among protein networks. *Science* 329 (5995):1038–43.

Lochmuller, H., A. Jani, J. Huard et al. 1994. Emergence of early region 1-containing replication-competent adenovirus in stocks of replication-defective adenovirus recombinants (delta E1 + delta E3) during multiple passages in 293 cells. *Hum Gene Ther* 5 (12):1485–91.

Louis, N., C. Evelegh, and F. L. Graham. 1997. Cloning and sequencing of the cellular-viral junctions from the human adenovirus type 5 transformed 293 cell line. *Virology* 233 (2):423–9.

Lusky, M., M. Christ, K. Rittner et al. 1998. *In vitro* and *in vivo* biology of recombinant adenovirus vectors with E1, E1/E2A, or E1/E4 deleted. *J Virol* 72 (3):2022–32.

Mack, C. A., W. R. Song, H. Carpenter et al. 1997. Circumvention of anti-adenovirus neutralizing immunity by administration of an adenoviral vector of an alternate serotype. *Hum Gene Ther* 8 (1):99–109.

Mastrangeli, A., B. G. Harvey, J. Yao et al. 1996. "Sero-switch" adenovirus-mediated *in vivo* gene transfer: Circumvention of anti-adenovirus humoral immune defenses against repeat adenovirus vector administration by changing the adenovirus serotype. *Hum Gene Ther* 7 (1):79–87.

Matsumoto, K., K. Nagata, M. Ui, and F. Hanaoka. 1993. Template activating factor I, a novel host factor required to stimulate the adenovirus core DNA replication. *J Biol Chem* 268 (14):10582–7.

Matyas, L., K. L. Schulte, J. A. Dormandy et al. 2005. Arteriogenic gene therapy in patients with unreconstructable critical limb ischemia: A randomized, placebo-controlled clinical trial of adenovirus 5-delivered fibroblast growth factor-4. *Hum Gene Ther* 16 (10):1202–11.

McConnell, M. J., and M. J. Imperiale. 2004. Biology of adenovirus and its use as a vector for gene therapy. *Hum Gene Ther* 15 (11):1022–33. doi: 10.1089/hum.2004.15.1022.

McCormick, F. 2003. Cancer-specific viruses and the development of ONYX-015. *Cancer Biol Ther* 2 (4 Suppl 1):S157–60.

McVie-Wylie, A. J., E. Y. Ding, T. Lawson et al. 2003. Multiple muscles in the AMD quail can be "cross-corrected" of pathologic glycogen accumulation after intravenous injection of an [E1-, polymerase-] adenovirus vector encoding human acid-alpha-glucosidase. *J Gene Med* 5 (5):399–406.

Mellman, I. 1992. The importance of being acidic: The role of acidification in intracellular membrane traffic. *J Exp Biol* 172 (Pt 1):39–45.

Michou, A. I., L. Santoro, M. Christ et al. 1997. Adenovirus-mediated gene transfer: Influence of transgene, mouse strain and type of immune response on persistence of transgene expression. *Gene Ther* 4 (5):473–82.

Mirza, M. A., and J. Weber. 1982. Structure of adenovirus chromatin. *Biochim Biophys Acta* 696 (1):76–86.

Mitani, K., F. L. Graham, C. T. Caskey, and S. Kochanek. 1995. Rescue, propagation, and partial purification of a helper virus-dependent adenovirus vector. *Proc Natl Acad Sci U S A* 92 (9):3854–8.

Mitchell, R. S., B. F. Beitzel, A. R. Schroder et al. 2004. Retroviral DNA integration: ASLV, HIV, and MLV show distinct target site preferences. *PLoS Biol* 2 (8):E234.

Mok, H., D. J. Palmer, P. Ng, and M. A. Barry. 2005. Evaluation of polyethylene glycol modification of first-generation and helper-dependent adenoviral vectors to reduce innate immune responses. *Mol Ther* 11 (1):66–79.

Morral, N., W. O'Neal, K. Rice et al. 1999. Administration of helper-dependent adenoviral vectors and sequential delivery of different vector serotype for long-term liver-directed gene transfer in baboons. *Proc Natl Acad Sci U S A* 96 (22):12816–21.

Morral, N., W. K. O'Neal, K. Rice et al. 2002. Lethal toxicity, severe endothelial injury, and a threshold effect with high doses of an adenoviral vector in baboons. *Hum Gene Ther* 13 (1):143–54.

Morris, S. J., and K. N. Leppard. 2009. Adenovirus serotype 5 L4-22K and L4-33K proteins have distinct functions in regulating late gene expression. *J Virol* 83 (7):3049–58. doi: 10.1128/JVI.02455-08 JVI.02455-08 [pii].

Morris, S. J., G. E. Scott, and K. N. Leppard. 2010. Adenovirus late-phase infection is controlled by a novel L4 promoter. *J Virol* 84 (14):7096–104. doi: 10.1128/JVI.00107-10 JVI.00107-10 [pii].

Morsy, M. A., M. Gu, S. Motzel et al. 1998. An adenoviral vector deleted for all viral coding sequences results in enhanced safety and extended expression of a leptin transgene. *Proc Natl Acad Sci U S A* 95 (14):7866–71.

Murakami, P., M. Havenga, F. Fawaz et al. 2004. Common structure of rare replication-deficient e1-positive particles in adenoviral vector batches. *J Virol* 78 (12):6200–8.

Murakami, P., E. Pungor, J. Files et al. 2002. A single short stretch of homology between adenoviral vector and packaging cell line can give rise to cytopathic effect-inducing, helper-dependent E1-positive particles. *Hum Gene Ther* 13 (8):909–20.

Muruve, D. A. 2004. The innate immune response to adenovirus vectors. *Hum Gene Ther* 15 (12):1157–66.

Muruve, D. A., M. J. Barnes, I. E. Stillman, and T. A. Libermann. 1999. Adenoviral gene therapy leads to rapid induction of multiple chemokines and acute neutrophil-dependent hepatic injury *in vivo*. *Hum Gene Ther* 10 (6):965–76.

Muruve, D. A., M. J. Cotter, A. K. Zaiss et al. 2004. Helper-dependent adenovirus vectors elicit intact innate but attenuated adaptive host immune responses *in vivo*. *J Virol* 78 (11):5966–72.

Muruve, D. A., V. Petrilli, A. K. Zaiss et al. 2008. The inflammasome recognizes cytosolic microbial and host DNA and triggers an innate immune response. *Nature* 452 (7183):103–7.

Nakai, H., E. Montini, S. Fuess et al. 2003. AAV serotype 2 vectors preferentially integrate into active genes in mice. *Nat Genet* 34 (3):297–302.

Nayak, S., and R. W. Herzog. 2010. Progress and prospects: Immune responses to viral vectors. *Gene Ther* 17 (3):295–304. doi: 10.1038/gt.2009.148 gt2009148 [pii].

Ng, P., C. Beauchamp, C. Evelegh, R. Parks, and F. L. Graham. 2001. Development of a FLP/frt system for generating helper-dependent adenoviral vectors. *Mol Ther* 3 (5 Pt 1): 809–15.

Ng, P., D. T. Cummings, C. M. Evelegh, and F. L. Graham. 2000. Yeast recombinase FLP functions effectively in human cells for construction of adenovirus vectors. *Biotechniques* 29 (3):524–6, 528.

Ng, P., R. J. Parks, D. T. Cummings et al. 1999. A high-efficiency Cre/loxP-based system for construction of adenoviral vectors. *Hum Gene Ther* 10 (16):2667–72.

Ng, P., R. J. Parks, D. T. Cummings, C. M. Evelegh, and F. L. Graham. 2000. An enhanced system for construction of adenoviral vectors by the two-plasmid rescue method. *Hum Gene Ther* 11 (5):693–9.

Nicol, C. G., D. Graham, W. H. Miller et al. 2004. Effect of adenovirus serotype 5 fiber and penton modifications on *in vivo* tropism in rats. *Mol Ther: J Am Soc Gene Ther* 10 (2):344–54. doi: 10.1016/j.ymthe.2004.05.020.

NIH. 1995. Report and recommendation of the panel to assess the NIH investment in research on gene therapy. http://osp.od.nih.gov/office-biotechnology-activities/orkin-motulsky-report.

NIH. 1999. Department of Health and Human Services, Recombinant DNA Advisory Committee, National Institutes of Health, Minutes of Symposium and Meeting, December 8–10, 1999. *Hum Gene Ther* 11:1591–1621.

NIH. 2002. Assessment of adenoviral vector safety and toxicity: Report of the national institutes of health recombinant DNA advisory committee. *Hum Gene Ther* 13:3–13.

Ohbayashi, F., M. A. Balamotis, A. Kishimoto et al. 2005. Correction of chromosomal mutation and random integration in embryonic stem cells with helper-dependent adenoviral vectors. *Proc Natl Acad Sci U S A* 102 (38):13628–33.

Okuwaki, M., and K. Nagata. 1998. Template activating factor-I remodels the chromatin structure and stimulates transcription from the chromatin template. *J Biol Chem* 273 (51):34511–8.

O'Malley, R. P., T. M. Mariano, J. Siekierka, and M. B. Mathews. 1986. A mechanism for the control of protein synthesis by adenovirus VA RNAI. *Cell* 44 (3):391–400. doi: 0092-8674(86)90460-5 [pii].

O'Reilly, M. S., T. Boehm, Y. Shing et al. 1997. Endostatin: An endogenous inhibitor of angiogenesis and tumor growth. *Cell* 88 (2):277–85. doi: S0092-8674(00)81848-6 [pii].

Ornelles, D. A., and T. Shenk. 1991. Localization of the adenovirus early region 1B 55-kilodalton protein during lytic infection: Association with nuclear viral inclusions requires the early region 4 34-kilodalton protein. *J Virol* 65 (1):424–9.

O'Shea, C. C., L. Johnson, B. Bagus et al. 2004. Late viral RNA export, rather than p53 inactivation, determines ONYX-015 tumor selectivity. *Cancer Cell* 6 (6):611–23.

Palmer, D., and P. Ng. 2003. Improved system for helper-dependent adenoviral vector production. *Mol Ther* 8 (5):846–52.

Palmer, D. J., and P. Ng. 2005. Helper-dependent adenoviral vectors for gene therapy. *Hum Gene Ther* 16 (1):1–16.

Parks, R. J., J. L. Bramson, Y. Wan, C. L. Addison, and F. L. Graham. 1999. Effects of stuffer DNA on transgene expression from helper-dependent adenovirus vectors. *J Virol* 73 (10):8027–34.

Parks, R. J., L. Chen, M. Anton et al. 1996. A helper-dependent adenovirus vector system: Removal of helper virus by Cre-mediated excision of the viral packaging signal. *Proc Natl Acad Sci U S A* 93 (24):13565–70.

Parks, R. J., C. M. Evelegh, and F. L. Graham. 1999. Use of helper-dependent adenoviral vectors of alternative serotypes permits repeat vector administration. *Gene Ther* 6 (9):1565–73.

Parks, R. J., and F. L. Graham. 1997. A helper-dependent system for adenovirus vector production helps define a lower limit for efficient DNA packaging. *J Virol* 71 (4):3293–8.

Pastore, L., N. Morral, H. Zhou et al. 1999. Use of a liver-specific promoter reduces immune response to the transgene in adenoviral vectors. *Hum Gene Ther* 10 (11):1773–81.

Peng, Z. 2005. Current status of gendicine in China: Recombinant human Ad-p53 agent for treatment of cancers. *Hum Gene Ther* 16 (9):1016–27.

Perez-Berna, A. J., R. Marabini, S. H. Scheres et al. 2009. Structure and uncoating of immature adenovirus. *J Mol Biol* 392 (2):547–57. doi: 10.1016/j.jmb.2009.06.057 S0022-2836(09)00787-6 [pii].

Perez-Berna, A. J., S. Marion, F. J. Chichon et al. 2015. Distribution of DNA-condensing protein complexes in the adenovirus core. *Nucl Acids Res* 43 (8):4274–83. doi: 10.1093/nar/gkv187 gkv187 [pii].

Pilder, S., M. Moore, J. Logan, and T. Shenk. 1986. The adenovirus E1B-55K transforming polypeptide modulates transport or cytoplasmic stabilization of viral and host cell mRNAs. *Mol Cell Biol* 6 (2):470–6.

Raper, S. E., N. Chirmule, F. S. Lee et al. 2003. Fatal systemic inflammatory response syndrome in a ornithine transcarbamylase deficient patient following adenoviral gene transfer. *Mol Genet Metab* 80 (1–2):148–58.

Raper, S. E., Z. J. Haskal, X. Ye et al. 1998. Selective gene transfer into the liver of nonhuman primates with E1-deleted, E2A-defective, or E1-E4 deleted recombinant adenoviruses. *Hum Gene Ther* 9 (5):671–9.

Raper, S. E., M. Yudkoff, N. Chirmule et al. 2002. A pilot study of *in vivo* liver-directed gene transfer with an adenoviral vector in partial ornithine transcarbamylase deficiency. *Hum Gene Ther* 13 (1):163–75.

Recchia, A., R. J. Parks, S. Lamartina et al. 1999. Site-specific integration mediated by a hybrid adenovirus/adeno-associated virus vector. *Proc Natl Acad Sci U S A* 96 (6):2615–20.

Reddy, V. S., S. K. Natchiar, P. L. Stewart, and G. R. Nemerow. 2010. Crystal structure of human adenovirus at 3.5 A resolution. *Science* 329 (5995):1071–5.

Reddy, V. S., and G. R. Nemerow. 2014. Structures and organization of adenovirus cement proteins provide insights into the role of capsid maturation in virus entry and infection. *Proc Natl Acad Sci U S A* 111 (32):11715–20. doi: 10.1073/pnas.1408462111 1408462111 [pii].

Rosenfeld, M. A., K. Yoshimura, B. C. Trapnell et al. 1992. In vivo transfer of the human cystic fibrosis transmembrane conductance regulator gene to the airway epithelium. *Cell* 68 (1):143–55.

Ross, P. J., and R. J. Parks. 2003. Oncolytic adenovirus: Getting there is half the battle. *Mol Ther* 8 (5):705–6.

Rowe, W. P., R. J. Huebner, L. K. Gilmore, R. H. Parrott, and T. G. Ward. 1953. Isolation of a cytopathogenic agent from human adenoids undergoing spontaneous degeneration in tissue culture. *Proc Soc Exp Biol Med* 84:570–3.

Ruben, M., S. Bacchetti, and F. Graham. 1983. Covalently closed circles of adenovirus 5 DNA. *Nature* 301 (5896):172–4.

Russell, W. C. 2009. Adenoviruses: Update on structure and function. *J Gen Virol* 90 (Pt 1): 1–20.

Saha, B., C. M. Wong, and R. J. Parks. 2014. The adenovirus genome contributes to the structural stability of the virion. *Viruses* 6 (9):3563–83. doi: 10.3390/v6093563 v6093563 [pii].

Sandig, V., R. Youil, A. J. Bett et al. 2000. Optimization of the helper-dependent adenovirus system for production and potency *in vivo*. *Proc Natl Acad Sci U S A* 97 (3):1002–7.

Sargent, K., R. A. Meulenbroek, and R. J. Parks. 2004. Activation of adenoviral gene expression by protein IX is not required for efficient virus replication. *J Virol* 78:5032–7.

Schiedner, G., S. Hertel, M. Johnston et al. 2002. Variables affecting *in vivo* performance of high-capacity adenovirus vectors. *J Virol* 76 (4):1600–9.

Schiedner, G., S. Hertel, and S. Kochanek. 2000. Efficient transformation of primary human amniocytes by E1 functions of Ad5: Generation of new cell lines for adenoviral vector production. *Hum Gene Ther* 11 (15):2105–16.

Schiedner, G., N. Morral, R. J. Parks et al. 1998. Genomic DNA transfer with a high-capacity adenovirus vector results in improved *in vivo* gene expression and decreased toxicity. *Nat Genet* 18 (2):180–3.

Schnell, M. A., Y. Zhang, J. Tazelaar et al. 2001. Activation of innate immunity in nonhuman primates following intraportal administration of adenoviral vectors. *Mol Ther* 3 (5 Pt 1): 708–22.

Schulte, M., M. Sorkin, S. Al-Benna et al. 2013. Innate immune response after adenoviral gene delivery into skin is mediated by AIM2, NALP3, DAI and mda5. *Springerplus* 2 (1):234. doi: 10.1186/2193-1801-2-234 302 [pii].

Shaw, G., S. Morse, M. Ararat, and F. L. Graham. 2002. Preferential transformation of human neuronal cells by human adenoviruses and the origin of HEK 293 cells. *Faseb J* 16 (8):869–71.

Shayakhmetov, D. M., A. Gaggar, S. Ni, Z. Y. Li, and A. Lieber. 2005. Adenovirus binding to blood factors results in liver cell infection and hepatotoxicity. *J Virol* 79 (12):7478–91.

Sherr, C. J., and F. McCormick. 2002. The RB and p53 pathways in cancer. *Cancer Cell* 2 (2):103–12. doi: S1535610802001022 [pii].

Shi, C. X., M. Hitt, P. Ng, and F. L. Graham. 2002. Superior tissue-specific expression from tyrosinase and prostate-specific antigen promoters/enhancers in helper-dependent compared with first-generation adenoviral vectors. *Hum Gene Ther* 13 (2):211–24.

Shi, C. X., M. A. Long, L. Liu et al. 2004. The human SCGB2A2 (mammaglobin-1) promoter/enhancer in a helper-dependent adenovirus vector directs high levels of transgene expression in mammary carcinoma cells but not in normal nonmammary cells. *Mol Ther* 10 (4):758–67.

Shi, J., and D. Zheng. 2009. An update on gene therapy in China. *Curr Opin Mol Ther* 11 (5):547–53.

Simon, R. H., J. F. Engelhardt, Y. Yang et al. 1993. Adenovirus-mediated transfer of the CFTR gene to lung of nonhuman primates: Toxicity study. *Hum Gene Ther* 4 (6):771–80.

Smith, J. S., Z. Xu, J. Tian, S. C. Stevenson, and A. P. Byrnes. 2008. Interaction of systemically delivered adenovirus vectors with Kupffer cells in mouse liver. *Hum Gene Ther* 19 (5):547–54. doi: 10.1089/hum.2008.004.

Smith, T. A., N. Idamakanti, M. L. Rollence et al. 2003. Adenovirus serotype 5 fiber shaft influences *in vivo* gene transfer in mice. *Hum Gene Ther* 14 (8):777–87.

Soifer, H., C. Higo, C. R. Logg et al. 2002. A novel, helper-dependent, adenovirus-retrovirus hybrid vector: Stable transduction by a two-stage mechanism. *Mol Ther* 5 (5 Pt 1): 599–608.

Steegenga, W. T., N. Riteco, A. G. Jochemsen, F. J. Fallaux, and J. L. Bos. 1998. The large E1B protein together with the E4orf6 protein target p53 for active degradation in adenovirus infected cells. *Oncogene* 16 (3):349–57. doi: 10.1038/sj.onc.1201540.

Stratford-Perricaudet, L. D., M. Levrero, J. F. Chasse, M. Perricaudet, and P. Briand. 1990. Evaluation of the transfer and expression in mice of an enzyme-encoding gene using a human adenovirus vector. *Hum Gene Ther* 1 (3):241–56.

Strunze, S., M. F. Engelke, I. H. Wang et al. 2011. Kinesin-1-mediated capsid disassembly and disruption of the nuclear pore complex promote virus infection. *Cell Host Microbe* 10 (3):210–23. doi: 10.1016/j.chom.2011.08.010 S1931-3128(11)00260-5 [pii].

Sullivan, D. E., S. Dash, H. Du et al. 1997. Liver-directed gene transfer in non-human primates. *Hum Gene Ther* 8 (10):1195–206.

Tamanini, A., E. Nicolis, A. Bonizzato et al. 2006. Interaction of adenovirus type 5 fiber with the coxsackievirus and adenovirus receptor activates inflammatory response in human respiratory cells. *J Virol* 80 (22):11241–54. doi: 10.1128/JVI.00721-06.

Tan, B. T., L. Wu, and A. J. Berk. 1999. An adenovirus-Epstein-Barr virus hybrid vector that stably transforms cultured cells with high efficiency. *J Virol* 73 (9):7582–9.

Tao, N., G. P. Gao, M. Parr et al. 2001. Sequestration of adenoviral vector by Kupffer cells leads to a nonlinear dose response of transduction in liver. *Mol Ther* 3 (1):28–35.

Teh, B. S., G. Ayala, L. Aguilar et al. 2004. Phase I-II trial evaluating combined intensity-modulated radiotherapy and *in situ* gene therapy with or without hormonal therapy in treatment of prostate cancer-interim report on PSA response and biopsy data. *Int J Radiat Oncol Biol Phys* 58 (5):1520–9. doi: 10.1016/j.ijrobp.2003.09.083 S0360301603020704 [pii].

Thaci, B., I. V. Ulasov, D. A. Wainwright, and M. S. Lesniak. 2011. The challenge for gene therapy: Innate immune response to adenoviruses. *Oncotarget* 2 (3):113–21.

Tjian, R. 1978. The binding site on SV40 DNA for a T antigen-related protein. *Cell* 13 (1):165–79.

Tollefson, A. E., B. Ying, K. Doronin, P. D. Sidor, and W. S. Wold. 2007. Identification of a new human adenovirus protein encoded by a novel late l-strand transcription unit. *J Virol* 81 (23):12918–26. doi: 10.1128/JVI.01531-07.

Tomko, R. P., R. Xu, and L. Philipson. 1997. HCAR and MCAR: The human and mouse cellular receptors for subgroup C adenoviruses and group B coxsackieviruses. *Proc Natl Acad Sci U S A* 94 (7):3352–6.

Toth, K., and W. S. Wold. 2002. HEK? No! *Mol Ther* 5 (6):654. doi: 10.1006/mthe.2002.0618 S1525001602906180 [pii].

Trentin, J. J., Y. Yabe, and G. Taylor. 1962. The quest for human cancer viruses. *Science* 137:835–41.

Triozzi, P. L., and E. C. Borden. 2011. VB-111 for cancer. *Expert Opin Biol Ther* 11 (12):1669–76. doi: 10.1517/14712598.2011.618122.

Umana, P., C. A. Gerdes, D. Stone et al. 2001. Efficient FLPe recombinase enables scalable production of helper-dependent adenoviral vectors with negligible helper-virus contamination. *Nat Biotechnol* 19 (6):582–5.

Uusi-Kerttula, H., S. Hulin-Curtis, J. Davies, and A. L. Parker. 2015. Oncolytic adenovirus: Strategies and insights for vector design and immuno-oncolytic applications. *Viruses* 7 (11):6009–42. doi: 10.3390/v7112923 v7112923 [pii].

Waddington, S. N., J. H. McVey, D. Bhella et al. 2008. Adenovirus serotype 5 hexon mediates liver gene transfer. *Cell* 132:397–409.

Wadsworth, S. C., H. Zhou, A. E. Smith, and J. M. Kaplan. 1997. Adenovirus vector-infected cells can escape adenovirus antigen-specific cytotoxic T-lymphocyte killing *in vivo*. *J Virol* 71 (7):5189–96.

Wang, Q., G. Greenburg, D. Bunch, D. Farson, and M. H. Finer. 1997. Persistent transgene expression in mouse liver following *in vivo* gene transfer with a delta E1/delta E4 adenovirus vector. *Gene Ther* 4 (5):393–400.

Wang, Q., X. C. Jia, and M. H. Finer. 1995. A packaging cell line for propagation of recombinant adenovirus vectors containing two lethal gene-region deletions. *Gene Ther* 2 (10):775–83.

Weitzman, M. D. 2005. Functions of the adenovirus E4 proteins and their impact on viral vectors. *Front Biosci: J Virtual Library* 10:1106–17.

Welsh, M. J., A. E. Smith, J. Zabner et al. 1994. Cystic fibrosis gene therapy using an adenovirus vector: *In vivo* safety and efficacy in nasal epithelium. *Hum Gene Ther* 5 (2):209–19.

Wickham, T. J., E. J. Filardo, D. A. Cheresh, and G. R. Nemerow. 1994. Integrin alpha v beta 5 selectively promotes adenovirus mediated cell membrane permeabilization. *J Cell Biol* 127 (1):257–64.

Wickham, T. J., P. Mathias, D. A. Cheresh, and G. R. Nemerow. 1993. Integrins alpha v beta 3 and alpha v beta 5 promote adenovirus internalization but not virus attachment. *Cell* 73 (2):309–19.

Wickham, T. J., D. M. Segal, P. W. Roelvink et al. 1996. Targeted adenovirus gene transfer to endothelial and smooth muscle cells by using bispecific antibodies. *J Virol* 70 (10):6831–8.

Wiethoff, C. M., and G. R. Nemerow. 2015. Adenovirus membrane penetration: Tickling the tail of a sleeping dragon. *Virology* 479–480:591–9. doi: 10.1016/j.virol.2015.03.006 S0042-6822(15)00142-7 [pii].

Wiethoff, C. M., H. Wodrich, L. Gerace, and G. R. Nemerow. 2005. Adenovirus protein VI mediates membrane disruption following capsid disassembly. *J Virol* 79 (4):1992–2000.

Wilson, J. M. 2005. Gendicine: The first commercial gene therapy product. *Hum Gene Ther* 16 (9):1014–5.

Wilson, J. M. 2009. Lessons learned from the gene therapy trial for ornithine transcarbamylase deficiency. *Mol Genet Metab* 96 (4):151–7. doi: 10.1016/j.ymgme.2008.12.016 S1096-7192(08)00499-X [pii].

Wilson, J. M., J. F. Engelhardt, M. Grossman, R. H. Simon, and Y. Yang. 1994. Gene therapy of cystic fibrosis lung disease using E1 deleted adenoviruses: A phase I trial. *Hum Gene Ther* 5 (4):501–19.

Wold, W. S., and M. S. Horwitz. 2007. Adenoviruses. In *Fields Virology*, edited by D. M. Knipe and P. M. Howley, 2396–2436. Philadelphia, PA: Lippincott Williams & Wilkins.

Worgall, S., G. Wolff, E. Falck-Pedersen, and R. G. Crystal. 1997. Innate immune mechanisms dominate elimination of adenoviral vectors following *in vivo* administration. *Hum Gene Ther* 8 (1):37–44.

Xiong, W., S. Goverdhana, S. A. Sciascia et al. 2006. Regulatable gutless adenovirus vectors sustain inducible transgene expression in the brain in the presence of an immune response against adenoviruses. *J Virol* 80 (1):27–37.

Xu, F., E. Ding, F. Migone et al. 2005. Glycogen storage in multiple muscles of old GSD-II mice can be rapidly cleared after a single intravenous injection with a modified adenoviral vector expressing hGAA. *J Gene Med* 7 (2):171–8.

Yabe, Y., J. J. Trentin, and G. Taylor. 1962. Cancer induction in hamsters by human type 12 adenovirus. Effect of age and of virus dose. *Proc Soc Exp Biol Med* 111:343–4.

Yang, Y., H. C. Ertl, and J. M. Wilson. 1994. MHC class I-restricted cytotoxic T lymphocytes to viral antigens destroy hepatocytes in mice infected with E1-deleted recombinant adenoviruses. *Immunity* 1 (5):433–42.

Yang, Y., F. A. Nunes, K. Berencsi et al. 1994. Cellular immunity to viral antigens limits E1-deleted adenoviruses for gene therapy. *Proc Natl Acad Sci U S A* 91 (10):4407–11.

Yant, S. R., A. Ehrhardt, J. G. Mikkelsen et al. 2002. Transposition from a gutless adeno-transposon vector stabilizes transgene expression *in vivo*. *Nat Biotechnol* 20 (10):999–1005.

Yarborough, M., and R. R. Sharp. 2009. Public trust and research a decade later: What have we learned since Jesse Gelsinger's death? *Mol Genet Metab* 97 (1):4–5. doi: 10.1016/j.ymgme.2009.02.002 S1096-7192(09)00041-9 [pii].

Ye, W., R. Liu, C. Pan et al. 2014. Multicenter randomized phase 2 clinical trial of a recombinant human endostatin adenovirus in patients with advanced head and neck carcinoma. *Mol Ther* 22 (6):1221–9. doi: 10.1038/mt.2014.53 mt201453 [pii].

Yeh, P., J. F. Dedieu, C. Orsini et al. 1996. Efficient dual transcomplementation of adenovirus E1 and E4 regions from a 293-derived cell line expressing a minimal E4 functional unit. *J Virol* 70 (1):559–65.

Yew, N. S., J. Marshall, M. Przybylska et al. 1999. Increased duration of transgene expression in the lung with plasmid DNA vectors harboring adenovirus E4 open reading frame 3. *Hum Gene Ther* 10 (11):1833–43.

Ying, B., A. E. Tollefson, and W. S. Wold. 2010. Identification of a previously unrecognized promoter that drives expression of the UXP transcription unit in the human adenovirus type 5 genome. *J Virol* 84 (21):11470–8. doi: 10.1128/JVI.01338-10.

Zabner, J., L. A. Couture, R. J. Gregory et al. 1993. Adenovirus-mediated gene transfer transiently corrects the chloride transport defect in nasal epithelia of patients with cystic fibrosis. *Cell* 75 (2):207–16.

Zabner, J., D. M. Petersen, A. P. Puga et al. 1994. Safety and efficacy of repetitive adenovirus-mediated transfer of CFTR cDNA to airway epithelia of primates and cotton rats. *Nat Genet* 6 (1):75–83.

Zhang, Y., N. Chirmule, G. P. Gao et al. 2001. Acute cytokine response to systemic adenoviral vectors in mice is mediated by dendritic cells and macrophages. *Mol Ther* 3 (5 Pt 1): 697–707.

Zhao, H., M. Chen, and U. Pettersson. 2014. A new look at adenovirus splicing. *Virology* 456–457:329–41. doi: 10.1016/j.virol.2014.04.006.

Zhou, H., W. O'Neal, N. Morral, and A. L. Beaudet. 1996. Development of a complementing cell line and a system for construction of adenovirus vectors with E1 and E2a deleted. *J Virol* 70 (10):7030–8.

Zhu, J., X. Huang, and Y. Yang. 2007. Innate immune response to adenoviral vectors is mediated by both Toll-like receptor-dependent and -independent pathways. *J Virol* 81 (7):3170–80. doi: 10.1128/JVI.02192-06.

Ziegler, R. J., C. Li, M. Cherry et al. 2002. Correction of the nonlinear dose response improves the viability of adenoviral vectors for gene therapy of Fabry disease. *Hum Gene Ther* 13 (8):935–45. doi: 10.1089/10430340252939041.

Zuckerman, J. B., C. B. Robinson, K. S. McCoy et al. 1999. A phase I study of adenovirus-mediated transfer of the human cystic fibrosis transmembrane conductance regulator gene to a lung segment of individuals with cystic fibrosis. *Hum Gene Ther* 10 (18):2973–85.

2 Innate and Adaptive Immune Responses to Adenovirus

Svetlana Atasheva and Dmitry M. Shayakhmetov

CONTENTS

ABSTRACT

Adenovirus vectors represent an attractive platform for the treatment of many important clinical diseases. However, numerous studies of adenovirus biodistribution in rodents and humans have shown that a large proportion of the administered vector is neutralized or sequestered by the host immune system. The mechanisms of adenovirus vector removal from the bloodstream are determined by an intricate interplay between the host factors, cells, and the virus. Macrophages in various organs are the first responders to the adenovirus vector, but require the assistance of various blood factors for successful elimination of the adenovirus. While there are various mechanisms of eradicating adenovirus vectors from the blood, they all culminate in the death and removal of virus-containing cells. For the successful application of adenovirus-based therapies, the negative influence of the innate and adaptive arms of the immune system should be taken into account when developing effective strategies of implementation of adenovirus vectors for clinical therapy.

Adenoviruses (Ads) possess a number of qualities proving them to be valuable tools for gene therapy. These qualities include the inability of the adenovirus genome to insert into the host genome, the large capacity of the vector, vector stability, and ease of manufacturing. To date, Ads are the most commonly used vectors in clinical trials for gene therapy applications. However, the results of these trials show that application of Ad vectors has considerable obstacles. The most significant limitations are (i) preexisting immunity to the vector in human populations due to exposure to naturally circulating Ads, and (ii) liver sequestration of a significant portion of the vector from the bloodstream during intravenous (i.v.) systemic delivery. Additionally, in some cases, the successful delivery of the vector to the target cells requires the injection of high doses of Ad vector, hence increasing the risks of toxicity. Dissecting the roles of the immune response to Ad administration will undoubtedly help in improving the Ad vector design for safe and effective therapies. Another important implication of understanding the immune response to Ad is the prediction of possible outcomes of the application of Ad-based therapies in patient cohorts with preexisting conditions, for example, liver cirrhosis. This chapter discusses the interactions between Ad and the host adaptive and innate immune systems and the outcomes of these interactions.

2.1 ADAPTIVE IMMUNITY

Human Adenovirus 5 (Ad5) is the most often used vector for Ad-based therapy (http://www.wiley.com//legacy/wileychi/genmed/clinical/). However, Ad5 is the causative agent of very common respiratory tract infections during childhood. The clinical trials of Ad5-based treatments allowed for thorough studies of the footprints of natural Ad infections in different human populations. These studies showed that the prevalence of Ad5 among countries varies dramatically. Approximately 50% of populations in North America and Western Europe have neutralizing antibodies to Ad5 in the serum.[1] The Sub-Saharan populations in Africa show prevalence of 80%–90%, while in Haiti about 82% of population have been previously exposed to Ad5.[2,3] In China the seroprevalence of Ad5 is 77%.[4] Natural infection with Ads leads to the development of immunity that includes both humoral and cellular adaptive responses. Preexisting immunity is one of the major obstacles of successful implementation of Ad-based therapies. Many studies have shown that the preexisting immunity can negatively affect vector effectivity or completely negate the treatment.[2] Both humoral and cellular preexisting immunity can be detrimental to successful Ad vector administration, however, their distinctive roles are not clear. The humoral response to Ad infection targets the serotype specific differences; neutralizing antibodies against one serotype fails to neutralize other serotypes.[5] The cellular immunity has a broad coverage and targets common, conserved antigens among different Ads, such as E1A or conserved epitopes within the hexon protein.[6]

2.1.1 HUMORAL IMMUNE RESPONSE

Localizing and pinpointing the epitopes for the neutralizing antibodies opens up opportunities for the design of Ad variants that can be effective regardless of the patient's immune status. Many studies have been conducted to understand which viral proteins are the targets for neutralization. Most of neutralizing antibodies that are produced in response to Ad infection target capsid proteins.[7,8] The Ad capsid consists of three major capsid proteins—hexon, penton, and the fiber. Together with minor capsid proteins they form an icosahedral viral particle. Within the icosahedral virion the hexon forms trimers that are connected with the penton base on the vertices of the particle. From the center of the penton base a trimer of the fiber protein is protruding outwards forming a shaft and a knob.[9] The structural relationships of these proteins are studied in detail and excellent cryoEM-based models of the whole viral particle are reconstructed with a resolution of 3.4 A.[10,11] These models allow for the identification of the surface-exposed domains that might serve as immunogenic epitopes for humoral immunity. Currently there are approximately 60 known types of human Ad. The basis for the major differences between serotypes is the variation of the fiber knob, length of the fiber shaft, exposed regions of the trimeric hexon, and variable loops on the penton base.[12–14] These protein determinants define the antigenic properties of each serotype. As a result, neutralizing antibodies are serotype specific and unable to neutralize other serotypes.

The molecular mechanisms of viral particle neutralization by antibodies are yet to be uncovered. Generally it is believed that neutralizing antibodies targeting fiber cause aggregation of viral particles thus forming immune complexes consisting of multiple virions bridged by antibodies.[5] These complexes can function as immune activating signals triggering viral clearance. Reports show that antibodies targeting another capsid protein—penton also play an important role in virus neutralization and synergistically enhance the neutralization effect of anti-fiber antibodies.[14] Penton possesses an Arg–Gly–Asp (RGD) motif that interacts with integrins on the cellular surface, this interaction mediates virus internalization and subsequent disassembly.[15] Interestingly, the Fab fragment of the antibody binding to the RGD motif on the penton protein has a neutralization activity; however, the full size antibody failed to neutralize the vector.[16] This phenomenon suggests that the trimer of the fiber protein protruding from the penton base produces a steric hindrance preventing the binding of the full size antibody to all five RGD copies on the penton base. These data show that the RGD loop can be an effective neutralization epitope for antibody targeting. Additional confirmation of the importance of the RGD loop as a target for neutralization is the ability of human alpha-defensins to bind to the adjacent region on the Ad penton and neutralize the virus by preventing the RGD loop function.[17] On the other side, some reports show that anti-penton antibodies do not play a significant role in neutralization, while antibodies against the hexon have the most neutralizing activity, particularly, antibodies against hypervariable regions (HVR).[18,19] By aligning the sequences of the hexon of different subtypes it was found that the Ad hexon has conservative regions that are intercalated by regions with high variability, named HVRs.[20] Structural studies have shown that these regions form exposed loops on the surface of the virion.[10] The HVRs vary greatly among different serotypes, thus significantly contributing to serotype specificity of neutralizing antibodies (nAb).[18] Therefore one of the current methods for avoiding preexisting immunity is replacing the HVRs from one serotype to another, for example, swapping HVRs of Ad5, a very common Ad to HVRs derived from less prevalent serotypes.[21,22] Another viable approach of avoiding Ad vector neutralization by antibodies is mutation of the immunogenic epitope. Recently, several monoclonal neutralizing antibodies were generated against the hexon protein of Ad3 and Ad7, the nAbs were bound to the HVR4 and HVR7, respectively.[23] In another screen for neutralizing Ab targeting chimpanzee Ad68 (AdC68) the researchers isolated seven different clones of Abs with neutralizing activity. However, the competition experiments suggested that all nAbs recognized the same hexon epitope. Mutational analysis showed that nAbs target the HVR1 of the hexon protein. Interestingly, mutation ETA to CDQ in the HVR1 was sufficient for virus escape of neutralization, however, the Abs were still capable of binding the capsid.[24] Nevertheless, the mechanism of neutralization of Ad infectivity is not very clear. The only known mechanism of neutralization was demonstrated for mouse monoclonal nAb 9C12. This nAb prevents hexon interactions with the microtubules or the microtubule motors and blocks virus trafficking to the nucleus.[25]

There is no consensus on which of the viral proteins elicits the strongest humoral response. Some reports show that the neutralizing activity of Ab in human blood with previous natural infections is directed against the hexon protein, while penton and fiber proteins do not play significant roles in neutralization.[18] On the other hand, a detailed study of cohorts participating in Ad5-based vaccine against HIV showed that about 50% of seropositive population has neutralizing antibody targeting fiber protein, and only 4% targeting hexon.[26] This study also analyzed the responses to Ad5 vector vaccination in naïve recipients and found that a non-replicating Ad5 vector elicits a response to hexon. In summary, the natural infection with Ad5 in humans leads to production of neutralizing antibodies that target all three major Ad capsid proteins.[26] Such broad response and existence of different antibodies can explain the variety of the reports that showed that modifying different capsid proteins can circumvent preexisting immunity.

2.1.2 Cellular Immunity

Ads have been known for their superb ability to elicit cellular immune response, particularly mediated by cytotoxic lymphocytes (CTLs).[27] This ability of Ad vectors has been extensively used for developing vaccine candidates for multiple pathogens, including HIV, Ebola virus, malaria, and others. In adult and pediatric patients with successfully resolved Ad infection it was found that both CD4 and CD8 CTLs can effectively lyse the infected cells.[28] Human Ads are represented by seven species (A–G), each species contain several types of Ads with designated numbers. Tests of different types of human and chimpanzee Ads showed that Ads belonging to group C elicit the most potent CTL response. Moreover, the minimal dose of the vector that is required for induction of a noticeable CTL response was significantly lower for group C Ads than for Ads belonging to other groups.[29] Historically, Ad5 belonging to group C has been the most commonly used vector for gene therapy applications. The ability of Ad5 to elicit very strong CTL response sparked works for understanding the mechanisms underlying this phenomenon. When applied to the vaccine development this knowledge will lead to the improvement of the immunogenicity and effectivity of vaccines. Immunization with Ad5 vector leads to antigen-specific

CD8 T cells expansion followed by a protracted contraction phase leaving a memory cell pool population.[30] Despite having no replication machinery, the Ad5 vectors are able to sustain antigen expression and presentation for more than a year, depending on the recombinant promoter driving antigen expression.[31] The duration of the antigen expression affected the quality of the immune response. Termination of expression of antigen before the peak of expansion CD8 T cells (day 3 post immunization) led only to a limited number of memory pool formation. Extension of antigen expression to 13–30 days allowed formation of a memory pool with a limited sustainability, while termination of antigen expression on day 90 did not have any negative effects on the quality of the memory pool.[30] The antigen presentation also plays a critical role for the quality of the CTL response. Bassett et al. showed that antigen presented by professional antigen-presenting cells (APC), for example, dendritic cells (DCs), is a critical step in memory formation. Surgical removal of draining lymph nodes (DLNs), the main sites for APC locale, prevented the development of CD8 T cell immunity.[27] The same effect was demonstrated in mice with impaired DLN homing mechanism, confirming that CTL priming occurs in DLNs.[32] Nevertheless, DLNs were required only during primary T cell activation, presenting of antigen by cells of non-hematopoietic origin was sufficient for a recall response.[27] The importance of fully functional DCs in priming was demonstrated in mice lacking a transcription factor involved in DC maturation, Baft3.[33] Additionally, mice deficient in MyD88, an adapter for toll-like receptor (TLR) signaling had significantly impaired CD8 T-cell responses, however, in mice lacking individual TLRs the CD8 T cell response was not diminished.[34]

Interestingly, the potency of the cellular immune response was independent of the route of Ad injection, the splenocytes from mice injected via intramuscular (i.m.), i.v., or intraperitoneal (i.p.) routes were able to lyse infected cells with the same efficiency.[35] Moreover, the non-replicating Ad vector was able to elicit CTL response, suggesting that activation of cellular immunity does not require *de novo* synthesis of viral proteins, the same results were achieved with injection of UV-inactivated Ad vector.[35] Immunogenic CTL epitopes in mice target very conserved proteins such as E1A and E2A, however, in humans the epitopes cover conservative regions of the hexon protein. Therefore the CTLs have high cross-reactivity, due to a high level of identity of the hexon among Ads. Trying to understand the nature of immunological Ad5-specific epitopes, Leen et al. ran a library of hexon-specific epitopes to elicit CTL response in human CTL-derived cell lines and found a number of potent immunogenic epitopes. Interestingly, only quarter of found epitopes were CD8-restricted with the rest being CD4-specific.

The role of cellular immunity in preexisting immunity is not very clear. In patients recruited in Merck clinical trial 016, Betts et al. measured CD8 CTL activity prior and post vaccination. The peripheral blood CD8 T cells population expanded after vaccination in both Ad-specific seronegative and seropositive cohorts. The kinetics of expansion was slightly different between cohorts, but there was no alteration in polyfunctionality, antigen expression, or memory phenotype of CD8 T cells.[36]

In a comprehensive study by Wohlleber et al. the authors investigated the mechanism of CD8 CTL-mediated death of Ad-infected hepatocytes. It is well known that the hepatocyte transduction is a very efficient process; it is possible to reach expression of virally encoded transgenes in nearly 90% of hepatocytes depending on the virus dose.[37,38] However, the level of transgene expression steeply declines with time and by 2 weeks after Ad administration no expression can be detected in immunocompetent hosts.[37,39,40] One of the reasons for transgene expression decline is elimination of infected cells by the CTLs.[41] Interestingly, the classical priming of CTLs by DCs was not required for elimination of the infected hepatocytes. Instead, liver sinusoid endothelial cells were able to process and cross-present viral antigens to the CTLs priming them to kill infected hepatocytes.[41] The killing mechanism was mediated by TNF-α, however, only infected cells were sensitized to TNF-α by expressing caspase 3. Expression of caspase 3 and enhanced TNF-α signaling resulted in elimination of infected hepatocytes.[41]

2.2   INNATE IMMUNITY

Effective gene therapy application of Ad vectors requires systemic administration of the vector at high doses resulting in the stimulation of innate and adaptive immune responses that can prove fatal for the host.[42–45] Experiments in rodents showed that upon i.v. Ad administration, the vector is rapidly sequestered from the bloodstream by different organs, predominantly the liver.[39,46–48] The removal of Ad from blood circulation temporally coincides with release of pro-inflammatory cytokines and chemokines, including IL-1α, IL-1β, IL-6, TNF-α, RANTES, IP-10, IL-8, MIP-1α, MIP-1β, and MIP-2.[49–53] The macrophages, including tissue residential macrophages (e.g., Kupffer cells in the liver), and DCs throughout the body are the principal sources of the released cytokines and chemokines.[54] In humans, i.v. injection of high doses of Ad vector (2×10^{12}–6×10^{13} viral particles) also results in elevated levels of IL-6 and IL-1 cytokines in the serum.[55–57] The local release of pro-inflammatory stimulus attracts leukocytes and neutrophils to the sites of infection resulting in tissues infiltration and local necrosis.[50,58] These processes have been demonstrated in the liver, lungs, and spleen, indicating that most, if not all, tissues in the body respond to i.v. Ad injection in a pro-inflammatory manner. The mechanisms of activation of the pro-inflammatory response to Ad injection have been extensively studied in rodents and humans to predict the immune response to vector administration in clinical settings. The examination in mouse and nonhuman primate preclinical animal models showed that transcriptionally defective Ad particles are capable of innate immune response induction.[51,59] Moreover, i.v. injection of a helper-dependent Ad vector that lacked all viral genes triggered a severe acute inflammatory response in a nonhuman primate model.[60] These data provide clear evidence that the activation of innate immune and inflammatory responses to the Ad vector does not require viral gene expression and depends on Ad particle interaction with the host. The induction of innate immunity prior to establishing the vector replication cycle or viral gene expression suggests that interactions of the Ad viral particle with host cells are critical events occurring immediately upon vector administration. Therefore, in this part of the chapter we aim to summarize *in vivo* data of the molecular mechanisms of Ad particle interactions with the host that lead to the activation of innate immunity.

2.2.1 Ad VECTOR INTERACTIONS WITH COMPONENTS OF THE BLOOD

Systemic delivery of Ad vectors for treatment of different inborn or acquired diseases requires using an i.v. route of administration. Blood, however, possesses multiple means of protecting the host from malicious pathogens. Different soluble factors that are found in the plasma, such as neutralizing antibodies, natural antibodies, complement, coagulation factors, and defensins can recognize viral invasion and either directly neutralize the Ad vector or activate an immune response against the administered virus.[61] Moreover, blood cells themselves can bind and sequester an Ad vector in the bloodstream targeting the vector for destruction. The interactions of the Ad vector with blood components play important roles in vector biodistribution[61] and understanding the detailed mechanisms of interactions will prove useful in directing the vector to the target organ or cell types.

2.2.1.1 Coagulation Factors

Many Ad vectors upon i.v. delivery demonstrate strong hepatotropism, however, under asanguineous conditions the perfused mouse livers were incapable for virus sequestration.[62] Testing the interacting partners of human Ad5 in the blood revealed that Ad5 directly interacts with coagulation factor IX and other plasma proteins (C4 binding protein, fibrinogen, pregnancy zone protein [PZP]).[62] Maturation of coagulation factor IX is regulated by vitamin K, which is also involved in post-translation modification of coagulation factors FVII, FX, and protein C. Although all of these factors can directly bind to HAd5, FX is the most effective co-factor mediating hepatocyte transduction both *in vivo* and *in vitro* compared to other coagulation factors.[63] Interestingly, these factors also increase Ad5 vector transduction efficiency in tissue culture (reviewed in Reference 64). Detailed structured studies of Ad5 interaction with FX pinpointed the FX binding site to the center of the hexon protein trimer on the Ad5 capsid. The cryo-EM studies showed that the glutamic acid domain of FX is bound at the top of each hexon trimer. The molecules of FX do not impose steric hindrance for other FX molecule binding to the capsid thus covering the entire virion with FX molecules.[38,65–67] Mutagenesis of the HVR5[68] and HVR7[66] reduced the binding capacities or completely ablated the FX-hexon interactions, respectively, suggesting their crucial roles in binding FX. The binding of coagulation factors to the Ad capsid has several implications for Ad biology. Binding of FX mediates attachment of the vector to heparan sulfate proteoglycans (HSPG) molecules that are scattered on the surface of hepatocytes and mediates hepatocytes transduction.[69] Another important role of FX binding to the surface of the virion is activation of signaling for triggering host immune response. The FX binding to Ad5 activates the NF-kB-dependent inflammatory pathway through the TLR4-TRIF/MyD88-TRAF6 signaling axis resulting in a release of pro-inflammatory cytokines and chemokines.[66] The Ad species that are capable of binding to FX activate a significantly broader spectrum of chemokines and cytokines than those whose hexon proteins cannot bind FX.[66] On the other hand, Xu et al.[70] showed that FX binding to the Ad capsid prevents recognition of the vector by natural IgM antibodies and subsequent complement-mediated degradation. The authors suggested that high concentration of FX in the plasma and the ability to shield the entire virion surface with FX helps the vector to escape detection by the immune system.

2.2.1.2   Complement

Intravenous administration of Ad activates both branches of the complement system, the classic and alternative pathways, which can be detected by measuring C3a concentration in the plasma. Complement activation is a very rapid process, within 10 minutes after injection of Ad the C3a levels in the plasma increases, reaches its peak at 30 minutes, and then declines to the normal level in 90 minutes after administration.[71,72] The mechanism of the complement activation in response to Ad vector administration is not understood in detail, but it requires cooperating of several components. Experiments with the ts1 mutant of Ad that can be internalized by the cells as the wild type vector, but is incapable for endosome rupture and escape, suggested that the complement system in unable to recognize and inactivate an Ad particle without assistance from the reticuloendothelial system.[72,73] Additionally, Xu et al.[70] showed the importance of the natural antibodies in complement activation in the absence of the FX. The *in vitro* studies supported dependence of complement activation on the presence of natural IgM antibodies.[61] However, Tian et al.[72] showed that there is no requirement for natural antibodies for complement activation *in vivo*. Clearly, natural antibodies and complement play significant roles in Ad vector neutralization in the bloodstream, however, to date, the exact molecular mechanisms that mediate *in vivo* virus neutralization remain unknown.

2.2.1.3   Defensins

Defensins are small amphiphilic peptides (18–45 amino acids) that are expressed by neutrophils, Paneth cells, and epithelial cells. They target membranous microorganisms and can directly disrupt the membrane resulting in inactivation of pathogens. These cationic peptides target a broad range of pathogens, including bacteria, viruses, fungi, and protozoa.[74] Surprisingly, Ad vectors are also sensitive to defensins albeit the absence of the membrane.[75] Smith et al.[17] showed that alpha-defensins HNP1 and HD5 have a very potent anti-Ad activity. The HD5 binds to the disordered region on the penton base and the negatively charged EDES sequence in the N-terminal fiber region.[76] The binding reinforces the penton–fiber vertex complex and stabilizes the virion structure resulting in an inability of capsid to release the penton and protein VI.[77] In the absence of protein VI release the Ad vector is unable to rupture and escape the endosome.

2.2.1.4   Blood Cells

The soluble factors in plasma mediate the effective means for recognition and inactivation of the Ad vector, however, the blood cells themselves are able to efficiently interact and disarm Ad. Erythrocytes, the largest fraction of the cells in the blood, express CAR, the coxsackie and Ad receptor[78,79] that interacts with the Ad5 fiber knob domain. Interestingly, murine erythrocytes, unlike human, lack CAR on the surface of the cells and Lyons et al. analyzed the Ad5 interaction with human and mouse blood cells. They found that after a short incubation with blood cells *ex vivo*, over 90% of the Ad vector was stably associated with human (but not murine) erythrocytes.[80] Moreover, in the presence of human or murine plasma, human erythrocytes can bind the Ad vector through the complement receptor (CRI).[81] Observations of clinical trials data involving cancer patients showed that most of the administered oncolytic Ad vector is associated with erythrocytes or granulocytes thus confirming the role of blood cells in Ad vector clearing in humans.[82] Binding to these receptors on the surface of erythrocytes is an efficient means for preventing Ad vector dissemination and infection of other cells.[78] Neutrophils are another cell type that can bind and subsequently internalize the Ad vector.[83] The interaction of Ad to neutrophils was reduced by blocking the CRI or the Fc receptors suggesting that the binding depends on the presence of complement and antibodies.[83]

Platelets, unlike neutrophils that are present in the circulation in high numbers only during times of acute host inflammatory responses, are constantly present in the blood in high numbers ($1.5–4.5 \times 10^5$ platelets/mL). Several reports demonstrated that Ad vector administration in mice leads to thrombocytopenia as a result of clearing of platelets.[84,85] Othman et al.[84] showed that platelets can actively internalize Ad without stimulation with ADP, which was required for internalization of inert latex particles.[86] Moreover, in mice Ad5 induced platelet–leukocyte aggregate formation, which was dependent on the von Willebrand factor. However, the inert latex particles failed to induce aggregates formation even in ADP-activated platelets.[84,86] Moreover, platelets interaction with the Ad vector in the blood and formation of aggregates can facilitate virus sequestration by the reticuloendothelial system. The importance of platelet–Ad interaction *in vivo* was demonstrated in experiments with mice that were treated to deplete platelets before i.v. Ad administration, the Ad vector sequestration to different organs was significantly reduced.[87] The mechanism of platelet recognition of Ad is unclear, however 78% of human platelets are positive for the Ad5 receptor, CAR.[61] It is plausible that platelet–Ad interaction is mediated by CAR, however Ad11 associates with platelets to a much higher degree than Ad5, and this interaction occurs in a fiber-independent manner.[88] Taken together these data suggest a novel and still uncharacterized mechanism of Ad interaction with circulating blood cells that involves targeted recognition of the viral capsid.

2.2.2 Ad INTERACTIONS WITH LIVER MACROPHAGES

Intravenously delivered Ad vectors based on Ad5, the most frequently used Ad vector for gene therapy purposes, demonstrate very strong hepatotropism.[46,48] The liver actively sequesters intravenously administrated Ad with remarkable speed and efficiency. The half-life of the injected HAd5 in the mouse bloodstream is 2 minutes.[46] Within 30 minutes after i.v. injection, the liver traps 99% of the injected Ad, thus eliminating it from the bloodstream.[46] Interactions between the Ad vector and liver cells lead to clinically significant hepatotoxicity and represent a major limitation for gene delivery to extrahepatic cells and tissues, such as disseminated metastatic tumors. Dissecting Ad interactions with host factors will further our understanding of the molecular mechanisms that dictate Ad biodistribution and identify targets for manipulation of Ad vectors for future clinical use.

Kupffer cells, the residential macrophages in the liver, reside in the lumen of sinusoids and filter the blood from foreign substances, for example, large particles. Kupffer cells uptake the majority (~90%) of the Ad vector from the blood. The liver sinusoids are lined with liver sinusoidal endothelial cells (LSECs) that have an extensive network of fenestrae allowing for transport of the nutrients from the bloodstream through the layer of endothelial cells to hepatocytes. LSECs and hepatocytes, the liver parenchyma cells, also are able to uptake the virus and account for uptake of 6% and 2% of the Ad vector, respectively.[89] The rest of the administered Ad vector is trapped in the space of Disse located between lining layer of LSECs and parenchymal hepatocytes.

The exact mechanism of Kupffer cell Ad uptake is being extensively studied but the details of vector entry are still not very clear. There are several investigations that describe the interaction of Ad vector and Kupffer cells, however, the data show that the interactions are very redundant and involve many players. Kupffer cells function as blood filters and sense foreign particles by recognizing certain molecular patterns. Kupffer cells can uptake inert latex particles, which suggests that they can discriminate foreign material based on size.[90] However, inactivated virus particles were sequestered by the liver at lower rates than a native virus, suggesting that capsid proteins interactions are important for Kupffer cells sequestration.[48] Another striking detail of Ad uptake by the liver is that it is independent of CAR, Ad receptor, which is required for viral transduction in tissue culture.[91,92] Moreover, liver sequestration of Ad5 and Ad11 that utilize different cellular receptors was comparable in the liver at 30 minutes post administration.[88] Additionally, the RGD motif within the penton base on the virus capsid that engages $\alpha_3\beta_1$, $\alpha_5\beta_1$, $\alpha_v\beta_1$, $\alpha_v\beta_3$, and $\alpha_v\beta_5$ integrin molecules on the cellular surface[93] was also unnecessary for Kupffer cells uptake.[89]

Interestingly, despite uptake of massive amounts of the virus, Kupffer cells do not support Ad replication or expression of an encoded transgene, and degrade 90% of the viral DNA virus in the first 24 hours,[39] suggesting that Kupffer cells represent a poor host for Ad propagation. Upon i.v. administration, Kupffer cells are the first line of cellular defense in the liver contributing to viral clearance. However, if the dose of the administered vector exceeds the threshold that Kupffer cells can uptake, the macrophages can no longer accumulate the incoming virus.[94] As a result, the excess of the virus can reach other liver cells and transduce hepatocytes.[89] Tao et al.[39] showed that the threshold can be reached with a single dose of the virus by injecting more than 1×10^{11} particles per mouse. Additionally, the saturation limit can be achieved by two separate administrations of the Ad vector. The first administration of the Ad vector saturates Kupffer cells and the second dose of the Ad vector can escape Kupffer cell sequestration if given just 5 minutes later.[39] The inability of Kupffer cells to uptake the second Ad dose leads to enhanced hepatocyte transduction efficiency and higher levels of transgene expression.[49] The same effect of higher levels of hepatocytes transduction can be achieved if Kupffer cells are depleted or functionally blocked.[95]

Uptake of the Ad vector by Kupffer cells depends on several critical components. In the report by Xu et al.[96] it was shown that scavenger receptors (SRs), natural antibodies, and complement play important roles in the process of virus uptake by Kupffer cells. SRs recognize negatively charged macromolecules including lipoproteins, proteoglycans, and carbohydrates.[97,98] It was shown that masking negatively charged areas of HVRs 1, 2, 5, and 7 of the hexon protein by PEGylation prevents sequestration by Kupffer cells. *In vitro* experiments in Chinese hamster ovary (CHO) cells (that naturally lack Ad receptor CAR) overexpressing different SRs, the

PEGylated vectors had impaired entry into CHO cells expressing SRA-II, suggesting that this receptor mediates recognition of HVRs 1, 2, 5, and 7.[99] However, mice lacking one of the major SRs, SRA-I, did not demonstrate any difference in Kupffer cells uptake of Ad compared to uptake by Kupffer cells expressing SRA-I.[100] Recently it was shown that SRs ES-1 in Kupffer cells can play a role in Ad uptake.[101] Peptides that target and inactivate these receptors increased hepatocytes transduction efficiency suggesting a blockage of Kupffer cell uptake of the Ad vector.[102] Another significant component that affects vector biodistribution are natural antibodies. The Rag1[-/-] mice lacking immunoglobulins had much higher hepatocytes transduction rates compared to wt mice due to reduced Kupffer cell uptake.[103,104] Complement importance was also supported by finding that CRig is the receptor on Kupffer cells that mediates early responses to Ad vector administration.[81] Taken together, these data show that the key components for Kupffer cells sequestration of Ad include native hexon protein, in particular, HVR 1, 2, 5, and 7; complement, natural antibodies, and SRs. However, the molecular details of how all of these seemingly distinct components of the virus–host interactions come into play remain unclear.

Following uptake, Kupffer cells release an array of pro-inflammatory cytokines and chemokines. Within 1 hour post Ad administration, IL-1α, IL-16, GM-CSF, CCL1, CCL4, CCL11, CXCL1, and CXCL2 are elevated in the plasma.[50] However, Kupffer cells respond to the Ad vector uptake rather drastically, they undergo severe cytosolic disorganization and die via necrotic type cell death.[105,106] Di Paolo et al. found that the mechanism of Kupffer cell death is independent of any known principal mediators of canonical necrotic or apoptotic cell death programs. By testing 35 different strains of mice the authors found that interferon regulated factor 3 (IRF3) is a non-redundant factor that is required for Ad-mediated Kupffer cell death. Interestingly, the cell death execution was independent of transcriptional activities of IRF3 or its upstream activators.[106] However, the Ad-mediated rupture of the endosomal membrane is required for Kupffer cells death, since the Ad ts1 mutant that fails to rupture and escape the endosome, did not induce Kupffer cell death.[106] The authors also showed that rapid Kupffer cells death is a suicidal mechanism for protective measure of the host defense against viral invasion. In IRF3[-/-] mice, Kupffer cells did not undergo cell death and had a higher viral load.[106] In summary, Kupffer cells in the liver represent the first line of host defense against blood-borne pathogens. They uptake the virus through various mechanisms including SRs, natural antibody, and complement system thus sequestering the pathogens and destroying them preventing virus dissemination.

Di Paolo et al.[89]. proposed a model for sequestration of blood-borne Ad in the liver. This model suggests that different interactions of blood components with a pathogen ensure the clearance of blood-borne Ad from the circulation in a redundant, synergistic, and orderly manner. The dominant mechanism is sequestering blood-borne Ad by Kupffer cells. When macrophages are overloaded and exceed their capacity, the hepatocytes absorb the virus in an FX-dependent manner, therefore, serving as a second dominant mechanism mediating sequestration of blood-borne Ad. However, when the Ad dose is very high, and both Kupffer cells and blood factor pathways are inactivated, LSECs and anatomical architecture of liver sinusoids become the third line of defense that sequesters Ad particles in an RGD motif-dependent manner. The independence of liver sequestration from fiber-specific receptors, rapid Kupffer cells response to Ad administration, and the engagement of host factors that recognize the Ad prior established replication cycle suggest that the mechanisms of virus clearance from the blood are likely decoy pathways that ensure the removal of pathogens from circulation to prevent their dissemination into other tissues.

2.2.3 Ad Interactions with Spleen Macrophages

The spleen is the largest secondary lymphoid organ in the body that is responsible for orchestrating immune responses to blood-borne antigens. Another function of the spleen is removal of old and damaged red blood cells. The spleen consists of two functionally and anatomically distinct compartments, the red pulp and white pulp. The red pulp functions as a blood filter and removes damaged erythrocytes and foreign material. The white pulp is comprised of three compartments: the periarteriolar lymphoid sheath (PALS), the follicles, and the marginal zone (MZ).[107] Blood that enters the marginal sinus percolates through the MZ where about 5% of the intravenously administered Ad is sequestered[108] by CD169+ and MARCO+ macrophages.[100] Within 10 minutes after Ad administration MARCO+ macrophages activate transcription of pro-inflammatory cytokines and chemokines IL-1α, IL-1β, and CXCL1.[100] The alacrity of the response suggested that MARCO+ macrophages activation is triggered by sensing the very early steps of viral invasion, most likely virus attachment or internalization. Indeed, administration of Ad5ΔRGD virus, which has a deletion of the RGD motif in the penton protein and is incapable of β3 integrin binding, did not trigger MARCO+ macrophages cytokine release.[93,100] Furthermore, the Ad ts1 mutant that can adhere to the cell surface and can be internalized but is incapable of escaping endosomes, failed to activate a full-scale immune response.[100] These data suggest that virus attachment to the integrins on the surface of MARCO+ macrophages activates an immune response. However, the immune response can be drastically amplified if the signal of virus attachment is coupled with endosomal membrane damage.[100] Interestingly, virus trapping led to drastic cytosol rearrangement and severe mitochondria distortions in MARCO+ macrophages that resulted in cell elimination within 24 hours.[109] Surprisingly, the death of MARCO+ macrophages was independent of IRF3, the key player in triggering Kupffer cell death in the liver.[106] Moreover, unlike in Kupffer cells, the cellular membranes of MARCO+ macrophages stayed intact and the necrotic cell death programs were not activated upon interaction with Ad.[109] The mechanism of the death of spleen macrophages was dependent on the release of IL-1α, which in its turn signals through IL-1R and activates production of both CXCL1 and CXCL2 in the spleen.[100] These chemokines are among the most potent chemoattractants stimulating neutrophil migration and activation.[110] Indeed, following Ad administration great numbers of polymorphonuclear leucocytes (PMNs) were released from the bone marrow into the bloodstream. Interestingly, about a third of all released PMNs were retained in the MZ of spleen in close proximity to MACRO+ macrophages.[100] However, sufficient retention of the PMNs in the spleen required cooperating of the complement component C3.[100] Only in the presence of C3 and IL-1R signaling neutrophils migrated to the splenic MZ and eliminated the virus-containing cells from the MZ.[111]

Another signaling that is involved in PMNs retention in the MZ is TLR4-dependent. It was found that mice that are deficient in TLR4 even with uncompromised IL-1α signaling cannot retain PMNs in the spleen MZ.[66] As a result, the PMNs are unable to clear the virus-containing cells, which leads to higher viral load.[66] The molecular mechanism of TLR4-dependent PMNs retention in the MZ is not clear and requires further investigation.

2.2.4 Ad INTERACTIONS WITH LUNG MACROPHAGES

Due to highly efficient virus sequestration by liver and spleen macrophages, only 1% of the injected virus can be recovered from the lungs after i.v. injection of Ad in mice.[108] However in certain conditions, like liver cirrhosis, the sequestration occurs predominantly in the lungs.[112] In rodents and humans, pulmonary i.v. macrophages are not constitutively present in the lungs, however, they can be induced during liver cirrhosis (reviewed in Reference [113]). Additionally, during liver cirrhosis the phagocytic macrophages migrate from the liver to other tissues.[114] This accumulation of the phagocytic macrophages in the lungs potentially can overtake some functions of the liver Kupffer cells and sequester the Ad vector. Interestingly, redistribution of sequestration of the virus to the lungs did not increase the transgene expression in the pulmonary tissue, suggesting that in the macrophages the Ad vector was degraded prior to transcription of the genome.[112] However, lung virus sequestration caused severe pulmonary hemorrhagic edema and increased levels of IL-6 and TNF-α, resulting in higher animal mortality compared to healthy rats.[115] The alveolar macrophages are constantly present in the alveoli and persistently survey respiratory epithelial surface.[116] The alveolar macrophages rapidly uptake intratracheally administrated Ad.[117,118] In response to Ad uptake, alveolar macrophages, but not airway epithelial or vascular endothelial cells, activated release of pro-inflammatory cytokines (TNF-α, IL-6, MIP-1α and MIP-2) as early as 30 minutes after administration.[117] The uptake of the Ad vector depends on the surfactant proteins, as the macrophages in mice deficient in major surfactant protein, SP-A, did not efficiently uptake the virus from the lumens.[119] Similar to macrophages in the liver, alveolar macrophages are not permissive for Ad vector replication and do not express a transgene encoded by the virus. Instead, they accumulate the virus and degrade viral DNA.[117] The degradation of viral DNA is a biphasic process. During the early phase, the alveolar macrophages degrade approximately 70% of the Ad DNA. This degradation is independent of the T-cell immune status of mice and occurs within 24 hours post virus administration.[118] The remaining Ad DNA degrades slowly during several days or weeks post virus administration. This second phase of degradation depends on the functions of the CD8 T cells that target the Ad infected cells.[120] Unlike alveolar macrophages, other lung cell types, such as airway epithelial, alveolar lining, and adventitial cells can be effectively transduced by Ad and express a transgene from 2 to 14 days post Ad delivery.[121] Interestingly, similar to the liver, the absence of alveolar macrophages allows for enhanced transduction of other lung cell types with an increase of viral DNA levels and amplified transgene expression.[118] Additionally, immune deficiencies in CD8+ T cells result in a prolonged expression of a transgene encoded by the Ad vector.[120]

In summary, i.v. Ad delivery causes a complex host immune response that is designed to sequester any pathogens to prevent their dissemination. The response involves multiple factors, including soluble blood factors, cells of hematopoietic and non-hematopoietic origin that all together form a cooperating network preventing possible invasion. Numerous studies of virus biodistribution following i.v. injection have shown that over 90% of the administered virus particles are accumulated in the liver and spleen. The mechanisms of Ad vector sequestration depend on an interplay of host factors that can promote vector uptake and transduction, or neutralize the virus. Different organs with their residential macrophages respond to Ad vector administration in a similar fashion that is aimed at eliminating invading pathogens. Although the molecular mechanisms of elimination are slightly variable, they all culminate in the death and removal of virus-containing cells. After trapping Ad particles from the bloodstream, liver residential macrophages, Kupffer cells, undergo IRF3-dependent necrosis. In the spleen, MARCO$^+$ MZ macrophages activate IL-1α-IL1RI-dependent pro-inflammatory cytokine and chemokine production. Cooperation of chemokines CXCL1 and CXCL2 with complement factor C3 results in recruiting and retaining PMNs in the splenic MZ in close proximity to virus-containing MARCO$^+$ macrophages. Activated PMNs eliminate both the virus and virus-containing cells. In the lungs, macrophages sequester Ad and release pro-inflammatory cytokines attracting and activating CD8$^+$ T cells that eliminate virus-infected cells. Collectively, cooperation of tissue macrophages, PMNs, and CTLs results in effective elimination of the virus from the blood and Ad-transduced cells from the tissues. If Ad-mediated therapy is to be successful, effective strategies need to be devised that would prevent Ad interaction with a network of circulating humoral factors and both innate and adaptive arms of the immune system.

REFERENCES

1. Mast, T. C. et al. International epidemiology of human pre-existing adenovirus (Ad) type-5, type-6, type-26 and type-36 neutralizing antibodies: Correlates of high Ad5 titers and implications for potential HIV vaccine trials. *Vaccine* **28**, 950–957, 2010. doi:10.1016/j.vaccine.2009.10.145.
2. Sumida, S. M. et al. Neutralizing antibodies to adenovirus serotype 5 vaccine vectors are directed primarily against the adenovirus hexon protein. *Journal of Immunology* **174**, 7179–7185, 2005.
3. Parker, A. L. et al. Effect of neutralizing sera on factor x-mediated adenovirus serotype 5 gene transfer. *Journal of Virology* **83**, 479–483, 2009. doi:10.1128/JVI.01878-08.
4. Sun, C. et al. Epidemiology of adenovirus type 5 neutralizing antibodies in healthy people and AIDS patients in Guangzhou, southern China. *Vaccine* **29**, 3837–3841, 2011. doi:10.1016/j.vaccine.2011.03.042.
5. Wohlfart, C. Neutralization of adenoviruses: Kinetics, stoichiometry, and mechanisms. *Journal of Virology* **62**, 2321–2328, 1988.
6. Leen, A. M. et al. Identification of hexon-specific CD4 and CD8 T-cell epitopes for vaccine and immunotherapy. *Journal of Virology* **82**, 546–554, 2008. doi:10.1128/JVI.01689-07.
7. Wohlfart, C. E., Svensson, U. K. and Everitt, E. Interaction between HeLa cells and adenovirus type 2 virions neutralized by different antisera. *Journal of Virology* **56**, 896–903, 1985.

8. Watson, G., Burdon, M. G. and Russell, W. C. An antigenic analysis of the adenovirus type 2 fibre polypeptide. *The Journal of General Virology* **69 (Pt 3)**, 525–535, 1988. doi:10.1099/0022-1317-69-3-525.

9. Nemerow, G. R., Stewart, P. L. and Reddy, V. S. Structure of human adenovirus. *Current Opinion in Virology* **2**, 115–121, 2012. doi:10.1016/j.coviro.2011.12.008.

10. Reddy, V. S., Natchiar, S. K., Stewart, P. L. and Nemerow, G. R. Crystal structure of human adenovirus at 3.5 A resolution. *Science* **329**, 1071–1075, 2010. doi:10.1126/science.1187292.

11. Liu, H. et al. Atomic structure of human adenovirus by cryo-EM reveals interactions among protein networks. *Science* **329**, 1038–1043, 2010. doi:10.1126/science.1187433.

12. Bradley, R. R., Lynch, D. M., Iampietro, M. J., Borducchi, E. N. and Barouch, D. H. Adenovirus serotype 5 neutralizing antibodies target both hexon and fiber following vaccination and natural infection. *Journal of Virology* **86**, 625–629, 2012. doi:10.1128/JVI.06254-11.

13. Sarkioja, M. et al. Changing the adenovirus fiber for retaining gene delivery efficacy in the presence of neutralizing antibodies. *Gene Therapy* **15**, 921–929, 2008. doi:10.1038/gt.2008.56.

14. Gahery-Segard, H. et al. Immune response to recombinant capsid proteins of adenovirus in humans: Antifiber and anti-penton base antibodies have a synergistic effect on neutralizing activity. *Journal of Virology* **72**, 2388–2397, 1998.

15. Nemerow, G. R. and Stewart, P. L. Role of alpha(v) integrins in adenovirus cell entry and gene delivery. *Microbiology and Molecular Biology Reviews: MMBR* **63**, 725–734, 1999.

16. Stewart, P. L. et al. Cryo-EM visualization of an exposed RGD epitope on adenovirus that escapes antibody neutralization. *EMBO J* **16**, 1189–1198, 1997. doi:10.1093/emboj/16.6.1189.

17. Smith, J. G. and Nemerow, G. R. Mechanism of adenovirus neutralization by Human alpha-defensins. *Cell Host & Microbe* **3**, 11–19, 2008. doi:10.1016/j.chom.2007.12.001.

18. Bradley, R. R. et al. Adenovirus serotype 5-specific neutralizing antibodies target multiple hexon hypervariable regions. *Journal of Virology* **86**, 1267–1272, 2012. doi:10.1128/JVI.06165-11.

19. Wu, H. et al. Construction and characterization of adenovirus serotype 5 packaged by serotype 3 hexon. *Journal of Virology* **76**, 12775–12782, 2002.

20. Crawford-Miksza, L. and Schnurr, D. P. Analysis of 15 adenovirus hexon proteins reveals the location and structure of seven hypervariable regions containing serotype-specific residues. *Journal of Virology* **70**, 1836–1844, 1996.

21. Roberts, D. M. et al. Hexon-chimaeric adenovirus serotype 5 vectors circumvent pre-existing anti-vector immunity. *Nature* **441**, 239–243, 2006. doi:10.1038/nature04721.

22. Ma, J. et al. Manipulating adenovirus hexon hypervariable loops dictates immune neutralisation and coagulation factor X-dependent cell interaction *in vitro* and *in vivo*. *PLoS Pathogens* **11**, e1004673, 2015. doi:10.1371/journal.ppat.1004673.

23. Tian, X. et al. Mapping the epitope of neutralizing monoclonal antibodies against human adenovirus type 3. *Virus Research* **208**, 66–72, 2015. doi:10.1016/j.virusres.2015.06.002.

24. Pichla-Gollon, S. L. et al. Structure-based identification of a major neutralizing site in an adenovirus hexon. *Journal of Virology* **81**, 1680–1689, 2007. doi:10.1128/JVI.02023-06.

25. Smith, J. G., Cassany, A., Gerace, L., Ralston, R. and Nemerow, G. R. Neutralizing antibody blocks adenovirus infection by arresting microtubule-dependent cytoplasmic transport. *Journal of Virology* **82**, 6492–6500, 2008. doi:10.1128/JVI.00557-08.

26. Cheng, C. et al. Differential specificity and immunogenicity of adenovirus type 5 neutralizing antibodies elicited by natural infection or immunization. *Journal of Virology* **84**, 630–638, 2010. doi:10.1128/JVI.00866-09.

27. Bassett, J. D. et al. CD8+ T-cell expansion and maintenance after recombinant adenovirus immunization rely upon cooperation between hematopoietic and non-hematopoietic antigen-presenting cells. *Blood* **117**, 1146–1155, 2011. doi:10.1182/blood-2010-03-272336.

28. Zandvliet, M. L. et al. Combined CD8+ and CD4+ adenovirus hexon-specific T cells associated with viral clearance after stem cell transplantation as treatment for adenovirus infection. *Haematologica* **95**, 1943–1951, 2010. doi:10.3324/haematol.2010.022947.

29. Colloca, S. et al. Vaccine vectors derived from a large collection of simian adenoviruses induce potent cellular immunity across multiple species. *Science Translational Medicine* **4**, 115ra112, 2012. doi:10.1126/scitranslmed.3002925.

30. Finn, J. D. et al. Persistence of transgene expression influences CD8+ T-cell expansion and maintenance following immunization with recombinant adenovirus. *Journal of Virology* **83**, 12027–12036, 2009. doi:10.1128/JVI.00593-09.

31. Holst, P. J. et al. MHC class II-associated invariant chain linkage of antigen dramatically improves cell-mediated immunity induced by adenovirus vaccines. *Journal of Immunology* **180**, 3339–3346, 2008.

32. Nielsen, K. N., Steffensen, M. A., Christensen, J. P. and Thomsen, A. R. Priming of CD8T cells by adenoviral vectors is critically dependent on B7 and dendritic cells but only partially dependent on CD28 ligation on CD8T cells. *Journal of Immunology* **193**, 1223–1232, 2014. doi:10.4049/jimmunol.1400197.

33. Quinn, K. M. et al. Antigen expression determines adenoviral vaccine potency independent of IFN and STING signaling. *Journal of Clinical Investigation* **125**, 1129–1146, 2015. doi:10.1172/JCI78280.

34. Rhee, E. G. et al. Multiple innate immune pathways contribute to the immunogenicity of recombinant adenovirus vaccine vectors. *Journal of Virology* **85**, 315–323, 2011. doi:10.1128/JVI.01597-10.

35. Kafri, T. et al. Cellular immune response to adenoviral vector infected cells does not require *de novo* viral gene expression: Implications for gene therapy. *Proceedings of the National Academy of Sciences of the United States of America* **95**, 11377–11382, 1998.

36. Hutnick, N. A. et al. Vaccination with Ad5 vectors expands Ad5-specific CD8T cells without altering memory phenotype or functionality. *PLoS One* **5**, e14385, 2010. doi:10.1371/journal.pone.0014385.

37. Shayakhmetov, D. M., Li, Z. Y., Ni, S. and Lieber, A. Analysis of adenovirus sequestration in the liver, transduction of hepatic cells, and innate toxicity after injection of fiber-modified vectors. *Journal of Virology* **78**, 5368–5381, 2004.

38. Kalyuzhniy, O. et al. Adenovirus serotype 5 hexon is critical for virus infection of hepatocytes *in vivo*. *Proceedings of the National Academy of Sciences of the United States of America* **105**, 5483–5488, 2008. doi:10.1073/pnas.0711757105.

39. Tao, N. et al. Sequestration of adenoviral vector by Kupffer cells leads to a nonlinear dose response of transduction in liver. *Molecular Therapy: The Journal of the American Society of Gene Therapy* **3**, 28–35, 2001. doi:10.1006/mthe.2000.0227.

40. Jover, R., Bort, R., Gomez-Lechon, M. J. and Castell, J. V. Cytochrome P450 regulation by hepatocyte nuclear factor 4 in human hepatocytes: A study using adenovirus-mediated antisense targeting. *Hepatology* **33**, 668–675, 2001. doi:10.1053/jhep.2001.22176.

41. Wohlleber, D. et al. TNF-induced target cell killing by CTL activated through cross-presentation. *Cell Reports* **2**, 478–487, 2012. doi:10.1016/j.celrep.2012.08.001.

42. Raper, S. E. et al. Fatal systemic inflammatory response syndrome in a ornithine transcarbamylase deficient patient following adenoviral gene transfer. *Molecular Genetics and Metabolism* **80**, 148–158, 2003.

43. Raper, S. E. et al. A pilot study of *in vivo* liver-directed gene transfer with an adenoviral vector in partial ornithine transcarbamylase deficiency. *Human Gene Therapy* **13**, 163–175, 2002. doi:10.1089/10430340152712719.

44. Morral, N. et al. Lethal toxicity, severe endothelial injury, and a threshold effect with high doses of an adenoviral vector in baboons. *Human Gene Therapy* **13**, 143–154, 2002. doi:10.1089/10430340152712692.
45. Lozier, J. N. et al. Toxicity of a first-generation adenoviral vector in rhesus macaques. *Human Gene Therapy* **13**, 113–124, 2002. doi:10.1089/10430340152712665.
46. Alemany, R., Suzuki, K. and Curiel, D. T. Blood clearance rates of adenovirus type 5 in mice. *The Journal of General Virology* **81**, 2605–2609, 2000.
47. Kirn, D. Clinical research results with dl1520 (Onyx-015), a replication-selective adenovirus for the treatment of cancer: What have we learned? *Gene Therapy* **8**, 89–98, 2001. doi:10.1038/sj.gt.3301377.
48. Worgall, S., Wolff, G., Falck-Pedersen, E. and Crystal, R. G. Innate immune mechanisms dominate elimination of adenoviral vectors following *in vivo* administration. *Human Gene Therapy* **8**, 37–44, 1997. doi:10.1089/hum.1997.8.1-37.
49. Lieber, A. et al. The role of Kupffer cell activation and viral gene expression in early liver toxicity after infusion of recombinant adenovirus vectors. *Journal of Virology* **71**, 8798–8807, 1997.
50. Muruve, D. A., Barnes, M. J., Stillman, I. E. and Libermann, T. A. Adenoviral gene therapy leads to rapid induction of multiple chemokines and acute neutrophil-dependent hepatic injury *in vivo*. *Human Gene Therapy* **10**, 965–976, 1999. doi:10.1089/10430349950018364.
51. Zhang, Y. et al. Acute cytokine response to systemic adenoviral vectors in mice is mediated by dendritic cells and macrophages. *Molecular Therapy: The Journal of the American Society of Gene Therapy* **3**, 697–707, 2001. doi:10.1006/mthe.2001.0329.
52. Borgland, S. L., Bowen, G. P., Wong, N. C., Libermann, T. A. and Muruve, D. A. Adenovirus vector-induced expression of the C-X-C chemokine IP-10 is mediated through capsid-dependent activation of NF-kappaB. *Journal of Virology* **74**, 3941–3947, 2000.
53. Bowen, G. P. et al. Adenovirus vector-induced inflammation: Capsid-dependent induction of the C–C chemokine RANTES requires NF-kappa B. *Human Gene Therapy* **13**, 367–379, 2002. doi:10.1089/10430340252792503.
54. Liu, Q. and Muruve, D. A. Molecular basis of the inflammatory response to adenovirus vectors. *Gene Therapy* **10**, 935–940, 2003. doi:10.1038/sj.gt.3302036.
55. Reid, T. et al. Intra-arterial administration of a replication-selective adenovirus (dl1520) in patients with colorectal carcinoma metastatic to the liver: A phase I trial. *Gene Therapy* **8**, 1618–1626, 2001. doi:10.1038/sj.gt.3301512.
56. Crystal, R. G. et al. Analysis of risk factors for local delivery of low- and intermediate-dose adenovirus gene transfer vectors to individuals with a spectrum of comorbid conditions. *Human Gene Therapy* **13**, 65–100, 2002. doi:10.1089/10430340152712647.
57. Ben-Gary, H., McKinney, R. L., Rosengart, T., Lesser, M. L. and Crystal, R. G. Systemic interleukin-6 responses following administration of adenovirus gene transfer vectors to humans by different routes. *Molecular Therapy: The journal of the American Society of Gene Therapy* **6**, 287–297, 2002.
58. McCoy, R. D. et al. Pulmonary inflammation induced by incomplete or inactivated adenoviral particles. *Human Gene Therapy* **6**, 1553–1560, 1995. doi:10.1089/hum.1995.6.12-1553.
59. Schnell, M. A. et al. Activation of innate immunity in nonhuman primates following intraportal administration of adenoviral vectors. *Molecular Therapy: The Journal of the American Society of Gene Therapy* **3**, 708–722, 2001. doi:10.1006/mthe.2001.0330.
60. Brunetti-Pierri, N. et al. Acute toxicity after high-dose systemic injection of helper-dependent adenoviral vectors into nonhuman primates. *Human Gene Therapy* **15**, 35–46, 2004. doi:10.1089/10430340460732445.

61. Baker, A. H., McVey, J. H., Waddington, S. N., Di Paolo, N. C. and Shayakhmetov, D. M. The influence of blood on *in vivo* adenovirus bio-distribution and transduction. *Molecular Therapy: The Journal of the American Society of Gene Therapy* **15**, 1410–1416, 2007. doi:10.1038/sj.mt.6300206.

62. Shayakhmetov, D. M., Gaggar, A., Ni, S., Li, Z. Y. and Lieber, A. Adenovirus binding to blood factors results in liver cell infection and hepatotoxicity. *Journal of Virology* **79**, 7478–7491, 2005. doi:10.1128/JVI.79.12.7478-7491.2005.

63. Parker, A. L. et al. Multiple vitamin K-dependent coagulation zymogens promote adenovirus-mediated gene delivery to hepatocytes. *Blood* **108**, 2554–2561, 2006. doi:10.1182/blood-2006-04-008532.

64. Lopez-Gordo, E., Denby, L., Nicklin, S. A. and Baker, A. H. The importance of coagulation factors binding to adenovirus: Historical perspectives and implications for gene delivery. *Expert Opinion on Drug Delivery* **11**, 1795–1813, 2014. doi:10.1517/17425247.2014.938637.

65. Irons, E. E. et al. Coagulation factor binding orientation and dimerization may influence infectivity of adenovirus-coagulation factor complexes. *Journal of Virology* **87**, 9610–9619, 2013. doi:10.1128/JVI.01070-13.

66. Doronin, K. et al. Coagulation factor X activates innate immunity to human species C adenovirus. *Science* **338**, 795–798, 2012. doi:10.1126/science.1226625.

67. Waddington, S. N. et al. Adenovirus serotype 5 hexon mediates liver gene transfer. *Cell* **132**, 397–409, 2008. doi:10.1016/j.cell.2008.01.016.

68. Alba, R. et al. Identification of coagulation factor (F)X binding sites on the adenovirus serotype 5 hexon: Effect of mutagenesis on FX interactions and gene transfer. *Blood* **114**, 965–971, 2009. doi:10.1182/blood-2009-03-208835.

69. Kritz, A. B. et al. Adenovirus 5 fibers mutated at the putative HSPG-binding site show restricted retargeting with targeting peptides in the HI loop. *Molecular Therapy: The Journal of the American Society of Gene Therapy* **15**, 741–749, 2007. doi:10.1038/sj.mt.6300094.

70. Xu, Z. et al. Coagulation factor X shields adenovirus type 5 from attack by natural antibodies and complement. *Nature Medicine* **19**, 452–457, 2013. doi:10.1038/nm.3107.

71. Baker, A. H., Nicklin, S. A. and Shayakhmetov, D. M. FX and host defense evasion tactics by adenovirus. *Molecular therapy: The Journal of the American Society of Gene Therapy* **21**, 1109–1111, 2013. doi:10.1038/mt.2013.100.

72. Tian, J. et al. Adenovirus activates complement by distinctly different mechanisms *in vitro* and *in vivo*: Indirect complement activation by virions *in vivo*. *Journal of Virology* **83**, 5648–5658, 2009. doi:10.1128/JVI.00082-09.

73. Shayakhmetov, D. M., Di Paolo, N. C. and Mossman, K. L. Recognition of virus infection and innate host responses to viral gene therapy vectors. *Molecular Therapy: The Journal of the American Society of Gene Therapy* **18**, 1422–1429, 2010. doi:10.1038/mt.2010.124.

74. Lehrer, R. I. and Lu, W. α-Defensins in human innate immunity. *Immunological Reviews* **245**, 84–112, 2012. doi:10.1111/j.1600-065X.2011.01082.x.

75. Bastian, A. and Schafer, H. Human alpha-defensin 1 (HNP-1) inhibits adenoviral infection *in vitro*. *Regulatory Peptides* **101**, 157–161, 2001.

76. Smith, J. G. et al. Insight into the mechanisms of adenovirus capsid disassembly from studies of defensin neutralization. *PLoS Pathogens* **6**, e1000959, 2010. doi:10.1371/journal.ppat.1000959.

77. Snijder, J. et al. Integrin and defensin modulate the mechanical properties of adenovirus. *Journal of Virology* **87**, 2756–2766, 2013. doi:10.1128/JVI.02516-12.

78. Carlisle, R. C. et al. Human erythrocytes bind and inactivate type 5 adenovirus by presenting Coxsackie virus-adenovirus receptor and complement receptor 1. *Blood* **113**, 1909–1918, 2009. doi:10.1182/blood-2008-09-178459.

79. Seiradake, E. et al. The cell adhesion molecule "CAR" and sialic acid on human eryth-rocytes influence adenovirus *in vivo* biodistribution. *PLoS Pathogens* **5**, e1000277, 2009. doi:10.1371/journal.ppat.1000277.

80. Lyons, M. et al. Adenovirus type 5 interactions with human blood cells may compro-mise systemic delivery. *Molecular Therapy: The Journal of the American Society of Gene Therapy* **14**, 118–128, 2006. doi:10.1016/j.ymthe.2006.01.003.

81. He, J. Q. et al. CRIg mediates early Kupffer cell responses to adenovirus. *Journal of leukocyte biology* **93**, 301–306, 2013. doi:10.1189/jlb.0612311.

82. Escutenaire, S. et al. *In vivo* and *in vitro* distribution of type 5 and fiber-modified onco-lytic adenoviruses in human blood compartments. *Annals of Medicine* **43**, 151–163, 2011. doi:10.3109/07853890.2010.538079.

83. Cotter, M. J., Zaiss, A. K. and Muruve, D. A. Neutrophils interact with adenovirus vec-tors via Fc receptors and complement receptor 1. *Journal of Virology* **79**, 14622–14631, 2005. doi:10.1128/JVI.79.23.14622-14631.2005.

84. Othman, M., Labelle, A., Mazzetti, I., Elbatarny, H. S. and Lillicrap, D. Adenovirus-induced thrombocytopenia: The role of von Willebrand factor and P-selectin in mediating accelerated platelet clearance. *Blood* **109**, 2832–2839, 2007. doi:10.1182/blood-2006-06-032524.

85. Wolins, N. et al. Intravenous administration of replication-incompetent adenovirus to rhesus monkeys induces thrombocytopenia by increasing *in vivo* platelet clearance. *British Journal of Haematology* **123**, 903–905, 2003.

86. Gupalo, E., Kuk, C., Qadura, M., Buriachkovskaia, L. and Othman, M. Platelet-adenovirus vs. inert particles interaction: Effect on aggregation and the role of platelet membrane receptors. *Platelets* **24**, 383–391, 2013. doi:10.3109/09537104.2012.703792.

87. Stone, D. et al. Adenovirus-platelet interaction in blood causes virus sequestration to the reticuloendothelial system of the liver. *Journal of Virology* **81**, 4866–4871, 2007. doi:10.1128/JVI.02819-06.

88. Stone, D. et al. Development and assessment of human adenovirus type 11 as a gene transfer vector. *Journal of Virology* **79**, 5090–5104, 2005. doi:10.1128/JVI.79.8.5090-5104.2005.

89. Di Paolo, N. C., van Rooijen, N. and Shayakhmetov, D. M. Redundant and synergistic mechanisms control the sequestration of blood-born adenovirus in the liver. *Molecular Therapy: The Journal of the American Society of Gene Therapy* **17**, 675–684, 2009. doi:10.1038/mt.2008.307.

90. Wheeler, M. D., Yamashina, S., Froh, M., Rusyn, I. and Thurman, R. G. Adenoviral gene delivery can inactivate Kupffer cells: Role of oxidants in NF-kappaB activation and cytokine production. *Journal of Leukocyte Biology* **69**, 622–630, 2001.

91. Alemany, R. and Curiel, D. T. CAR-binding ablation does not change biodistribution and toxicity of adenoviral vectors. *Gene Therapy* **8**, 1347–1353, 2001. doi:10.1038/sj.gt.3301515.

92. Einfeld, D. A. et al. Reducing the native tropism of adenovirus vectors requires removal of both CAR and integrin interactions. *Journal of Virology* **75**, 11284–11291, 2001. doi:10.1128/JVI.75.23.11284-11291.2001.

93. Wickham, T. J., Mathias, P., Cheresh, D. A. and Nemerow, G. R. Integrins alpha v beta 3 and alpha v beta 5 promote adenovirus internalization but not virus attachment. *Cell* **73**, 309–319, 1993.

94. Shashkova, E. V., Doronin, K., Senac, J. S. and Barry, M. A. Macrophage depletion com-bined with anticoagulant therapy increases therapeutic window of systemic treatment with oncolytic adenovirus. *Cancer Research* **68**, 5896–5904, 2008. doi:10.1158/0008-5472.CAN-08-0488.

95. Schiedner, G. et al. Selective depletion or blockade of Kupffer cells leads to enhanced and prolonged hepatic transgene expression using high-capacity adenoviral vectors. *Molecular Therapy: The Journal of the American Society of Gene Therapy* **7**, 35–43, 2003.

96. Xu, Z., Tian, J., Smith, J. S. and Byrnes, A. P. Clearance of adenovirus by Kupffer cells is mediated by scavenger receptors, natural antibodies, and complement. *Journal of Virology* **82**, 11705–11713, 2008. doi:10.1128/JVI.01320-08.

97. Zani, I. A. et al. Scavenger receptor structure and function in health and disease. *Cells* **4**, 178–201, 2015. doi:10.3390/cells4020178.

98. Canton, J., Neculai, D. and Grinstein, S. Scavenger receptors in homeostasis and immunity. *Nature Reviews Immunology* **13**, 621–634, 2013. doi:10.1038/nri3515.

99. Khare, R., Reddy, V. S., Nemerow, G. R. and Barry, M. A. Identification of adenovirus serotype 5 hexon regions that interact with scavenger receptors. *Journal of Virology* **86**, 2293–2301, 2012. doi:10.1128/JVI.05760-11.

100. Di Paolo, N. C. et al. Virus binding to a plasma membrane receptor triggers interleukin-1 alpha-mediated proinflammatory macrophage response *in vivo. Immunity* **31**, 110–121, 2009. doi:10.1016/j.immuni.2009.04.015.

101. Piccolo, P. et al. SR-A and SREC-I are Kupffer and endothelial cell receptors for helper-dependent adenoviral vectors. *Molecular Therapy: The Journal of the American Society of Gene Therapy* **21**, 767–774, 2013. doi:10.1038/mt.2012.287.

102. Piccolo, P., Annunziata, P., Mithbaokar, P. and Brunetti-Pierri, N. SR-A and SREC-I binding peptides increase HDAd-mediated liver transduction. *Gene Therapy* **21**, 950–957, 2014. doi:10.1038/gt.2014.71.

103. Khare, R., Hillestad, M. L., Xu, Z., Byrnes, A. P. and Barry, M. A. Circulating antibodies and macrophages as modulators of adenovirus pharmacology. *Journal of Virology* **87**, 3678–3686, 2013. doi:10.1128/JVI.01392-12.

104. Khare, R. et al. Generation of a Kupffer cell-evading adenovirus for systemic and liver-directed gene transfer. *Molecular Therapy: The Journal of the American Society of Gene Therapy* **19**, 1254–1262, 2011. doi:10.1038/mt.2011.71.

105. Manickan, E. et al. Rapid Kupffer cell death after intravenous injection of adenovirus vectors. *Molecular Therapy: The Journal of the American Society of Gene Therapy* **13**, 108–117, 2006. doi:10.1016/j.ymthe.2005.08.007.

106. Di Paolo, N. C., Doronin, K., Baldwin, L. K., Papayannopoulou, T. and Shayakhmetov, D. M. The transcription factor IRF3 triggers "defensive suicide" necrosis in response to viral and bacterial pathogens. *Cell Reports* **3**, 1840–1846, 2013. doi:10.1016/j.celrep.2013.05.025.

107. Cesta, M. F. Normal structure, function, and histology of the spleen. *Toxicologic Pathology* **34**, 455–465, 2006. doi:10.1080/01926230600867743.

108. Johnson, M., Huyn, S., Burton, J., Sato, M. and Wu, L. Differential biodistribution of adenoviral vector *in vivo* as monitored by bioluminescence imaging and quantitative polymerase chain reaction. *Human Gene Therapy* **17**, 1262–1269, 2006. doi:10.1089/hum.2006.17.1262.

109. Di Paolo, N. C. et al. IL-1 alpha and complement cooperate in triggering local neutrophilic inflammation in response to adenovirus and eliminating virus-containing cells. *PLoS Pathogens* **10**, e1004035, 2014. doi:10.1371/journal.ppat.1004035.

110. Kuida, K. et al. Altered cytokine export and apoptosis in mice deficient in interleukin-1 beta converting enzyme. *Science* **267**, 2000–2003, 1995.

111. Silva, M. T. When two is better than one: Macrophages and neutrophils work in concert in innate immunity as complementary and cooperative partners of a myeloid phagocyte system. *Journal of Leukocyte Biology* **87**, 93–106, 2010.

112. Smith, J. S., Tian, J., Muller, J. and Byrnes, A. P. Unexpected pulmonary uptake of adenovirus vectors in animals with chronic liver disease. *Gene Therapy* **11**, 431–438, 2004. doi:10.1038/sj.gt.3302149.

113. Schneberger, D., Aharonson-Raz, K. and Singh, B. Pulmonary intravascular macrophages and lung health: What are we missing? *American Journal of Physiology. Lung Cellular and Molecular Physiology* **302**, L498–503, 2012. doi:10.1152/ajplung.00322.2011.

114. Holmberg, J. T. et al. Radiolabelled colloid uptake distribution and pulmonary contents and localization of lysosomal enzymes in cholestatic rats. *Scandinavian Journal of Gastroenterology* **21**, 291–299, 1986.

115. Smith, J. S., Tian, J., Lozier, J. N. and Byrnes, A. P. Severe pulmonary pathology after intravenous administration of vectors in cirrhotic rats. *Molecular Therapy: The Journal of the American Society of Gene Therapy* **9**, 932–941, 2004. doi:10.1016/j.ymthe.2004.03.010.

116. Hussell, T. and Bell, T. J. Alveolar macrophages: Plasticity in a tissue-specific context. *Nature reviews. Immunology* **14**, 81–93, 2014. doi:10.1038/nri3600.

117. Zsengeller, Z., Otake, K., Hossain, S. A., Berclaz, P. Y. and Trapnell, B. C. Internalization of adenovirus by alveolar macrophages initiates early proinflammatory signaling during acute respiratory tract infection. *Journal of Virology* **74**, 9655–9667, 2000.

118. Worgall, S. et al. Role of alveolar macrophages in rapid elimination of adenovirus vectors administered to the epithelial surface of the respiratory tract. *Human Gene Therapy* **8**, 1675–1684, 1997. doi:10.1089/hum.1997.8.14-1675.

119. Harrod, K. S., Trapnell, B. C., Otake, K., Korfhagen, T. R. and Whitsett, J. A. SP-A enhances viral clearance and inhibits inflammation after pulmonary adenoviral infection. *The American Journal of Physiology* **277**, L580–588, 1999.

120. Yang, Y., Li, Q., Ertl, H. C. and Wilson, J. M. Cellular and humoral immune responses to viral antigens create barriers to lung-directed gene therapy with recombinant adenoviruses. *Journal of Virology* **69**, 2004–2015, 1995.

121. Rodman, D. M. et al. In vivo gene delivery to the pulmonary circulation in rats: Transgene distribution and vascular inflammatory response. *American Journal of Respiratory Cell and Molecular Biology* **16**, 640–649, 1997. doi:10.1165/ajrcmb.16.6.9191465.

3 Helper-Dependent Adenoviral Vectors for Cell and Gene Therapy

Nicola Brunetti-Pierri and Philip Ng

CONTENTS

ABSTRACT

Helper-dependent adenoviral vectors (HDAds) that are deleted of all viral genes can efficiently transduce a wide variety of dividing and nondividing cell types to mediate high levels of transgene expression. In contrast to early generation adenoviral vectors, the lack of viral genes in HDAds results in long-term transgene expression without chronic toxicity. Moreover, deletion of viral genes permits a large cloning capacity of 37 kb. Because HDAd genomes exist in transduced cells as non-integrating episomes, the risks of germ line transmission and insertional mutagenesis are negligible. This chapter will review studies of particular significance and describe the most recent advancements in this vector technology.

3.1   INTRODUCTION

Helper-dependent adenoviral vectors (HDAds) (also referred to as gutless, gutted, mini, fully-deleted, high-capacity, Δ, pseudo, encapsidated adenovirus mini-chromosome) are deleted of all viral coding sequences. Like early generation adenoviral (Ad) vectors, HDAds can efficiently transduce a wide variety of dividing and non-dividing cells to mediate high-level transgene expression. In contrast to early generation Ad vectors, the lack of viral genes enables HDAds to mediate long-term transgene expression without chronic toxicity. Moreover, deletion of viral genes permits a large cloning capacity of 37 kb. Because HDAd genomes exist in episomal states in the nuclei of transduced cells, the risks of germline transmission and insertional mutagenesis are negligible. Many methods for producing HDAds and many *in vitro*, *ex vivo*, and *in vivo* applications of HDAd-mediated gene transfer for gene and cell therapy have been reported. Rather than providing a comprehensive review of all of these studies, this chapter describes examples of particular significance or interest.

3.2 HDAd PRODUCTION

The most efficient and widely used method for producing HDAds is based on the Cre/loxP system (Parks et al. 1996) (Figure 3.1). Although other strategies have been developed (Alba et al. 2007; Ng et al. 2001; Parks 2005; Sargent et al. 2004; Umana et al. 2001), the Cre/loxP system remains the method of choice for HDAd production based on vector yield and purity. In the Cre/loxP system the HDAd genome is first constructed in a bacterial plasmid. Minimally, the HDAd genome includes the expression cassette of interest and ~500 bp of *cis*-acting Ad sequences required for vector DNA replication: inverted terminal repeats (ITRs) and packaging signal (ψ). Moreover, a small segment of noncoding Ad sequence from the E4 region adjacent to the right ITR can be included to increase vector yield, possibly by enhancing packaging of the HDAd genome (Sandig et al. 2000). Stuffer DNA is often required to bring the size of the HDAd genome up to the packaging requirements of the viral capsid which is between 27.7 and 37.8 kb (Bett et al. 1993; Parks and Graham 1997).

To convert the plasmid form of the HDAd genome into the viral form, the plasmid is first digested with the appropriate restriction enzyme to release the HDAd genome from the bacterial plasmid sequences. 293 cells expressing the Cre recombinase (293Cre) are then transfected with the linearized HDAd genome and subsequently infected with a helper virus (HV). The HV is an E1-deleted Ad that bears a packaging signal flanked by loxP sites. Therefore, following infection of 293Cre cells, the packaging signal is excised from the HV genome by Cre-mediated site-specific recombination between the loxP sites (Figure 3.1) rendering the HV genome unpackagable but still able to replicate and *trans*-complement the replication and encapsidation of the HDAd genome. The titer of the HDAd is increased by serial co-infections of 293Cre cells with the HDAd and the HV. Improved reagents and optimized methods have permitted rapid and robust large-scale production of high-quality HDAd with very low HV contamination levels (Palmer and Ng 2003). Detailed methodologies for producing HDAd are described elsewhere (Palmer and Ng 2007, 2008). Because the HV is an E1-deleted Ad, the generation of replication-competent Ad (RCA; E1[+]) as a consequence of homologous recombination between the HV and the Ad sequences present in 293-derived cells can occur (Parks et al. 1996; Reddy et al. 2002). To prevent the formation of RCA, a "stuffer" sequence is inserted into the E3

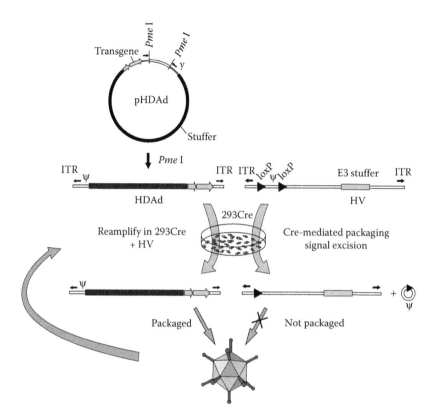

FIGURE 3.1 The Cre/loxP system for generating HDAd vectors. The HDAd contains only ~500 bp of *cis*-acting Ad sequences required for DNA replication (ITRs) and packaging (ψ), the remainder of the genome consists of the desired transgene and non-Ad "stuffer" sequences. The HDAd genome is constructed as a bacterial plasmid (pHDAd) and is liberated by restriction enzyme digestion (e.g., *Pme*I). To rescue the HDAd, the liberated genome is transfected into 293 cells expressing Cre and infected with an HV bearing a packaging signal (ψ) flanked by loxP sites. The HV genome also contains a stuffer sequence in E3 to prevent the formation of RCA in 293-derived cells. Cre-mediated excision of ψ renders the HV genome unpackagable, but still able to replicate and provide all of the necessary *trans*-acting factors for propagation of the HDAd. The titer of the HDAd vector is increased by serial co-infections of 293Cre cells with the HDAd and the HV.

region of the HV to render any E1+ recombinants too large to be packaged (Parks et al. 1996; Suzuki et al. 2011; Umana et al. 2001).

Although current systems cannot produce HDAd free of HV, the HV is an E1-deleted Ad and is present in HDAd preparation at low contaminating amounts which are far below the much higher amounts of E1-deleted Ad that have been administered to several patients in clinical trials without adverse events. Moreover, in mouse models, intravenous (Reddy et al. 2002) or intramuscular (Maione et al. 2001) injections of HDAd with up to 10% HV contamination did not reduce the duration of transgene expression or result in significantly higher toxicity compared to preparations with only 0.1%–0.5% contamination.

The large cloning capacity of HDAd permits inclusion of transgenes in their native genomic context that results in superior level and duration of expression of transgenes compared to their cDNA counterparts (Kim et al. 2001; Schiedner et al. 1998, 2002). This is likely due to a more physiological regulation of gene expression. HDAds offer the advantage of transferring multiple genes in their genomic context in contrast to other vectors (e.g., early generation Ad, retroviral, lentiviral, or adeno-associated viral [AAV] vectors) that have limited cloning capacity.

Because HDAd genomes exist episomally, high-level transgene expression will likely not be permanent, especially in dividing transduced cells. Unfortunately, simple vector re-administration to reestablish transgene expression is not possible because the initial administration elicits neutralizing anti-Ad antibodies. Fortunately, serotype switching has overcome this problem. In addition to the Ad serotype 5-based HV, other HV based on serotypes 1, 2, and 6 have been generated (Parks et al. 1999; Weaver et al. 2009a,b). Therefore, genetically identical HDAds of different serotypes can be generated simply by changing the serotype of the HV used for vector production. There are ~50 human serotypes of Ad. Therefore, it may be possible to create a panel of different serotype HVs for producing different serotypes but genetically identical HDAds. These HDAds could then be given sequentially when transgene expression wanes from the previous vector administration, as shown by several *in vivo* studies, especially when transduced cells are dividing (Kim et al. 2001; Parks et al. 1999; Weaver et al. 2009a,b). Given the large number of Ad serotypes, this could be theoretically repeated for the lifetime of the patient.

A number of serotype 5 HVs containing genetic elements from other serotype Ads have been developed for the production of chimeric HDAds with novel and useful properties. For example, the fiber gene of a serotype 5 HV was replaced with the fiber gene from serotype 35 and this chimeric HV was used to produce a chimeric HDAd that utilized CD46 as the cellular receptor instead of coxsackie and adenovirus receptor (CAR) (Shayakhmetov et al. 2004) to transduce cells that do not express CAR. Similarly, serotype 5 HV bearing the fiber knob domain from serotype 3 was used to produce a chimeric HDAd that could more efficiently transduce adult muscle in mice (Guse et al. 2012). Replacement of the capsid hexon protein hypervariable regions (HVRs) in the serotype 5 HV with the HVRs from serotype 6 permitted production of a chimeric HDAd able to evade Kupffer cell uptake following intravenous injection into mice (Khare et al. 2011). Moreover, desired chimeric capsid-modified Ad can be generated from different sources and switching between commonly used early generation Ad and HDAd vectors allowing the generation of tailored Ad vectors with distinct features (Muck-Hausl et al. 2015).

3.3 INTRACELLULAR STATUS OF HDAd VECTORS

HDAd genomes appear to exist in the nucleus of transduced cells as replication-deficient linear monomers both in cell culture and mouse livers (Jager and Ehrhardt 2009). However, one study found that approximately 1%–3% of HDAd genomes have circularized and contain end-to-end joining of the Ad termini, at least in cell culture (Kreppel and Kochanek 2004). The intracellular HDAd genome is assembled into chromatin through association with cellular histones that promotes efficient transgene expression (Ross et al. 2009, 2011).

Although Ad and Ad vectors are episomal, it is reasonable to assume that integration of vector DNA into the host genome occurs sporadically. Several studies in cell culture have investigated the frequency of HDAd genome integration and found random integration frequencies to be 10^{-3} to 10^{-5} per cell, depending on the experimental conditions (Harui et al. 1999; Hillgenberg et al. 2001; Ohbayashi et al. 2005; Stephen et al. 2008; Suzuki et al. 2008). Further, these studies revealed that vector genomes appear to integrate as intact monomers with little or no loss of sequences at the vector ends, and an apparent preference for integration into genes (Hillgenberg et al. 2001; Stephen et al. 2008). However, artificial conditions of cell culture and inherent genetic instability of cultured cells might have overestimated the frequency of HDAd genomic integration (Stephen et al. 2010). Indeed, a lower *in vivo* integration frequency of 6.72×10^{-5} per hepatocyte was detected in mouse livers compared to culture cells, and the vector appeared to integrate through its termini (Stephen et al. 2010).

In non-dividing cells, the HDAd genome exists as a stable episome thus conferring stable transgene expression. However, in dividing cells, the HDAd genome is lost because it does not integrate into the host DNA, and it lacks any replicative or nuclear retention mechanisms. Therefore, HDAds are not useful for long-term expression in actively dividing cells. Nevertheless, episomal persistence of HDAd genome in murine liver is greater than plasmid DNA by a mechanism that has not been yet identified (Ehrhardt et al. 2003). To overcome loss in dividing cells, several hybrid vectors based on the HDAd platform have been developed which are reviewed in depth in Chapter 5 of this book.

3.4   *EX VIVO* GENE TRANSFER

3.4.1   HDAd AND STEM CELLS

Embryonic stem cells (ESCs) and induced pluripotent stem cells (iPSCs) are promising for cell-based therapies and regenerative medicine. Targeted gene repair/editing of stem cells offer the potential for autologous cell therapy for treatment of a wide spectrum of human diseases. High-efficiency HDAd-mediated gene targeting has been accomplished in mouse ESCs (Ohbayashi et al. 2005), monkey ESCs (Suzuki et al. 2008), and into a wide variety of loci in different human ESCs and iPSCs (Aizawa et al. 2012; Li et al. 2011; Liu et al. 2011; Saydaminova et al. 2015; Umeda et al. 2013). Collectively, these studies showed that HDAd can mediate efficient knockin and knockout at transcriptionally active or inactive loci in human ESCs and iPSCs with no effect on undifferentiated state and pluripotency, no ectopic random integration/off-target effects and/or introduction of additional mutations, while maintaining genetic and epigenetic integrity. Indeed, targeted gene correction of human iPSCs by HDAd minimally impacts whole-genome mutational load as determined by whole-genome sequencing (Suzuki et al. 2014). Importantly, the disease phenotype was reversed in patient-derived cells after HDAd-mediated gene targeting (Li et al. 2011; Liu et al. 2011).

In summary, as a gene targeting vector for stem cells, HDAds offer several advantages: (i) they can transduce stem cells with high efficiency and low cytotoxicity; (ii) they can accommodate very long regions of homology to the target chromosome locus to enhance homologous recombination and permit a wide range of choices for promoters and selectable markers; (iii) they may result in correction of multiple mutations at the target locus; (iv) their high efficient transduction contributes to high targeting efficiency thus allowing for smaller numbers of starting cells. Moreover, the introduction of artificial double-strand breaks at the target loci, which are potentially mutagenic and may result in unpredictable and undetectable off-target effects, is not required.

3.4.2 Ex vivo Gene Transfer in Dermal Fibroblasts

HDAds have been used in an *ex vivo* clinical trial to treat patients with anemia secondary to chronic kidney failure (Mitrani et al. 2011). In this phase I–II study, a small number of dermal fibroblasts removed from the patients' skin were transduced *ex vivo* with an HDAd expressing erythropoietin (EPO) and implanted autologously in the subcutaneous tissue under local anesthesia. Following transduction, the amount of EPO produced by transduced cells was measured so that a precise number of transduced cells could be implanted to achieve pre-determined blood EPO levels. No adverse events were reported in this trial and importantly, increased hemoglobin levels were sustained for up to 1 year after a single treatment with the HDAd transduced cells (Mitrani et al. 2011). Although it remains to be established whether this approach is effective for diseases requiring blood levels of therapeutic proteins higher than EPO, this safe and simple approach has potential for applications in several diseases.

3.5   IN VIVO GENE THERAPY

3.5.1 Liver-Directed Gene Therapy

The liver is an attractive target for gene therapy because it is affected in several acquired and genetic diseases, and can be used as a factory for secretion of vector-encoded therapeutic proteins into the blood circulation. Ad vectors, including HDAds, have been used extensively to investigate liver-directed gene therapy because simple intravenous injection of Ad-based vectors can result in efficient hepatocyte transduction. Many examples of *in vivo* HDAd-mediated, liver-directed gene therapy have been reported. In general, all these studies show that HDAd can mediate long-term phenotypic correction without chronic toxicity (Brunetti-Pierri and Ng 2011). For example, the correction of hypercholesterolemia in apolipoprotein E (apoE)-deficient mice (Kim et al. 2001) and hyperbilirubinemia in the rat model of Crigler–Najjar syndrome (Toietta et al. 2005) are paradigmatic; these studies showed that a single intravenous injection of HDAd resulted in lifelong expression of the therapeutic transgene and permanent phenotypic correction of a genetic disease with negligible toxicity (Kim et al. 2001; Toietta et al. 2005). Importantly, long-term expression by HDAd has also been recapitulated in clinically relevant large animal models (Brunetti-Pierri et al. 2005a, 2006, 2007, 2012, 2013; McCormack et al. 2006; Morral et al. 1999).

Unfortunately, high HDAd doses are required to achieve efficient hepatic transduction following systemic intravascular administration. This is because there is a nonlinear dose response to hepatic transduction following intravenous Ad injection; low doses yield very low to undetectable levels of transgene expression, but high doses resulting in disproportionately high levels of transgene expression. The Kupffer cells of the liver are responsible for this nonlinear dose response by avidly sequestering blood-borne Ad particles (Schiedner et al. 2003b; Tao et al. 2001). Natural or preexisting neutralizing antibodies are involved in vector uptake by the Kupffer cells (Khare et al. 2013; Xu et al. 2013) through opsonization of vector particles that enhance Fc-receptor-mediated vector uptake. Antibodies can also bind indirectly to viral particles through binding to complement factor C3 (Xu et al. 2008). Antibody-virus complexes activate the classical complement proteins C1, C2, and C4 (Tian et al. 2009; Xu et al. 2013). Complement activation results in covalent binding of C3 fragments to the viral capsid and Ad particle uptake by the Kupffer cells via complement receptor Ig-superfamily (CRIg) that regulate death of these cells in the liver (He et al. 2013). Following uptake of Ad, the Kupffer cells undergo rapid pro-inflammatory necrotic death that is controlled by interferon-regulatory factor 3 (IRF3) (Di Paolo et al. 2013; Nociari et al. 2007; Schiedner et al. 2003a; Smith et al. 2008). Ad uptake by the Kupffer cells and their necrosis appear to play a protective role and may represent a defensive suicide strategy preventing disseminated virus infection (Doronin et al. 2012). Consequently, for the host that lacks this macrophage population, even a sub-lethal virus infection may lead to compromised resistance and be detrimental for survival. Mice depleted of tissue macrophages by clodronate liposomes indeed showed higher virus DNA burden, greater hepatotoxicity, and increased lethality (Doronin et al. 2012). The observation that administration of polyinosine, as well as other polyanionic ligands, into mice prior to intravenous Ad injection drastically reduces Ad accumulation in the Kupffer cells and increases hepatocyte gene transfer (Haisma et al. 2008; Xu et al. 2008) has led to the recognition of scavenger receptor-A (SR-A) and scavenger receptor expressed on endothelial cells type I (SREC-I) on both Kupffer cells and endothelial cells as mediators of Ad vector uptake (Haisma et al. 2009; Khare et al. 2012; Piccolo et al. 2013, 2014).

In addition, Ad-mediated hepatocyte transduction is hampered by the physical barrier of liver endothelial fenestrations (Lievens et al. 2004; Snoeys et al. 2007; Wisse et al. 2008): Ad5 particles have a diameter of 93 nm with protruding fibers of 30 nm (Snoeys et al. 2007) whereas the diameter of human liver fenestration is ~107 nm and thus, the relative smaller size of liver fenestrations may be an obstacle for hepatocyte transduction in humans (Wisse et al. 2008).

Systemic administration of high doses of Ad vectors likely results in widespread transduction of a large number of various extrahepatic cell types (e.g., blood cells, endothelium, spleen, and lung) which are important barriers to efficient hepatocyte transduction. Over 90% of Ad vectors bind to human erythrocytes *ex vivo* (Lyons et al. 2006) through the CAR that is expressed on erythrocytes from humans but not from mice or rhesus macaques (Carlisle et al. 2009). Furthermore, erythrocytes from humans but not from mice bear the complement receptor 1 (CR1) which binds

Ad5 in the presence of antibodies and complement (Carlisle et al. 2009). Although the liver takes up more vector than any other organ following systemic injection, the total amount of vector, on a vector genome copy number per μg of genomic DNA basis, is abundantly distributed throughout the body in mice (Zhang et al. 2001), in nonhuman primates (Schnell et al. 2001; Varnavski et al. 2002), and in a human patient (Raper et al. 2003).

Unfortunately, the systemic intravascular administration of high vector doses needed to achieve efficient hepatic transduction results in activation of the innate inflammatory response and acute toxicity. This acute toxic response occurs immediately after vector administration and its severity is dose-dependent as a consequence of activation of the innate immune response by the viral capsid proteins (Liu and Muruve 2003; Muruve et al. 1999; Schnell et al. 2001; Zhang et al. 2001). Indeed, dose-dependent acute toxicity, consistent with activation of the innate inflammatory immune response, was observed following systemic administration of HDAd into nonhuman primates (Brunetti-Pierri et al. 2004). The role of the viral capsid in causing acute toxicity was confirmed by studies that have shown induction of acute expression of several inflammatory cytokine and chemokines following intravenous injection of either FGAd or HDAd vectors into mice (Muruve et al. 2004). However, FGAd but not HDAd vectors, also induced a second phase of liver inflammation at 7 days post-injection. These studies demonstrate that HDAds induce intact innate but attenuated adaptive immune responses *in vivo*. A detailed review regarding activation of the innate inflammatory response to Ad vectors can be found in Chapter 10 of this book.

However, it is clear that a threshold of innate immune activation must first be attained, as a consequence of high dose and systemic exposure of the vector to many cell types and blood-borne components, before severe and lethal acute toxicity manifests. Evidence of robust activation of the acute inflammatory response is observed in both rodents and nonhuman primates given a comparable systemic high dose Ad (on a per kg basis). However, there are differences in the activation of the acute response and rodents can tolerate higher doses compared to primates. In addition, there are differences even among mouse strains in the degree of activation of the acute toxicity. This may reflect intra-species and inter-species differences in the quality of the innate immune response or sensitivities of the end organs to pathologic sequelae (Schnell et al. 2001; Zhang et al. 2001). Differences in natural IgM between BALB/c and C57BL/6 mice have been found to contribute to lower efficiency of Ad5 gene transfer in BALB/c mice (Qiu et al. 2015) and are possibly involved in the difference in activation of the acute toxicity.

There has been a single case of intravascular administration of HDAd into a human patient. In this clinical trial, 4.3×10^{11} vp/kg of an HDAd expressing factor VIII (FVIII) was intravenously injected into a hemophilia A patient (White and Monahan 2007). This subject developed grade 3 liver toxicity, marked increase in IL-6, thrombocytopenia, and laboratory signs of disseminated intravascular coagulopathy, but all these values returned to baseline by day 19 post-infusion. Unfortunately, no evidence of FVIII expression was detected (White and Monahan 2007) and much detail remains unknown about this unpublished study.

Different strategies to overcome the threshold to hepatocyte transduction and the obstacle of acute toxicity have been reported. Because the severity of the acute response is dose-dependent and appears to correlate with extrahepatic systemic vector dissemination, one such approach involved preferential targeting of the vector to the liver, thereby allowing for lower vector doses. One strategy involved injection of HDAd directly into the surgically isolated liver in nonhuman primates and was shown to achieve higher efficiency hepatic transduction with reduced systemic vector dissemination compared to systemic injection (Brunetti-Pierri et al. 2006). However, this approach is invasive and consequently, minimally invasive percutaneous balloon occlusion catheter-based methods were developed to achieve preferential hepatocyte transduction. One such strategy mimics hydrodynamic injection but without rapid, large-volume injection (Brunetti-Pierri et al. 2007) and was based on the observation that hydrodynamic injection of HDAd into mice resulted in increased hepatocyte transduction with reduced markers of acute toxicity and reduced systemic vector dissemination (Brunetti-Pierri et al. 2005b). This so-called pseudo-hydrodynamic injection, developed in nonhuman primates, involved transient occlusion of hepatic venous outflow using two balloon occlusion catheters percutaneously placed in the inferior vena cava (IVC), above and below the hepatic veins (HV). Because blood entering the liver from the hepatic artery (HA) and portal vein (PV) remains unobstructed, an increase in intrahepatic pressure is achieved, mimicking the high pressures achieved by systemic hydrodynamic injection in mice. Following balloon deflation, the vector is then administered by peripheral intravenous injection in small volume which resulted in high-efficiency hepatic transduction with minimal toxicity in nonhuman primates (Brunetti-Pierri et al. 2007). In a subsequent refined method, a balloon occlusion catheter was percutaneously positioned in the IVC to occlude hepatic venous outflow and HDAd was injected directly into the occluded liver via a percutaneously placed HA catheter (Figure 3.2a). This resulted in up to 80-fold improvement in hepatic transduction compared to systemic vector injection with negligible toxicity in nonhuman primates (Figure 3.2b) (Brunetti-Pierri et al. 2009). This balloon catheter method was used to deliver a low dose of HDAd expressing human factor IX (hFIX) into rhesus macaques which resulted in plasma hFIX levels within the therapeutic range for up to 2.8 years post-injection (Brunetti-Pierri et al. 2012). However, a similar approach to deliver the low-density lipoprotein receptor (LDLR) gene in LDLR +/− monkeys resulted in short-term reduction of the hypercholesterolemia (Oka et al. 2015). The reasons for this different outcome are unclear although differences in the amount of LDLR need to lower blood cholesterol versus the amount of FIX needed for detection by ELISA may be responsible; LDLR is a membrane protein and high LDLR levels are required to achieve measurable effects on plasma cholesterol, whereas FIX is a secreted protein and sensitive assays are available to measure even very low plasma levels.

A follow-up of the three methods above reported that transgene expression had persisted for up to 7 years which is more than half the life span of most captive baboons (Martin et al. 2002), for all injected nonhuman primates without long-term adverse effects. However, in all cases, transgene expression levels slowly declined over time to less than 10% of peak values by the end of the observation period but remained 2.3–111-fold above baseline values (Figure 3.3) (Brunetti-Pierri et al. 2013). The slow, steady decline in transgene expression over time is likely dependent

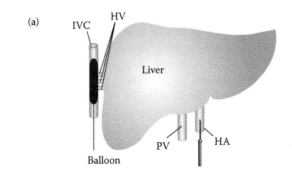

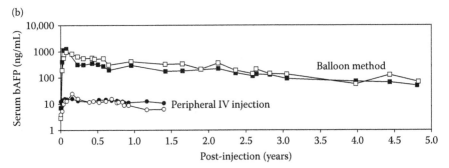

FIGURE 3.2   (a) A sausage-shaped balloon catheter is positioned in the IVC under fluoro-scopic guidance. Inflation of the balloon results in hepatic venous outflow occlusion from the HV. The HDAd is administered by injection through a percutaneously positioned HA cath-eter. (b) Serum baboon alpha-fetoprotein (bAFP) levels following administration of HDAd expressing bAFP by either the balloon method or by peripherial intravenous (IV) injection. (Adapted from Brunetti-Pierri N et al. 2009. *Mol Ther* 17(2):327–333.)

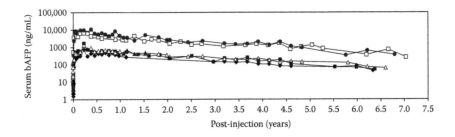

FIGURE 3.3   Long-term transgene expression in nonhuman primates following injection of HDAd directly into the isolated liver. Duration of transgene expression following administra-tion of HDAd expressing bAFP in baboons by the method described previously (Brunetti-Pierri et al. 2006). (Adapted from Brunetti-Pierri N et al. 2009. *Mol Ther* 17(2):327–333.)

upon the gradual loss of transduced hepatocytes due to physiologic hepatocyte turn-over, loss of the extrachromosomal vector genome, or a combination of both.

Physical or chemical methods to enlarge fenestration diameter could increase hepatocyte transduction with lower vector doses thereby reducing the acute toxicity. For example, *N*-acetylcysteine combined with transient liver ischemia (Snoeys et al. 2007) and Na-decanoate (Lievens et al. 2004) have been shown to increase the size of sinusoidal fenestrae and augment Ad-mediated hepatocyte transduction. Moreover, pretreatment with vasoactive intestinal peptide (VIP) increases hepatic transduction and reduces innate immune response following administration of HDAd (Vetrini et al. 2010).

Specific coupling of 5 K polyethylene glycol (PEG) or transferrin to the hexon capsid protein of FGAd and HDAd vectors can improve liver transduction (Prill et al. 2011). The mechanism for this improvement appears to be the evasion of Kupffer cells. Avoiding Kupffer cell uptake using a chimeric vector, in which the hypervari-able region of Ad5 is replaced with that of Ad6, increases liver transduction approxi-mately 10-fold in BALB/c mice. Additionally, ALT levels were significantly lower in mice given the Ad5/6 chimeric vector than mice receiving the Ad5 vector (Khare et al. 2011). Scavenger receptors on Kupffer cells bind Ad particles and remove them from the circulation, thus preventing hepatocyte transduction (Haisma et al. 2008; Xu et al. 2008). HDAd particles interact *in vitro* and *in vivo* with SR-A and with SREC-I and by blocking these receptors with specific antigen-binding fragments or small peptides, the efficiency of hepatocyte transduction *in vivo* by HDAd can be increased (Piccolo et al. 2013, 2014).

"Masking" of the viral capsid by encapsidation within bilamellar cationic lipo-somes (Yotnda et al. 2002) and PEGylation (Croyle et al. 2005; Leggiero et al. 2013; Mok et al. 2005) attenuates the severity of the innate inflammatory response with-out compromising hepatic transduction efficiency. IL-6 elevation is further inhibited by PEGylated Ad in combination with methylprednisolone, an anti-inflammatory glucocorticoid (De Geest et al. 2005). Similarly, a single administration of dexa-methasone, another anti-inflammatory glucocorticoid, prior to Ad administration significantly reduced both the innate and adaptive immune responses (Seregin et al. 2009). These approaches, individually or in combination, may improve the therapeu-tic index of HDAd. Although the above studies showed that hepatic transduction in mice was not compromised by PEGylated HDAd, this was not the case in nonhuman primates where hepatic transduction was reduced (Wonganan et al. 2010), empha-sizing that caution should be taken in extrapolating results from rodents to larger animals and humans.

3.5.2 Lung Gene Therapy

Ad vectors have been extensively used for pulmonary gene transfer with the goal of treating cystic fibrosis (CF) due to mutations in the cystic fibrosis conductance regulator (*CFTR*) gene. However, shortcomings limiting applications of Ad vectors for lung-directed gene transfer include: (i) a dose-dependent inflammation and pneu-monia induced by pulmonary delivery of FGAd vectors (Harvey et al. 2002; Joseph et al. 2001; Simon et al. 1993; Wilmott et al. 1996; Yei et al. 1994), and (ii) inefficient pulmonary delivery in small and large animals, and humans (Grubb et al. 1994; Harvey et al. 1999; Joseph et al. 2001; Perricone et al. 2001; Zuckerman et al. 1999) due to localization of cellular receptor for Ad (and other viral vectors) on the baso-lateral surface of airway epithelial cells (Pickles et al. 2000). Transient disruption of the tight junctions of airway epithelial cells by various chemicals such as EGTA (Chu et al. 2001), EDTA (Wang et al. 2000), polycations (Kaplan et al. 1998), sodium caprate (Johnson et al. 2003), L-α-lysophosphatidylcholine (LPC) (Koehler et al. 2005; Limberis et al. 2002) and other agents, allow vector access to their basolateral receptors to permit efficient transduction.

Pulmonary inflammation following delivery of FGAd has been attributed to expression of viral genes present in the vector backbone which is directly cyto-toxic and also provokes an adaptive cellular immune response against the trans-duced cells consequently resulting in transient transgene expression and toxicity (Dai et al. 1995; Morral et al. 1997; O'Neal et al. 1998; Yang et al. 1994, 1995a,b). Therefore, this toxicity is not expected to occur with HDAds. Indeed, in contrast to FGAd, lungs of mice receiving HDAd following EGTA to disrupt tight junc-tions by nasal instillations were free of inflammation and indistinguishable from saline-treated controls (Toietta et al. 2003). Moreover, HDAd resulted in exten-sive and long-term transduction of proximal and distal airways, from the trachea to bronchiolar epithelium and submucosal glands, which are considered important targets for CF gene therapy, and little transduction of alveolar cells (Koehler et al. 2001; O'Neal et al. 2000). Using the relatively large (4.1 kb) human cytokeratin 18 (K18) control elements, that can be easily accommodated into HDAd, the trans-gene expression pattern was restricted to cells that physiologically express CFTR, including the epithelium of large airways, bronchioles, and submucosal glands with the exclusion of alveoli (Koehler et al. 2001). Taken together, these results indi-cated that intranasal administration of HDAd following disruption of tight junctions resulted in high-efficiency pulmonary transduction in cell types relevant for CF. Moreover, HDAd can express properly localized CFTR in the appropriate target cell types for CF gene therapy in mice (Koehler et al. 2003) and in nasal epithelial cells obtained from CF patients (Cao et al. 2015), and protects the airways from infections by opportunistic pathogens (Koehler et al. 2003). Finally, HDAd vec-tors can be re-administered to the airways through transient immunosuppression in mice (Cao et al. 2011).

The intranasal delivery as performed in mice (spontaneous liquid inhalation) is not applicable to larger animals. A clinically relevant method of vector delivery was developed using an intracorporeal nebulizing catheter (AeroProbe, Trudell Medical International) which aerosolizes material directly into the trachea and lungs. Using the AeroProbe, to aerosolize 0.1% LPC containing the HDAd vectors carrying an K18 promoter driving LacZ (HDAd-K18LacZ) into the lungs of rabbits (Koehler et al. 2005), exceedingly high and unprecedented levels gene transfer were achieved in all cell types of the proximal and distal airway epithelium, from the trachea to terminal bronchioles. All rabbits, including those given LPC only as controls, exhib-ited a transient decrease in dynamic lung compliance immediately following aerosol delivery. Fever and mild-to-moderate patchy pneumonia without edema likely due also to LPC were also observed indicating that further optimization of the dose of LPC and/or vector is needed to eliminate/minimize this toxicity.

Moreover, HDAds have been investigated for airway delivery in pigs and ferrets. The HDAd vectors carrying the K18 promoter driving LacZ or human CFTR cDNA formulated with LPC delivered as an aerosol under bronchoscopic guidance to 25- to 30-kg pigs resulted in strong transgene expression in the airway epithelium including submucosal glands. Although acute inflammatory cytokine and chemokine production, and infiltration of neutrophils into the pig airway epithelium were transiently observed after HDAd administration, there was no evidence of systemic toxicity and no significant difference in inflammatory cell infiltration in the bronchi and alveolar regions were detected 1 week after vector delivery (Cao et al. 2013). Intratracheal delivery of HDAd-K18LacZ formulated with LPC and DEAE-dextran in newborn ferrets also resulted in transgene expression in epithelial cells of intralobar conducting airways with little expression in the alveolar regions of the lung (Yan et al. 2015). Taken together, these studies demonstrate that high efficiency transduction of the airway epithelium in large animal models and are promising for CF gene therapy.

3.5.3 GENE THERAPY FOR BRAIN, EYE, AND MUSCLE DISORDERS

Because of their ability to infect post-mitotic cells and to mediate long-term transgene expression, Ad-based vectors are attractive for gene therapy of brain, eye, and muscle disorders. For brain gene transfer intravascular delivery of Ad vectors is ineffective because the blood-brain barrier prevents the access of the blood-borne vector to the brain. Therefore, Ad vectors for brain gene therapy have been investigated with direct intracranial injection. In contrast to liver delivery following intravenous injections, FGAd-mediated transduction of adult brain cells leads to stable transgene expression (Davidson et al. 1993; Le Gal La Salle et al. 1993) because the brain is relatively protected from the effects of the immune response (Byrnes et al. 1996). However, loss of transgene expression and chronic inflammation are observed when a peripheral immune response against Ad is elicited after natural infection or following vector re-administration (Thomas et al. 2000). Interestingly, these negative effects are not seen with HDAds (Thomas et al. 2000; Xiong et al. 2006) that mediate significantly higher transgene expression levels and induce a substantially reduced inflammatory and immune response (Dindot et al. 2010; Zou et al. 2001).

FGAds have been used in clinical trials for brain cancer (glioblastoma multiforme) and intratumoral injection has been associated with increased survival in two different trials (Germano et al. 2003; Immonen et al. 2004). Given the high risk that FGAd treatment of glioblastoma multiforme can be compromised by exposure to natural Ad infection, HDAds encoding regulated therapeutic genes could offer a significantly safer and more effective treatment for patients with this type and potentially other brain cancers (Candolfi et al. 2006, 2007; Muhammad et al. 2012; Puntel et al. 2013; VanderVeen et al. 2013).

For brain directed gene therapy, helper-dependent canine adenovirus (CAV-2) vectors have also been developed and investigated in lysosomal storage disorders, such as MPSIIIA and MPSVII (Ariza et al. 2014; Cubizolle et al. 2014; Lau et al. 2012). These vectors preferentially transduce neurons resulting in stable, high-level expression, and efficiently traffic via axonal retrograde transport (Bru et al. 2010; Simao et al. 2016). The main advantages of helper-dependent CAV-2 vectors over serotype 5 HDAd vectors include the lower prevalence of preexisting humoral immunity because 98% of subjects are negative for neutralizing antibodies against CAV-2 (Kremer et al. 2000), reduced activation of the innate response (Keriel et al. 2006) and dendritic cells (Perreau et al. 2007), and neuronal retrograde transport (Hnasko et al. 2006).

Eye gene therapy is attractive for safety reasons because affected patients are only exposed to low vector doses which remain confined in an immunologically privileged site. Intravitreal injections of HDAd expressing anti-angiogenic factors have been investigated to counteract ocular neovascularization occurring in diabetic retinopathy and age-related macular degeneration (Chen et al. 2008; Lamartina et al. 2007). These studies showed long-term therapeutic transgene expression, inhibition of retinal neovascularization, and substantially reduced inflammatory response with HDAd compared to FGAd.

Following subretinal injections, Ad-based vectors preferentially target the retinal pigment epithelium (RPE) and HDAd can mediate long-term expression of therapeutic genes in the RPE without evidence of adverse immune reactions or significant toxicity (Kreppel et al. 2002). Because of its large cloning capacity HDAds were particularly attractive for several inherited retinopathies due to large therapeutic genes that are beyond the cargo capacity of other vectors commonly used for eye gene therapy, such as the AAV. For example, Stargardt disease, the most common form of juvenile onset macular degeneration, is caused by mutations in *ABCA4* gene, whose cDNA is 6.8 kb, and Leber's Congenital Amaurosis (LCA10) is due to mutations in *CEP290*, whose cDNA is 7.4 kb. Unfortunately, none of the Ad serotypes investigated so far has shown efficiency greater than AAV8 vectors for transduction of photoreceptors that are the target cells for treatment of inherited retinopathies (Lam et al. 2014; Puppo et al. 2014).

The muscle is the target tissue for therapy of Duchenne muscular dystrophy (DMD) due to mutations in the dystrophin gene, one of the most common genetic diseases that lacks effective treatments. The length of dystrophin cDNA (14 kb) precluded its inclusion into most gene therapy viral vectors but HDAds with their large cloning capacity can accommodate not only one full-length dystrophin cDNA but also two copies of the gene (Dudley et al. 2004; Gilbert et al. 2003). HDAds expressing full-length dystrophin have been shown to restore the full dystrophin–glycoprotein complex in the skeletal muscle, resulting in decreased muscle degeneration and amelioration of muscle function (Clemens et al. 1996; DelloRusso et al. 2002; Dudley et al. 2004; Gilbert et al. 2003; Gilchrist et al. 2002). However, in contrast to newborn mice, adult mice require higher doses of vector for efficient transduction of mature muscle (DelloRusso et al. 2002). The inefficient transduction of mature muscle is presumably due to low CAR levels. Incorporation of polylysine into the H-I loop of the Ad fiber protein can improve HDAd transduction of mature muscle cells, resulting in increased transduction compared to the unmodified counterpart (Bramson et al. 2004). In addition, HDAd bearing the fiber knob domain from Ad serotype 3 significantly improved skeletal muscle transduction following intramuscular injections in adult mice and mediated long-term transgene expression (Guse et al. 2012). However, with the development of microdystrophin that can be accommodated by AAV (Yoshimura et al. 2004), the advantage of HDAd great cloning capacity may be diminished with regards to DMD gene therapy.

3.6 CONCLUSIONS

Dose-dependent activation of the innate inflammatory response by the viral capsid following intravascular delivery remains an important obstacle, particularly for liver-directed applications requiring high vector doses to achieve clinically relevant phenotypic improvements (Piccolo and Brunetti-Pierri 2015). Although important knowledge has been gained on Ad–host interactions occurring following systemic intravascular administrations, effective strategies still need to be developed to minimize, if not eliminate, this innate inflammatory response. In the meantime, *in vivo* applications that require very low and/or localized vector doses, or *ex vivo* gene and cell-based strategies that do not provoke an innate inflammatory response hold potential for clinical translation.

In all animal models studied so far, HDAd-transduced hepatocytes (as well as all other target cell types examined) are not destroyed by an adaptive cellular immune response, thus leading to multi-year transgene expression. However, whether this holds true for humans is not known, especially considering the outcomes of recent liver-directed clinical trials for FIX-deficiency with AAV vectors (Manno et al. 2006; Nathwani et al. 2011, 2014). Like HDAd, AAV vectors do not contain any viral genes and mediate long-term transgene expression following hepatocyte transduction in several disease animal models. However, in humans, AAV-mediated transgene expression from transduced hepatocytes is subject to killing by AAV-specific cytotoxic T lymphocytes (CTLs). The source of immunogen has been attributed to AAV capsid peptides derived directly from the injected vector particles (Manno et al. 2006; Mingozzi et al. 2007; Pien et al. 2009). Similar to AAV, HDAd capsid proteins derived from the administered particles may be a source of immunogen (Molinier-Frenkel et al. 2000; Smith et al. 1996). HDAd transduction of dendritic cells *in vitro* can stimulate activation of anti-Ad-specific CTLs (Roth et al. 2002), which are also generated following intravascular (Muruve et al. 2004) and intranasal administrations of HDAd in mice (Kushwah et al. 2008). Collectively, these studies show that HDAd can indeed induce a CTL response against viral proteins derived from capsid independent *de novo* viral protein synthesis. However, whether these Ad-specific CTLs will eliminate HDAd transduced cells *in vivo* remains to be shown and animal modeling may not be useful to address this important issue as it was not in the case of AAV vectors (Li et al. 2007, 2009; Wang et al. 2007).

The safety of AAV has been challenged by a few studies that documented hepatocellular carcinoma and vector genomic integration after AAV gene delivery in mice (Chandler et al. 2015; Donsante et al. 2007; Reiss and Hahnewald 2011). Moreover, a recent study reports that natural infections in humans with AAV serotype 2 result in chromosomal insertions activating proto-oncogenes in the liver and suggests that the AAV integrations cause the tumors (Nault et al. 2015). Although the risk of hepatocellular carcinoma development by AAV vectors remains to be fully understood, it is important to continue investigations on other vector systems, particularly those, such as HDAd vectors, that have a very low frequency of genomic integration.

ACKNOWLEDGMENTS

PN is supported by the National Institutes of Health (R01 DK067324 and P51OD011133). NB-P is supported by the Italian Telethon Foundation (TCBMT3TELD) and the European Research Council (IEMTx).

REFERENCES

Aizawa E, Hirabayashi Y, Iwanaga Y, Suzuki K, Sakurai K, Shimoji M, Aiba K et al. 2012. Efficient and accurate homologous recombination in hESCs and hiPSCs using helper-dependent adenoviral vectors. *Mol Ther* 20(2):424–431.

Alba R, Hearing P, Bosch A, Chillon M. 2007. Differential amplification of adenovirus vectors by flanking the packaging signal with attB/attP-PhiC31 sequences: Implications for helper-dependent adenovirus production. *Virology* 367(1):51–58.

Ariza L, Gimenez-Llort L, Cubizolle A, Pages G, Garcia-Lareu B, Serratrice N, Cots D et al. 2014. Central nervous system delivery of helper-dependent canine adenovirus corrects neuropathology and behavior in mucopolysaccharidosis type VII mice. *Hum Gene Ther* 25(3):199–211.

Bett AJ, Prevec L, Graham FL. 1993. Packaging capacity and stability of human adenovirus type 5 vectors. *J Virol* 67(10):5911–5921.

Bramson JL, Grinshtein N, Meulenbroek RA, Lunde J, Kottachchi D, Lorimer IA, Jasmin BJ, Parks RJ. 2004. Helper-dependent adenoviral vectors containing modified fiber for improved transduction of developing and mature muscle cells. *Hum Gene Ther* 15(2):179–188.

Bru T, Salinas S, Kremer EJ. 2010. An update on canine adenovirus type 2 and its vectors. *Viruses* 2(9):2134–2153.

Brunetti-Pierri N, Palmer DJ, Beaudet AL, Carey KD, Finegold M, Ng P. 2004. Acute toxicity after high-dose systemic injection of helper-dependent adenoviral vectors into non-human primates. *Hum Gene Ther* 15(1):35–46.

Brunetti-Pierri N, Nichols TC, McCorquodale S, Merricks E, Palmer DJ, Beaudet AL, Ng P. 2005a. Sustained phenotypic correction of canine hemophilia B after systemic administration of helper-dependent adenoviral vector. *Hum Gene Ther* 16(7):811–820.

Brunetti-Pierri N, Palmer DJ, Mane V, Finegold M, Beaudet AL, Ng P. 2005b. Increased hepatic transduction with reduced systemic dissemination and proinflammatory cytokines following hydrodynamic injection of helper-dependent adenoviral vectors. *Mol Ther* 12(1):99–106.

Brunetti-Pierri N, Ng T, Iannitti DA, Palmer DJ, Beaudet AL, Finegold MJ, Carey KD, Cioffi WG, Ng P. 2006. Improved hepatic transduction, reduced systemic vector dissemination, and long-term transgene expression by delivering helper-dependent adenoviral vectors into the surgically isolated liver of nonhuman primates. *Hum Gene Ther* 17(4):391–404.

Brunetti-Pierri N, Stapleton GE, Palmer DJ, Zuo Y, Mane VP, Finegold MJ, Beaudet AL, Leland MM, MullinsCE, Ng P. 2007. Pseudo-hydrodynamic delivery of helper-dependent adenoviral vectors into non-human primates for liver-directed gene therapy. *Mol Ther* 15(4):732–740.

Brunetti-Pierri N, Stapleton GE, Law M, Breinholt J, Palmer DJ, Zuo Y, Grove NC et al. 2009. Efficient, long-term hepatic gene transfer using clinically relevant HDAd doses by balloon occlusion catheter delivery in nonhuman primates. *Mol Ther* 17(2):327–333.

Brunetti-Pierri N, Ng P. 2011. Helper-dependent adenoviral vectors for liver-directed gene therapy. *Hum Mol Genet* 20(R1):R7–R13.

Brunetti-Pierri N, Liou A, Patel P, Palmer D, Grove N, Finegold M, Piccolo P et al. 2012. Balloon catheter delivery of helper-dependent adenoviral vector results in sustained, therapeutic hFIX expression in rhesus macaques. *Mol Ther* 20(10):1863–1870.

Brunetti-Pierri N, Ng T, Iannitti D, Cioffi W, Stapleton G, Law M, Breinholt J et al. 2013. Transgene expression up to 7 years in nonhuman primates following hepatic transduction with helper-dependent adenoviral vectors. *Hum Gene Ther* 24(8):761–765.

Byrnes AP, Wood MJ, Charlton HM. 1996. Role of T cells in inflammation caused by adenovirus vectors in the brain. *Gene Ther* 3(7):644–651.

Candolfi M, Curtin JF, Xiong WD, Kroeger KM, Liu C, Rentsendorj A, Agadjanian H et al. 2006. Effective high-capacity gutless adenoviral vectors mediate transgene expression in human glioma cells. *Mol Ther* 14(3):371–381.

Candolfi M, Pluhar GE, Kroeger K, Puntel M, Curtin J, Barcia C, Muhammad A K et al. 2007. Optimization of adenoviral vector-mediated transgene expression in the canine brain in vivo, and in canine glioma cells *in vitro*. *Neuro-oncol* 9(3):245–258.

Cao H, Yang T, Li XF, Wu J, Duan C, Coates AL, Hu J. 2011. Readministration of helper-dependent adenoviral vectors to mouse airway mediated via transient immunosuppression. *Gene Ther* 18(2):173–181.

Cao H, Machuca TN, Yeung JC, Wu J, Du K, Duan C, Hashimoto K et al. 2013. Efficient gene delivery to pig airway epithelia and submucosal glands using helper-dependent adenoviral vectors. *Mol Ther Nucl Acids* 2:e127.

Cao H, Ouyang H, Ip W, Du K, Duan W, Avolio J, Wu J et al. 2015. Testing gene therapy vectors in human primary nasal epithelial cultures. *Mol Ther Methods Clin Dev* 2:15034.

Carlisle RC, Di Y, Cerny AM, Sonnen AF, Sim RB, Green NK, Subr V et al. 2009. Human erythrocytes bind and inactivate type 5 adenovirus by presenting coxsackievirus-adenovirus receptor and complement receptor 1. *Blood* 113(9):1909–1918.

Chandler RJ, LaFave MC, Varshney GK, Trivedi NS, Carrillo-Carrasco N, Senac JS, Wu W et al. 2015. Vector design influences hepatic genotoxicity after adeno-associated virus gene therapy. *J Clin Invest* 125(2):870–880.

Chen P, Hamilton M, Thomas CA, Kroeger K, Carrion M, Macgill RS, Gehlbach P et al. 2008. Persistent expression of PEDF in the eye using high-capacity adenovectors. *Mol Ther* 16(12):1986–1994.

Chu Q, St George JA, Lukason M, Cheng SH, Scheule R K, Eastman SJ. 2001. EGTA enhancement of adenovirus-mediated gene transfer to mouse tracheal epithelium *in vivo*. *Hum Gene Ther* 12(5):455–467.

Clemens PR, Kochanek S, Sunada Y, Chan S, Chen HH, Campbell KP, Caskey CT. 1996. *In vivo* muscle gene transfer of full-length dystrophin with an adenoviral vector that lacks all viral genes. *Gene Ther* 3(11):965–972.

Croyle MA, Le HT, Linse KD, Cerullo V, Toietta G, Beaudet A, Pastore L. 2005. PEGylated helper-dependent adenoviral vectors: Highly efficient vectors with an enhanced safety profile. *Gene Ther* 12(7):579–587.

Cubizolle A, Serratrice N, Skander N, Colle MA, Ibanes S, Gennetier A, Bayo-Puxan N et al. 2014. Corrective GUSB transfer to the canine mucopolysaccharidosis VII brain. *Mol Ther* 22(4):762–773.

Dai Y, Schwarz EM, Gu D, Zhang WW, Sarvetnick N, Verma IM. 1995. Cellular and humoral immune responses to adenoviral vectors containing factor IX gene: Tolerization of factor IX and vector antigens allows for long-term expression. *Proc Natl Acad Sci U S A* 92(5):1401–1405.

Davidson BL, Allen ED, Kozarsky KF, Wilson JM, Roessler BJ. 1993. A model system for *in vivo* gene transfer into the central nervous system using an adenoviral vector. *Nat Genet* 3(3):219–223.

De Geest B, Snoeys J, Van Linthout S, Lievens J, Collen D. 2005. Elimination of innate immune responses and liver inflammation by PEGylation of adenoviral vectors and methylprednisolone. *Hum Gene Ther* 16(12):1439–1451.

DelloRusso C, Scott JM, Hartigan-O'Connor D, Salvatori G, Barjot C, Robinson AS, Crawford RW, Brooks SV, Chamberlain JS. 2002. Functional correction of adult mdx mouse muscle using gutted adenoviral vectors expressing full-length dystrophin. *Proc Natl Acad Sci U S A* 99(20):12979–12984.

Di Paolo NC, Doronin K, Baldwin LK, Papayannopoulou T, Shayakhmetov DM. 2013. The transcription factor IRF3 triggers "defensive suicide" necrosis in response to viral and bacterial pathogens. *Cell Rep* 3(6):1840–1846.

Dindot S, Piccolo P, Grove NC, Palmer D, Brunetti-Pierri N. 2010. Intrathecal injection of HDAd vectors results in long-term transgene expression in neuroependymal cells and neurons. *Hum Gene Ther* 22(6):745–751.

Donsante A, Miller DG, Li Y, Vogler C, Brunt EM, Russell DW, Sands MS. 2007. AAV vector integration sites in mouse hepatocellular carcinoma. *Science* 317(5837):477.

Doronin K, Flatt JW, Di Paolo NC, Khare R, Kalyuzhniy O, Acchione M, Sumida JP et al. 2012. Coagulation factor X activates innate immunity to human species C adenovirus. *Science* 338(6108):795–798.

Dudley RW, Lu Y, Gilbert R, Matecki S, Nalbantoglu J, Petrof BJ, Karpati G. 2004. Sustained improvement of muscle function one year after full-length dystrophin gene transfer into mdx mice by a gutted helper-dependent adenoviral vector. *Hum Gene Ther* 15(2):145–156.

Ehrhardt A, Xu H, Kay MA. 2003. Episomal persistence of recombinant adenoviral vector genomes during the cell cycle *in vivo*. *J Virol* 77(13):7689–7695.

Germano IM, Fable J, Gultekin SH, Silvers A. 2003. Adenovirus/herpes simplex-thymidine kinase/ganciclovir complex: Preliminary results of a phase I trial in patients with recurrent malignant gliomas. *J Neurooncol* 65(3):279–289.

Gilbert R, Dudley RW, Liu AB, Petrof BJ, Nalbantoglu J, Karpati G. 2003. Prolonged dystrophin expression and functional correction of mdx mouse muscle following gene transfer with a helper-dependent (gutted) adenovirus-encoding murine dystrophin. *Hum Mol Genet* 12(11):1287–1299.

Gilchrist SC, Ontell MP, Kochanek S, Clemens PR. 2002. Immune response to full-length dystrophin delivered to Dmd muscle by a high-capacity adenoviral vector. *Mol Ther* 6(3):359–368.

Grubb BR, Pickles RJ, Ye H, Yankaskas JR, Vick RN, Engelhardt JF, Wilson JM, Johnson LG, Boucher RC. 1994. Inefficient gene transfer by adenovirus vector to cystic fibrosis airway epithelia of mice and humans. *Nature* 371(6500):802–806.

Guse K, Suzuki M, Sule G, Bertin TK, Tyynismaa H, Ahola-Erkkila S, Palmer D et al. 2012. Capsid-modified adenoviral vectors for improved muscle-directed gene therapy. *Hum Gene Ther* 23(10):1065–1070.

Haisma HJ, Kamps JA, Kamps GK, Plantinga JA, Rots MG, Bellu AR. 2008. Polyinosinic acid enhances delivery of adenovirus vectors *in vivo* by preventing sequestration in liver macrophages. *J Gen Virol* 89(Pt 5):1097–1105.

Haisma HJ, Boesjes M, Beerens AM, van der Strate BW, Curiel DT, Pluddemann A, Gordon S, Bellu AR. 2009. Scavenger receptor A: A new route for adenovirus 5. *Mol Pharm* 6(2):366–374.

Harui A, Suzuki S, Kochanek S, Mitani K. 1999. Frequency and stability of chromosomal integration of adenovirus vectors. *J Virol* 73(7):6141–6146.

Harvey BG, Leopold PL, Hackett NR, Grasso TM, Williams PM, Tucker AL, Kaner RJ et al. 1999. Airway epithelial CFTR mRNA expression in cystic fibrosis patients after repetitive administration of a recombinant adenovirus. *J Clin Invest* 104(9):1245–1255.

Harvey BG, Maroni J, O'Donoghue KA, Chu KW, Muscat JC, Pippo AL, Wright CE et al. 2002. Safety of local delivery of low- and intermediate-dose adenovirus gene transfer vectors to individuals with a spectrum of morbid conditions. *Hum Gene Ther* 13(1):15–63.

He JQ, Katschke KJ, Jr., Gribling P, Suto E, Lee WP, Diehl L, Eastham-Anderson J et al. 2013. CRIg mediates early Kupffer cell responses to adenovirus. *J Leukocyte Biol* 93(2):301–306.

Hillgenberg M, Tonnies H, Strauss M. 2001. Chromosomal integration pattern of a helper-dependent minimal adenovirus vector with a selectable marker inserted into a 27.4-kilobase genomic stuffer. *J Virol* 75(20):9896–9908.

Hnasko TS, Perez FA, Scouras AD, Stoll EA, Gale SD, Luquet S, Phillips PE, Kremer EJ, Palmiter RD. 2006. Cre recombinase-mediated restoration of nigrostriatal dopamine in dopamine-deficient mice reverses hypophagia and bradykinesia. *Proc Natl Acad Sci U S A* 103(23):8858–8863.

Immonen A, Vapalahti M, Tyynela K, Hurskainen H, Sandmair A, Vanninen R, Langford G, Murray N, Yla-Herttuala S. 2004. AdvHSV-tk gene therapy with intravenous ganciclovir improves survival in human malignant glioma: A randomised, controlled study. *Mol Ther* 10(5):967–972.

Jager L, Ehrhardt A. 2009. Persistence of high-capacity adenoviral vectors as replication-defective monomeric genomes *in vitro* and in murine liver. *Hum Gene Ther* 20(8):883–896.

Johnson LG, Vanhook MK, Coyne CB, Haykal-Coates N, Gavett SH. 2003. Safety and efficiency of modulating paracellular permeability to enhance airway epithelial gene transfer *in vivo*. *Hum Gene Ther* 14(8):729–747.

Joseph PM, O'Sullivan BP, Lapey A, Dorkin H, Oren J, Balfour R, Perricone MA et al. 2001. Aerosol and lobar administration of a recombinant adenovirus to individuals with cystic fibrosis. I. Methods, safety, and clinical implications. *Hum Gene Ther* 12(11):1369–1382.

Kaplan JM, Pennington SE, St George JA, Woodworth LA, Fasbender A, Marshall J, Cheng SH, Wadsworth SC, Gregory RJ, Smith AE. 1998. Potentiation of gene transfer to the mouse lung by complexes of adenovirus vector and polycations improves therapeutic potential. *Hum Gene Ther* 9(10):1469–1479.

Keriel A, Rene C, Galer C, Zabner J, Kremer EJ. 2006. Canine adenovirus vectors for lung-directed gene transfer: Efficacy, immune response, and duration of transgene expression using helper-dependent vectors. *J Virol* 80(3):1487–1496.

Khare R, May SM, Vetrini F, Weaver EA, Palmer D, Rosewell A, Grove N, Ng P, Barry MA. 2011. Generation of a Kupffer cell-evading adenovirus for systemic and liver-directed gene transfer. *Mol Ther* 19(7):1254–1262.

Khare R, Reddy VS, Nemerow GR, Barry MA. 2012. Identification of adenovirus serotype 5 hexon regions that interact with scavenger receptors. *J Virol* 86(4):2293–2301.

Khare R, Hillestad ML, Xu Z, Byrnes AP, Barry MA. 2013. Circulating antibodies and macrophages as modulators of adenovirus pharmacology. *J Virol* 87(7):3678–3686.

Kim IH, Jozkowicz A, Piedra PA, Oka K, Chan L. 2001. Lifetime correction of genetic deficiency in mice with a single injection of helper-dependent adenoviral vector. *Proc Natl Acad Sci U S A* 98(23):13282–13287.

Koehler DR, Hannam V, Belcastro R, Steer B, Wen Y, Post M, Downey G, Tanswell AK, Hu J. 2001. Targeting transgene expression for cystic fibrosis gene therapy. *Mol Ther* 4(1):58–65.

Koehler DR, Sajjan U, Chow YH, Martin B, Kent G, Tanswell AK, McKerlie C, Forstner JF, Hu J. 2003. Protection of Cftr knockout mice from acute lung infection by a helper-dependent adenoviral vector expressing Cftr in airway epithelia. *Proc Natl Acad Sci U S A* 100(26):15364–15369.

Koehler DR, Frndova H, Leung K, Louca E, Palmer D, Ng P, McKerlie C, Cox P, Coates AL, Hu J. 2005. Aerosol delivery of an enhanced helper-dependent adenovirus formulation to rabbit lung using an intratracheal catheter. *J Gene Med* 7(11):1409–1420.

Kremer EJ, Boutin S, Chillon M, Danos O. 2000. Canine adenovirus vectors: An alternative for adenovirus-mediated gene transfer. *J Virol* 74(1):505–512.

Kreppel F, Luther TT, Semkova I, Schraermeyer U, Kochanek S. 2002. Long-term transgene expression in the RPE after gene transfer with a high-capacity adenoviral vector. *Invest Ophthalmol Vis Sci* 43(6):1965–1970.

Kreppel F, Kochanek S. 2004. Long-term transgene expression in proliferating cells mediated by episomally maintained high-capacity adenovirus vectors. *J Virol* 78(1):9–22.

Kushwah R, Cao H, Hu J. 2008. Characterization of pulmonary T cell response to helper-dependent adenoviral vectors following intranasal delivery. *J Immunol* 180(6):4098–4108.

Lam S, Cao H, Wu J, Duan R, Hu J. 2014. Highly efficient retinal gene delivery with helper-dependent adenoviral vectors. *Genes Dis* 1(2):227–237.

Lamartina S, Cimino M, Roscilli G, Dammassa E, Lazzaro D, Rota R, Ciliberto G, Toniatti C. 2007. Helper-dependent adenovirus for the gene therapy of proliferative retinopathies: Stable gene transfer, regulated gene expression and therapeutic efficacy. *J Gene Med* 9(10):862–874.

Lau AA, Rozaklis T, Ibanes S, Luck AJ, Beard H, Hassiotis S, Mazouni K, Hopwood JJ, Kremer EJ, Hemsley KM. 2012. Helper-dependent canine adenovirus vector-mediated transgene expression in a neurodegenerative lysosomal storage disorder. *Gene* 491(1):53–57.

Le Gal La Salle G, Robert JJ, Berrard S, Ridoux V, Stratford-Perricaudet LD, Perricaudet M, Mallet J. 1993. An adenovirus vector for gene transfer into neurons and glia in the brain. *Science* 259(5097):988–990.

Leggiero E, Astone D, Cerullo V, Lombardo B, Mazzaccara C, Labruna G, Sacchetti L, Salvatore F, Croyle M, Pastore L. 2013. PEGylated helper-dependent adenoviral vector expressing human Apo A-I for gene therapy in LDLR-deficient mice. *Gene Ther* 20:1124–1130.

Li H, Murphy SL, Giles-Davis W, Edmonson S, Xiang Z, Li Y, Lasaro MO, High KA, Ertl HC. 2007. Pre-existing AAV capsid-specific CD8+ T cells are unable to eliminate AAV-transduced hepatocytes. *Mol Ther* 15(4):792–800.

Li H, Lin SW, Giles-Davis W, Li Y, Zhou D, Xiang ZQ, High KA, Ertl HC. 2009. A preclinical animal model to assess the effect of pre-existing immunity on AAV-mediated gene transfer. *Mol Ther* 17(7):1215–1224.

Li M, Suzuki K, Qu J, Saini P, Dubova I, Yi F, Lee J, Sancho-Martinez I, Liu GH, Izpisua Belmonte JC. 2011. Efficient correction of hemoglobinopathy-causing mutations by homologous recombination in integration-free patient iPSCs. *Cell Res* 21(12):1740–1744.

Lievens J, Snoeys J, Vekemans K, Van Linthout S, de Zanger R, Collen D, Wisse E, De Geest B. 2004. The size of sinusoidal fenestrae is a critical determinant of hepatocyte transduction after adenoviral gene transfer. *Gene Ther* 11(20):1523–1531.

Limberis M, Anson DS, Fuller M, Parsons DW. 2002. Recovery of airway cystic fibrosis transmembrane conductance regulator function in mice with cystic fibrosis after single-dose lentivirus-mediated gene transfer. *Hum Gene Ther* 13(16):1961–1970.

Liu GH, Suzuki K, Qu J, Sancho-Martinez I, Yi F, Li M, Kumar S et al. 2011. Targeted gene correction of laminopathy-associated LMNA mutations in patient-specific iPSCs. *Cell Stem Cell* 8(6):688–694.

Liu Q, Muruve DA. 2003. Molecular basis of the inflammatory response to adenovirus vectors. *Gene Ther* 10(11):935–940.

Lyons M, Onion D, Green NK, Aslan K, Rajaratnam R, Bazan-Peregrino M, Phipps S et al. 2006. Adenovirus type 5 interactions with human blood cells may compromise systemic delivery. *Mol Ther* 14(1):118–128.

Maione D, Della Rocca C, Giannetti P, D'Arrigo R, Liberatoscioli L, Franlin LL, Sandig V, Ciliberto G, La Monica N, Savino R. 2001. An improved helper-dependent adenoviral vector allows persistent gene expression after intramuscular delivery and overcomes preexisting immunity to adenovirus. *Proc Natl Acad Sci U S A* 98(11):5986–5991.

Manno CS, Pierce GF, Arruda VR, Glader B, Ragni M, Rasko JJ, Ozelo MC et al. 2006. Successful transduction of liver in hemophilia by AAV-Factor IX and limitations imposed by the host immune response. *Nat Med* 12(3):342–347.

Martin LJ, Mahaney MC, Bronikowski AM, Dee Carey K, Dyke B, Comuzzie AG. 2002. Lifespan in captive baboons is heritable. *Mech Ageing Dev* 123(11):1461–1467.

McCormack WM, Jr., Seiler MP, Bertin TK, Ubhayakar K, Palmer DJ, Ng P, Nichols TC, Lee B. 2006. Helper-dependent adenoviral gene therapy mediates long-term correction of the clotting defect in the canine hemophilia A model. *J Thromb Haemost* 4(6):1218–1225.

Mingozzi F, Maus MV, Hui DJ, Sabatino DE, Murphy SL, Rasko JE, Ragni MV et al. 2007. CD8(+) T-cell responses to adeno-associated virus capsid in humans. *Nat Med* 13(4):419–422.

Mitrani E, Pearlman A, Stern B, Miari R, Goltsman H, Kunicher N, Panet A. 2011. Biopump: Autologous skin-derived micro-organ genetically engineered to provide sustained continuous secretion of therapeutic proteins. *Dermatol Ther* 24(5):489–497.

Mok H, Palmer DJ, Ng P, Barry MA. 2005. Evaluation of polyethylene glycol modification of first-generation and helper-dependent adenoviral vectors to reduce innate immune responses. *Mol Ther* 11(1):66–79.

Molinier-Frenkel V, Gahery-Segard H, Mehtali M, Le Boulaire C, Ribault S, Boulanger P, Tursz T, Guillet JG, Farace F. 2000. Immune response to recombinant adenovirus in humans: Capsid components from viral input are targets for vector-specific cytotoxic T lymphocytes. *J Virol* 74(16):7678–7682.

Morral N, O'Neal W, Zhou H, Langston C, Beaudet A. 1997. Immune responses to reporter proteins and high viral dose limit duration of expression with adenoviral vectors: Comparison of E2a wild type and E2a deleted vectors. *Hum Gene Ther* 8(10):1275–1286.

Morral N, O'Neal W, Rice K, Leland M, Kaplan J, Piedra PA, Zhou H et al. 1999. Administration of helper-dependent adenoviral vectors and sequential delivery of different vector serotype for long-term liver-directed gene transfer in baboons. *Proc Natl Acad Sci U S A* 96(22):12816–12821.

Muck-Hausl M, Solanki M, Zhang W, Ruzsics Z, Ehrhardt A. 2015. Ad 2.0: A novel recombineering platform for high-throughput generation of tailored adenoviruses. *Nucl Acids Res* 43(8):e50.

Muhammad AK, Xiong W, Puntel M, Farrokhi C, Kroeger KM, Salem A, Lacayo L et al. 2012. Safety profile of gutless adenovirus vectors delivered into the normal brain parenchyma: Implications for a glioma phase 1 clinical trial. *Hum Gene Ther Methods* 23(4):271–284.

Muruve DA, Barnes MJ, Stillman IE, Libermann TA. 1999. Adenoviral gene therapy leads to rapid induction of multiple chemokines and acute neutrophil-dependent hepatic injury *in vivo*. *Hum Gene Ther* 10(6):965–976.

Muruve DA, Cotter MJ, Zaiss AK, White LR, Liu Q, Chan T, Clark SA et al. 2004. Helper-dependent adenovirus vectors elicit intact innate but attenuated adaptive host immune responses *in vivo*. *J Virol* 78(11):5966–5972.

Nathwani AC, Tuddenham EG, Rangarajan S, Rosales C, McIntosh J, Linch DC, Chowdary P et al. 2011. Adenovirus-associated virus vector-mediated gene transfer in hemophilia B. *N Engl J Med* 365(25):2357–2365.

Nathwani AC, Reiss UM, Tuddenham EG, Rosales C, Chowdary P, McIntosh J, Della Peruta M et al. 2014. Long-term safety and efficacy of factor IX gene therapy in hemophilia B. *N Engl J Med* 371(21):1994–2004.

Nault JC, Datta S, Imbeaud S, Franconi A, Mallet M, Couchy G, Letouze E et al. 2015. Recurrent AAV2-related insertional mutagenesis in human hepatocellular carcinomas. *Nat Genet* 47(10):1187–1193.

Ng P, Beauchamp C, Evelegh C, Parks R, Graham FL. 2001. Development of a FLP/frt system for generating helper-dependent adenoviral vectors. *Mol Ther* 3(5 Pt 1):809–815.

Nociari M, Ocheretina O, Schoggins JW, Falck-Pedersen E. 2007. Sensing infection by adenovirus: Toll-like receptor-independent viral DNA recognition signals activation of the interferon regulatory factor 3 master regulator. *J Virol* 81(8):4145–4157.

O'Neal WK, Zhou H, Morral N, Aguilar-Cordova E, Pestaner J, Langston C, Mull B, Wang Y, Beaudet AL, Lee B. 1998. Toxicological comparison of E2a-deleted and first-generation adenoviral vectors expressing alpha1-antitrypsin after systemic delivery. *Hum Gene Ther* 9(11):1587–1598.

O'Neal WK, Rose E, Zhou H, Langston C, Rice K, Carey D, Beaudet AL. 2000. Multiple advantages of alpha-fetoprotein as a marker for *in vivo* gene transfer. *Mol Ther* 2(6):640–648.

Ohbayashi F, Balamotis MA, Kishimoto A, Aizawa E, Diaz A, Hasty P, Graham FL, Caskey CT, Mitani K. 2005. Correction of chromosomal mutation and random integration in embryonic stem cells with helper-dependent adenoviral vectors. *Proc Natl Acad Sci U S A* 102(38):13628–13633.

Oka K, Mullins CE, Kushwaha RS, Leen AM, Chan L. 2015. Gene therapy for rhesus monkeys heterozygous for LDL receptor deficiency by balloon catheter hepatic delivery of helper-dependent adenoviral vector. *Gene Ther* 22(1):87–95.

Palmer D, Ng P. 2003. Improved system for helper-dependent adenoviral vector production. *Mol Ther* 8(5):846–852.

Palmer D, Ng P. 2007. *Methods for the Production and Characterization of Helper-Dependent Adenoviral Vectors*. Cold Spring Harbor Press, Cold Spring Harbor, NY.

Palmer D, Ng P. 2008. Methods for the preparation of helper-dependent adenoviral vectors. *Methods Mol Med* 433:33–53.

Parks R, Evelegh C, Graham F. 1999. Use of helper-dependent adenoviral vectors of alternative serotypes permits repeat vector administration. *Gene Ther* 6(9):1565–1573.

Parks RJ, Chen L, Anton M, Sankar U, Rudnicki MA, Graham FL. 1996. A helper-dependent adenovirus vector system: Removal of helper virus by Cre-mediated excision of the viral packaging signal. *Proc Natl Acad Sci U S A* 93(24):13565–13570.

Parks RJ, Graham FL. 1997. A helper-dependent system for adenovirus vector production helps define a lower limit for efficient DNA packaging. *J Virol* 71(4):3293–3298.

Parks RJ. 2005. Adenovirus protein IX: A new look at an old protein. *Mol Ther* 11(1):19–25.

Perreau M, Mennechet F, Serratrice N, Glasgow JN, Curiel DT, Wodrich H, Kremer EJ. 2007. Contrasting effects of human, canine, and hybrid adenovirus vectors on the phenotypical and functional maturation of human dendritic cells: Implications for clinical efficacy. *J Virol* 81(7):3272–3284.

Perricone MA, Morris JE, Pavelka K, Plog MS, O'Sullivan BP, Joseph PM, Dorkin H et al. 2001. Aerosol and lobar administration of a recombinant adenovirus to individuals with cystic fibrosis. II. Transfection efficiency in airway epithelium. *Hum Gene Ther* 12(11):1383–1394.

Piccolo P, Vetrini F, Mithbaokar P, Grove NC, Bertin T, Palmer D, Ng P, Brunetti-Pierri N. 2013. SR-A and SREC-I are Kupffer and endothelial cell receptors for helper-dependent adenoviral vectors. *Mol Ther* 21(4):767–774.

Piccolo P, Annunziata P, Mithbaokar P, Brunetti-Pierri N. 2014. SR-A and SREC-I binding peptides increase HDAd-mediated liver transduction. *Gene Ther* 21(11):950–957.

Piccolo P, Brunetti-Pierri N. 2015. Gene therapy for inherited diseases of liver metabolism. *Hum Gene Ther* 26(4):186–192.

Pickles RJ, Fahrner JA, Petrella JM, Boucher RC, Bergelson JM. 2000. Retargeting the coxsackievirus and adenovirus receptor to the apical surface of polarized epithelial cells reveals the glycocalyx as a barrier to adenovirus-mediated gene transfer. *J Virol* 74(13):6050–6057.

Pien GC, Basner-Tschakarjan E, Hui DJ, Mentlik AN, Finn JD, Hasbrouck NC, Zhou S et al. 2009. Capsid antigen presentation flags human hepatocytes for destruction after transduction by adeno-associated viral vectors. *J Clin Invest* 119(6):1688–1695.

Prill JM, Espenlaub S, Samen U, Engler T, Schmidt E, Vetrini F, Rosewell A et al. 2011. Modifications of adenovirus hexon allow for either hepatocyte detargeting or targeting with potential evasion from Kupffer cells. *Mol Ther* 19(1):83–92.

Puntel M, KMG A, Farrokhi C, Vanderveen N, Paran C, Appelhans A, Kroeger KM et al. 2013. Safety profile, efficacy, and biodistribution of a bicistronic high-capacity adenovirus vector encoding a combined immunostimulation and cytotoxic gene therapy as a prelude to a phase I clinical trial for glioblastoma. *Toxicol Appl Pharmacol* 268(3):318–330.

Puppo A, Cesi G, Marrocco E, Piccolo P, Jacca S, Shayakhmetov DM, Parks RJ et al. 2014. Retinal transduction profiles by high-capacity viral vectors. *Gene Ther* 21(10):855–865.

Qiu Q, Xu Z, Tian J, Moitra R, Gunti S, Notkins AL, Byrnes AP. 2015. Impact of natural IgM concentration on gene therapy with adenovirus type 5 vectors. *J Virol* 89(6):3412–3416.

Raper SE, Chirmule N, Lee FS, Wivel NA, Bagg A, Gao GP, Wilson JM, Batshaw ML. 2003. Fatal systemic inflammatory response syndrome in a ornithine transcarbamylase deficient patient following adenoviral gene transfer. *Mol Genet Metab* 80(1–2):148–158.

Reddy PS, Sakhuja K, Ganesh S, Yang L, Kayda D, Brann T, Pattison S et al. 2002. Sustained human factor VIII expression in hemophilia A mice following systemic delivery of a gutless adenoviral vector. *Mol Ther* 5(1):63–73.

Reiss J, Hahnewald R. 2011. Molybdenum cofactor deficiency: Mutations in GPHN, MOCS1, and MOCS2. *Hum Mutat* 32(1):10–18.

Ross PJ, Kennedy MA, Parks RJ. 2009. Host cell detection of noncoding stuffer DNA contained in helper-dependent adenovirus vectors leads to epigenetic repression of transgene expression. *J Virol* 83(17):8409–8417.

Ross PJ, Kennedy MA, Christou C, Risco Quiroz M, Poulin KL, Parks RJ. 2011. Assembly of helper-dependent adenovirus DNA into chromatin promotes efficient gene expression. *J Virol* 85(8):3950–3958.

Roth MD, Cheng Q, Harui A, Basak SK, Mitani K, Low TA, Kiertscher SM. 2002. Helper-dependent adenoviral vectors efficiently express transgenes in human dendritic cells but still stimulate antiviral immune responses. *J Immunol* 169(8):4651–4656.

Sandig V, Youil R, Bett AJ, Franlin LL, Oshima M, Maione D, Wang F, Metzker ML, Savino R, Caskey CT. 2000. Optimization of the helper-dependent adenovirus system for production and potency *in vivo*. *Proc Natl Acad Sci U S A* 97(3):1002–1007.

Sargent KL, Ng P, Evelegh C, Graham FL, Parks RJ. 2004. Development of a size-restricted pIX-deleted helper virus for amplification of helper-dependent adenovirus vectors. *Gene Ther* 11(6):504–511.

Saydaminova K, Ye X, Wang H, Richter M, Ho M, Chen H, Xu N et al. 2015. Efficient genome editing in hematopoietic stem cells with helper-dependent Ad5/35 vectors expressing site-specific endonucleases under microRNA regulation. *Mol Ther Methods Clin Dev* 1:14057.

Schiedner G, Morral N, Parks RJ, Wu Y, Koopmans SC, Langston C, Graham FL, Beaudet AL, Kochanek S. 1998. Genomic DNA transfer with a high-capacity adenovirus vector results in improved *in vivo* gene expression and decreased toxicity. *Nat Genet* 18(2):180–183.

Schiedner G, Hertel S, Johnston M, Biermann V, Dries V, Kochanek S. 2002. Variables affecting *in vivo* performance of high-capacity adenovirus vectors. *J Virol* 76(4):1600–1609.

Schiedner G, Bloch W, Hertel S, Johnston M, Molojavyi A, Dries V, Varga G, Van Rooijen N, Kochanek S. 2003a. A hemodynamic response to intravenous adenovirus vector particles is caused by systemic Kupffer cell-mediated activation of endothelial cells. *Hum Gene Ther* 14(17):1631–1641.

Schiedner G, Hertel S, Johnston M, Dries V, van Rooijen N, Kochanek S. 2003b. Selective depletion or blockade of Kupffer cells leads to enhanced and prolonged hepatic transgene expression using high-capacity adenoviral vectors. *Mol Ther* 7(1):35–43.

Schnell MA, Zhang Y, Tazelaar J, Gao GP, Yu QC, Qian R, Chen SJ. 2001. Activation of innate immunity in nonhuman primates following intraportal administration of adenoviral vectors. *Mol Ther* 3(5 Pt 1):708–722.

Seregin SS, Appledorn DM, McBride AJ, Schuldt NJ, Aldhamen YA, Voss T, Wei J et al. 2009. Transient pretreatment with glucocorticoid ablates innate toxicity of systemically delivered adenoviral vectors without reducing efficacy. *Mol Ther* 17(4):685–696.

Shayakhmetov DM, Li ZY, Gaggar A, Gharwan H, Ternovoi V, Sandig V, Lieber A. 2004. Genome size and structure determine efficiency of postinternalization steps and gene transfer of capsid-modified adenovirus vectors in a cell-type-specific manner. *J Virol* 78(18):10009–10022.

Simao D, Pinto C, Fernandes P, Peddie CJ, Piersanti S, Collinson LM, Salinas S et al. 2016. Evaluation of helper-dependent canine adenovirus vectors in a 3D human CNS model. *Gene Ther* 23(1):86–94.

Simon RH, Engelhardt JF, Yang Y, Zepeda M, Weber-Pendleton S, Grossman M, Wilson JM. 1993. Adenovirus-mediated transfer of the CFTR gene to lung of nonhuman primates: Toxicity study. *Hum Gene Ther* 4(6):771–780.

Smith CA, Woodruff LS, Kitchingman GR, Rooney CM. 1996. Adenovirus-pulsed dendritic cells stimulate human virus-specific T-cell responses *in vitro*. *J Virol* 70(10):6733–6740.

Smith JS, Xu Z, Tian J, Stevenson SC, Byrnes AP. 2008. Interaction of systemically delivered adenovirus vectors with Kupffer cells in mouse liver. *Hum Gene Ther* 19(5):547–554.

Snoeys J, Lievens J, Wisse E, Jacobs F, Duimel H, Collen D, Frederik P, De Geest B. 2007. Species differences in transgene DNA uptake in hepatocytes after adenoviral transfer correlate with the size of endothelial fenestrae. *Gene Ther* 14(7):604–612.

Stephen SL, Sivanandam VG, Kochanek S. 2008. Homologous and heterologous recombination between adenovirus vector DNA and chromosomal DNA. *J Gene Med* 10(11):1176–1189.

Stephen SL, Montini E, Sivanandam VG, Al-Dhalimy M, Kestler HA, Finegold M, Grompe M, Kochanek S. 2010. Chromosomal integration of adenoviral vector DNA *in vivo*. *J Virol* 84(19):9987–9994.

Suzuki K, Mitsui K, Aizawa E, Hasegawa K, Kawase E, Yamagishi T, Shimizu Y, Suemori H, Nakatsuji N, Mitani K. 2008. Highly efficient transient gene expression and gene targeting in primate embryonic stem cells with helper-dependent adenoviral vectors. *Proc Natl Acad Sci U S A* 105(37):13781–13786.

Suzuki K, Yu C, Qu J, Li M, Yao X, Yuan T, Goebl A et al. 2014. Targeted gene correction minimally impacts whole-genome mutational load in human-disease-specific induced pluripotent stem cell clones. *Cell Stem Cell* 15(1):31–36.

Suzuki T, Sasaki T, Yano K, Sakurai F, Kawabata K, Kondoh M, Hayakawa T, Yagi K, Mizuguchi H. 2011. Development of a recombinant adenovirus vector production system free of replication-competent adenovirus by utilizing a packaging size limit of the viral genome. *Virus Res* 158(1–2):154–160.

Tao N, Gao GP, Parr M, Johnston J, Baradet T, Wilson JM, Barsoum J, Fawell SE. 2001. Sequestration of adenoviral vector by Kupffer cells leads to a nonlinear dose response of transduction in liver. *Mol Ther* 3(1):28–35.

Thomas CE, Schiedner G, Kochanek S, Castro MG, Lowenstein PR. 2000. Peripheral infection with adenovirus causes unexpected long-term brain inflammation in animals injected intracranially with first-generation, but not with high-capacity, adenovirus vectors: Toward realistic long-term neurological gene therapy for chronic diseases. *Proc Natl Acad Sci U S A* 97(13):7482–7487.

Tian J, Xu Z, Smith JS, Hofherr SE, Barry MA, Byrnes AP. 2009. Adenovirus activates complement by distinctly different mechanisms *in vitro* and *in vivo*: Indirect complement activation by virions *in vivo*. *J Virol* 83(11):5648–5658.

Toietta G, Koehler DR, Finegold MJ, Lee B, Hu J, Beaudet AL. 2003. Reduced inflammation and improved airway expression using helper-dependent adenoviral vectors with a K18 promoter. *Mol Ther* 7(5 Pt 1):649–658.

Toietta G, Mane VP, Norona WS, Finegold MJ, Ng P, McDonagh AF, Beaudet AL, Lee B. 2005. Lifelong elimination of hyperbilirubinemia in the Gunn rat with a single injection of helper-dependent adenoviral vector. *Proc Natl Acad Sci U S A* 102(11):3930–3935.

Umana P, Gerdes CA, Stone D, Davis JR, Ward D, Castro MG, Lowenstein PR. 2001. Efficient FLPe recombinase enables scalable production of helper-dependent adenoviral vectors with negligible helper-virus contamination. *Nat Biotechnol* 19(6):582–585.

Umeda K, Suzuki K, Yamazoe T, Shiraki N, Higuchi Y, Tokieda K, Kume K, Mitani K, Kume S. 2013. Albumin gene targeting in human embryonic stem cells and induced pluripotent stem cells with helper-dependent adenoviral vector to monitor hepatic differentiation. *Stem Cell Res* 10(2):179–194.

VanderVeen N, Paran C, Krasinkiewicz J, Zhao L, Palmer D, Hervey-Jumper S, Ng P, Lowenstein PR, Castro MG. 2013. Effectiveness and preclinical safety profile of doxycycline to be used "off-label" to induce therapeutic transgene expression in a phase I clinical trial for glioma. *Hum Gene Ther Clin Dev* 24(3):116–126.

Varnavski AN, Zhang Y, Schnell M, Tazelaar J, Louboutin JP, Yu QC, Bagg A, Gao GP, Wilson JM. 2002. Preexisting immunity to adenovirus in rhesus monkeys fails to prevent vector-induced toxicity. *J Virol* 76(11):5711–5719.

Vetrini F, Brunetti-Pierri N, Palmer DJ, Bertin T, Grove NC, Finegold MJ, Ng P. 2010. Vasoactive intestinal peptide increases hepatic transduction and reduces innate immune response following administration of helper-dependent Ad. *Mol Ther* 18(7):1339–1345.

Wang G, Zabner J, Deering C, Launspach J, Shao J, Bodner M, Jolly DJ, Davidson BL, McCray PB, Jr. 2000. Increasing epithelial junction permeability enhances gene transfer to airway epithelia *in vivo*. *Am J Respir Cell Mol Biol* 22(2):129–138.

Wang L, Figueredo J, Calcedo R, Lin J, Wilson JM. 2007. Cross-presentation of adeno-associated virus serotype 2 capsids activates cytotoxic T cells but does not render hepatocytes effective cytolytic targets. *Hum Gene Ther* 18(3):185–194.

Weaver EA, Nehete PN, Buchl SS, Senac JS, Palmer D, Ng P, Sastry KJ, Barry MA. 2009a. Comparison of replication-competent, first generation, and helper-dependent adenoviral vaccines. *PLoS One* 4(3):e5059.

Weaver EA, Nehete PN, Nehete BP, Buchl SJ, Palmer D, Montefiori DC, Ng P, Sastry KJ, Barry MA. 2009b. Protection against Mucosal SHIV Challenge by Peptide and Helper-Dependent Adenovirus Vaccines. *Viruses* 1(3):920.

White GI, Monahan PE. 2007. *Gene Therapy for Hemophilia A*. Oxford, UK: Blackwell Publishing Ltd.

Wilmott RW, Amin RS, Perez CR, Wert SE, Keller G, Boivin GP, Hirsch R et al. 1996. Safety of adenovirus-mediated transfer of the human cystic fibrosis transmembrane conductance regulator cDNA to the lungs of nonhuman primates. *Hum Gene Ther* 7(3):301–318.

Wisse E, Jacobs F, Topal B, Frederik P, De Geest B. 2008. The size of endothelial fenestrae in human liver sinusoids: Implications for hepatocyte-directed gene transfer. *Gene Ther* 15(17):1193–1199.

Wonganan P, Clemens CC, Brasky K, Pastore L, Croyle MA. 2010. Species differences in the pharmacology and toxicology of PEGylated helper-dependent adenovirus. *Mol Pharm* 8(1):78–92.

Xiong W, Goverdhana S, Sciascia SA, Candolfi M, Zirger JM, Barcia C, Curtin JF et al. 2006. Regulatable gutless adenovirus vectors sustain inducible transgene expression in the brain in the presence of an immune response against adenoviruses. *J Virol* 80(1):27–37.

Xu Z, Tian J, Smith JS, Byrnes AP. 2008. Clearance of adenovirus by Kupffer cells is mediated by scavenger receptors, natural antibodies, and complement. *J Virol* 82(23):11705–11713.

Xu Z, Qiu Q, Tian J, Smith JS, Conenello GM, Morita T, Byrnes AP. 2013. Coagulation factor X shields adenovirus type 5 from attack by natural antibodies and complement. *Nat Med* 19(4):452–457.

Yan Z, Stewart ZA, Sinn PL, Olsen JC, Hu J, McCray PB, Jr., Engelhardt JF. 2015. Ferret and pig models of cystic fibrosis: Prospects and promise for gene therapy. *Hum Gene Ther Clin Dev* 26(1):38–49.

Yang Y, Nunes FA, Berencsi K, Furth EE, Gonczol E, Wilson JM. 1994. Cellular immunity to viral antigens limits E1-deleted adenoviruses for gene therapy. *Proc Natl Acad Sci U S A* 91(10):4407–4411.

Yang Y, Li Q, Ertl HC, Wilson JM. 1995a. Cellular and humoral immune responses to viral antigens create barriers to lung-directed gene therapy with recombinant adenoviruses. *J Virol* 69(4):2004–2015.

Yang Y, Xiang Z, Ertl HC, Wilson JM. 1995b. Upregulation of class I major histocompatibility complex antigens by interferon gamma is necessary for T-cell-mediated elimination of recombinant adenovirus-infected hepatocytes *in vivo*. *Proc Natl Acad Sci U S A* 92(16):7257–7261.

Yei S, Mittereder N, Wert S, Whitsett JA, Wilmott RW, Trapnell BC. 1994. *In vivo* evaluation of the safety of adenovirus-mediated transfer of the human cystic fibrosis transmembrane conductance regulator cDNA to the lung. *Hum Gene Ther* 5(6):731–744.

Yoshimura M, Sakamoto M, Ikemoto M, Mochizuki Y, Yuasa K, Miyagoe-Suzuki Y, Takeda S. 2004. AAV vector-mediated microdystrophin expression in a relatively small percentage of mdx myofibers improved the mdx phenotype. *Mol Ther* 10(5):821–828.

Yotnda P, Chen DH, Chiu W, Piedra PA, Davis A, Templeton NS, Brenner MK. 2002. Bilamellar cationic liposomes protect adenovectors from preexisting humoral immune responses. *Mol Ther* 5(3):233–241.

Zhang Y, Chirmule N, Gao GP, Qian R, Croyle M, Joshi B, Tazelaar J, Wilson JM. 2001. Acute cytokine response to systemic adenoviral vectors in mice is mediated by dendritic cells and macrophages. *Mol Ther* 3(5 Pt 1):697–707.

Zou L, Yuan X, Zhou H, Lu H, Yang K. 2001. Helper-dependent adenoviral vector-mediated gene transfer in aged rat brain. *Hum Gene Ther* 12(2):181–191.

Zuckerman JB, Robinson CB, McCoy KS, Shell R, Sferra TJ, Chirmule N, Magosin SA et al. 1999. A phase I study of adenovirus-mediated transfer of the human cystic fibrosis transmembrane conductance regulator gene to a lung segment of individuals with cystic fibrosis. *Hum Gene Ther* 10(18):2973–2985.

4 Chemical and Combined Genetic and Chemical Modifications of Adenovirus Vector Capsids to Overcome Barriers for *In Vivo* Vector Delivery

Florian Kreppel

CONTENTS

ABSTRACT

Adenovirus (Ad)-based vectors are the most frequently used vector type in clinical trials up to date. However, a multitude of vector-patient interactions that occur after in vivo delivery of the vectors into patients imposes barriers for efficacious clinical application of Ad-based vectors in gene therapy, vaccination, or oncolysis.

This chapter outlines the efficacy-limiting barriers and describes chemical vector capsid modifications as a potential solution.

4.1   INTRODUCTION

4.1.1   Ad Vectors and Gene Therapy

More than 2000 clinical gene therapy trials have been approved to date (see http://www.abedia.com/wiley/years.php) and several of them have successfully demonstrated that the delivery of genes can be a causative treatment with significant clinical benefit for patients suffering from genetic disorders. Among the most successful trials were those addressing the genetic immunodeficiencies severe combined immunodeficiency-adenosine deaminase deficiency (SCID-ADA) (Aiuti et al. 2002, 2009; Gaspar et al. 2006, 2011), X-linked severe combined immunodeficiency 1 (SCID-X1), X-linked chronic granulomatous disease (X-CGD) (Ott et al. 2006, 2007; Kang et al. 2011), Wiskott–Aldrich syndrome (Aiuti et al. PMID 23845947), the brain demyelinating disease X-linked adrenoleukodystrophy (ALD) (Cartier et al. 2009, 2012) and metachromatic leukodystrophy (Biffi et al. PMID 23845948), the genetic retinal disorder Leber's congenital amaurosis type 2 (LCA2) (Bainbridge et al. 2008, 2015; Hauswirth et al. 2008; Maguire et al. 2008; Testa et al. 2013), and lipoprotein lipase deficiency (Gaudet et al. 2010, 2012, 2013).

In the light of these very remarkable recent successes it is important to note that a wide variety of different diseases may be treated by gene therapy protocols. Such diseases include malignant tumors and disseminated metastases, genetic diseases of lung, skeletal muscle, and liver, diseases that could be treated by using secretory organs such as liver or muscle as production sites for therapeutic proteins, and infectious diseases that may be prevented or cured or whose progression may be halted by genetic vaccination. However, the treatment of all of these diseases either requires or least benefits from systemic vector delivery to exploit the enormous potential of gene therapy. A comprehensive vector and disease-specific understanding of vector–host interactions is of paramount importance prior to safe and efficacious systemic vector delivery.

Ad-based vectors are the most frequently used vector type in clinical trials (see http://www.abedia.com/wiley/vectors.php). They have been used in more than 490 clinical trials and have been delivered to thousands of patients. With one tragic exception (Raper et al. 2003), Ad vectors have been shown to be well-tolerated and safe. In addition, the world's first nationally licensed gene therapy products Gendicine and Oncorine are based on Ad (Peng 2005; Wilson 2005; Räty et al. 2008; Liang 2012).

However, data from clinical and preclinical trials have also pointed out that the clinical applicability of Ad vectors is currently limited by numerous vector–host interactions. These interactions lead to rapid vector neutralization, acute toxicity, and vector sequestration and mistargeting. From a virological perspective the wild type human Ad type 5 (Ad5) may be considered one of the best-characterized viruses. Yet it is an enormous challenge to describe and understand the complexity of interactions between the Ad vector and a patient's organism after *in vivo* vector delivery on a molecular level. Such comprehensive understanding is mandatory before Ad vectors can become clinical routine for a large number of patients and a wide variety of diseases.

The multitude of interactions between Ad vectors and patients after *in vivo* vector delivery imposes a large number of barriers that negatively influence efficient and specific gene delivery and determine the toxicity of the vectors to a large degree. While clinical studies and data from animal models have already revealed a large number of such barriers, only few of them are understood in great detail. Furthermore, in particular with the ongoing development of Ad vectors based on different types and species, it appears very likely that numerous barriers have not even been identified yet.

The aim of this chapter is to outline the complexity of Ad vector–host interactions as far as possible today and to figure out the utility of chemical and combined genetic and chemical Ad vector modifications as tools to analyze and overcome efficacy-limiting barriers for Ad vector-mediated gene transfer. Since most data in the past were generated with Ad5-based vectors, a strong focus will be on this Ad type, which can serve as an excellent paradigm to also understand interactions of other Ad types. Most if not all of the barriers described below impact on the efficacy of *in vivo* delivered Ad vectors independent of the route of delivery.

In the early years of Ad vector development numerous approaches tried to manipulate the vector tropism and vector–host interactions by genetic means. However, the introduction of small ligand motifs (for targeting) or point mutations (for detargeting) was typically not sufficient to overcome the multitude of biological *in vivo* barriers and to enable successful clinical vector application. Therefore, this chapter focuses on chemical and combined genetic and chemical capsid modifications, which hold great promise for efficacious *in vivo* delivery of Ad vectors. The reader should note that in parallel to chemical capsid modifications, type switching strategies, chimeric vectors, and approaches of directed evolution (Bauzon and Hermiston 2012) have currently been developed with great success. While these are not a direct subject of this chapter it will become obvious that such strategies can beneficially be combined with chemical approaches to create improved Ad-based vectors now (Nguyen et al. 2016) and in the future.

4.2 BASIC Ad BIOLOGY

4.2.1 CLASSIFICATION

Ads are non-enveloped viruses with an icosahedral protein capsid that harbors a double-stranded linear DNA genome with a size of 36–40 kbp. They were first isolated in 1953 from adenoid tissue (Rowe et al. 1953). The family of Adenoviridae comprises the five genera AtAd, AviAd, IchtAd, Siadenovirus, and MastAd with the latter including human Ads. Human Ads were divided into species A–G based on their ability to agglutinate erythrocytes (Rosen 1960). Based on sequence analysis at least 57 types can be distinguished. The most frequently used and best-characterized Ad gene transfer vectors are derived from types 5 and 2 of human species C. Importantly, during the past years other types and chimera have been characterized and generated (e.g., Mack et al. 1997; Wickham et al. 1997; Shayakhmetov et al. 2000; Mei et al. 2002; Lu et al. 2009). Also non-human Ads have been vectorized (see e.g., Roy et al. 2006, 2007, 2009; Peruzzi et al. 2009; Colloca et al. 2012; Capone et al. 2013), studied *in vitro* in animal models and in humans (Green et al. 2015; Kelly et al. 2015).

4.2.2 STRUCTURE

The capsid of human Ad5 is 90 nm in diameter and packages a genome of 36 kbp (crystal structure data can be found in Reddy et al. (2010a,b), a general structural description in Knipe and Howley (2013)). The capsid consists of 11 structural proteins, which are called proteins II–IX, IIIa, terminal protein, and μ (protein X). The most abundant capsomere, which is present in 240 trimers per virion is the hexon capsomere. Twelve hexon trimers comprise each facet of the icosahedral capsid. At each capsid vertex there is one protein complex called penton. Penton is comprised of five penton monomers, which form the so-called penton-base and one fiber trimer, which is non-covalently associated to the penton base via its N-terminal domain and protrudes from the capsid as a rod-like structure. Penton, that is, the trimeric fiber protein and loop motifs in the pentameric penton-base protein are involved in receptor-binding and cellular uptake of virions at least *in vitro* (Wickham et al. 1993; Bergelson et al. 1997; Roelvink et al. 1998). The capsid protein IX stabilizes the capsid and is located at the central nine hexon trimers of each facet. Protein VI contributes to capsid stabilization but is located beneath the peripentoneal hexons in the inner capsid. It plays an important role for the endosomal escape of Ad species C

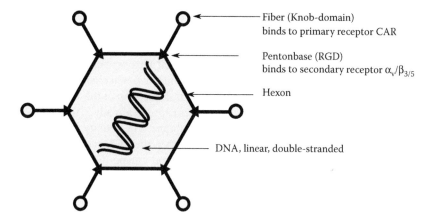

Fiber (Knob-domain)
binds to primary receptor CAR

Pentonbase (RGD)
binds to secondary receptor $\alpha_v/\beta_{3/5}$

Hexon

DNA, linear, double-stranded

FIGURE 4.1 Schematic illustration of the Ad capsid. Ad is a non-enveloped icosahedral virus with a linear double-stranded DNA genome. The capsomeres fiber and penton interact with the primary and secondary receptors. The capsomere hexon is the most abundant capsomere and involved in the *in vivo* tropism and various virus–host interactions.

(Wiethoff et al. 2005). Besides the genome the core of the virions contains the proteins/peptides V, VII, and μ, all of which are basic, arginine-rich proteins. Protein VII condenses the virus DNA genome into a chromatin-like structure. The 5′-end of each DNA strand is covalently attached to the terminal protein, which is important for the initiation of genome replication (Tamanoi and Stillman 1982). The capsid core harbors the viral protease p23, which is important for virus particle assembly, maturation, disassembly, and endosomal escape of the virions (Greber et al. 1996; Greber 1998). Figure 4.1 shows a schematic outline of the Ad capsid. Most early interactions of Ad vectors with cellular and non-cellular host components occur at the capsid surface and involve the capsomeres hexon, fiber, and penton base.

4.2.3 Cell Entry *In Vitro*

The entry of Ad vectors into host cells is a highly dynamic multistep process, which has been described to great detail for Ad species C type 5 (Meier and Greber 2004). This knowledge has recently been expanded to types 3, 7, 11, 14 (Wang et al. 2011), and 37 (Nilsson et al. 2011).

Members of all subgroups except subgroup B (Roelvink et al. 1998) bind with the globular head domains (called knob) of their fiber proteins to the coxsackie and Ad receptor, CAR (Bergelson et al. 1997; Tomko et al. 1997). After binding to CAR the fiber protein bends to give access to an arginine–glycine–aspartic acid (RGD) motif in penton base. This flexible bending depends on a lysine–lysine–threonine–lysine (KKTK) motif located in the shaft of the fiber protein (Wu and Nemerow 2004). After binding to the primary receptor CAR and fiber bending, the RGD motif in penton base binds to the secondary receptor, an integrin αvβ3 or αvβ5 (Wickham et al. 1993; Bai et al. 1994). The binding to integrins triggers dynamin-dependent endocytosis of the particles via clathrin-coated pits and vesicles. At the time of internalization the capsid starts to disassemble (Nakano et al. 2000). Already upon binding to CAR fibers are partially shed from the virions and released to the extracellular space. In the early endosome fiber dissociation is completed and the endosomal acidification triggers the dissociation of peripentonal hexons, penton base, cement protein pIIIa and pVI (Greber et al. 1993). These conformational changes are required for endosomal escape and allow for the cytosolic translocation of the virions. Protein VI is mandatory for endosomal escape. It becomes exposed already at the cell surface and disrupts the endosomal membrane by its membrane-lytic activity (Wiethoff et al. 2005; Maier et al. 2012; Martinez et al. 2013). Further, the RGD motif in the penton base and the protease p23 take part in the endosomal escape (Greber et al. 1996). After endosomal escape, the partially disassembled virions are transported to the nuclear pore by microtubuli and the hexon-associated cement protein IX is detached in the cytosol by the time the virus localizes at the nuclear membrane (Greber et al. 1993; Suomalainen et al. 1999; Greber and Way 2006; Gazzola et al. 2009).

The integrin activation and internalization of Ad5 triggers a series of signaling cascades, which prepare the cell for virus arrival (Greber 2002; Wolfrum and Greber 2013). These signaling cascades induce cytoskeletal changes, regulate the endosomal transport, and influence the endosomal escape of the virions to promote transport of the particles via the dynein/dynactin motor complex (Suomalainen et al. 2001).

The partially disassembled particles, which still harbor the condensed DNA arrive at the perinuclear space and bind to the nuclear pore complex. For Ad2 it has been shown that these particles bind to the nuclear pore complex filament protein CAN/Nup214 (Trotman et al. 2001). At this position histone H1 binds to the hexon capsomere of the capsid and attracts the import factors Imp7 and Impβ. Nuclear import signals on the viral chromatin proteins or histone H1 in addition to other factors mediate translocation of virus DNA into the nucleus (Cassany et al. 2015).

The processes of endosomal escape and nuclear transport are highly efficient and the vast majority of Ad5 particles that enter a cell do reach the nucleus within 20–30 min. Importantly, any modifications of the viral capsid (or core) with the aim to improve Ad vector delivery in terms of targeting, shielding, or both should not interfere with these highly dynamic intracellular processes in order to maintain gene transfer efficiency.

4.2.4 VIRUS GENOME AND LIFE CYCLE

The doubled-stranded DNA genome of Ad is flanked by inverted terminal repeats (ITRs) (for detailed description of the genome organization see (Knipe and Howley 2013)). These ITRs serve as replication origins. A packaging signal sequence Ψ is located between the left ITR and the genome's first transcription unit. It is required for packaging the adenoviral genome into capsids. The genome is divided into 100 equally sized "map units" and according to transcription analyses at different time points after virus entry into cells, the gene regions of Ad are classified as early ("E") and late ("L") gene regions, relative to the onset of DNA replication. The early genes are encoded by the four gene regions *E1–E4*. E1A gene products are essential for the expression of all other virus genes. Consistently, a deletion of the *E1* gene region renders the virus a replication defective vector.

During the infectious cycle, replication of the viral genomes starts 4–6 hours after the onset of $E1$ expression. At the same time five families of late mRNA ($L1$–$L5$) coding for the different capsid proteins are transcribed from the so-called major late promoter. The life cycle from the arrival of the vector genome inside the nucleus until the release of progeny virus is a highly regulated process in time and space (Knipe and Howley 2013).

4.2.5 TYPES OF Ad VECTORS

Until today Ad-based gene transfer vectors are the most frequently used vector type in gene therapy clinical trials. They are characterized by a relative ease of production to high titers, their ability to transduce dividing as well as non-dividing cells, a high genome stability, and a very low degree of vector genome integration into the host cell. Based on modifications of the virus genomes different types of Ad vectors have been developed. For detailed reviews on the different vector types see (Imperiale and Kochanek 2004; Volpers and Kochanek 2004). The different vector types exhibit different characteristics with respect to longevity of transgene expression and immune stimulation and can be used for diverse applications such as oncolytic virotherapy, genetic vaccination, or classic gene therapy. Table 4.1 outlines the different vector types and their potential applications.

"First-generation vectors" ($\Delta E1$ vectors) have the $E1$ and optionally the $E3$ gene region deleted. They have a capacity for foreign transgenes of up to 8 kbp. Since the majority of viral genes are retained in this vector type, low-level expression of these genes can occur after transduction and that triggers potent cytotoxic T-lymphocytes (CTL) responses against transduced cells. Therefore, this vector type is preferable when strong short-term transgene expression is required and immunostimulatory side effects neither matter nor are required (Yang et al. 1994, 1995). Consistently, first-generation vectors hold out promise for genetic vaccination (Tatsis and Ertl 2004; Lasaro and Ertl 2009; Small and Ertl 2011). Of note, the world's first

TABLE 4.1
Different Types of Ad Vectors and Their Potential Applications

Vector Type	Transgene Capacity (kbp)	Immunogenicity	Application
First-generation $\Delta E1/\Delta E3$	8	High	Genetic vaccination, tumor therapy
Second-generation $\Delta E1/\Delta E2/\Delta E4$	12	High	Genetic vaccination, tumor therapy
High-capacity "gutless": devoid of all viral coding sequences	36	Low	Classic gene therapy, genetic vaccination is also possible
Oncolytic/ armed oncolytic	1.2	High	Tumor therapy: oncolysis

Note: The extent of genome deletions influences the immunogenicity of the Ad vector type. This determines the potential application.

commercialized anti-cancer gene therapy drug Gendicine licensed in China is based on this vector type (Peng 2005; Wilson 2005).

"Second-generation" Ad vectors have, in addition to the $E1$ deletion, deletions or inactivations of $E2$ and/or $E4$ genes (Engelhardt et al. 1994; Yang et al. 1994; Armentano et al. 1995; Gao et al. 1996; Wang et al. 1997). Compared to first-generation vectors, such second-generation vectors may allow for prolonged transgene expression and decreased Ad gene expression-induced inflammatory responses, although this is subject of some controversy (Yang et al. 1994; Wang et al. 1997). Importantly, these vectors also have an increased capacity for the uptake of transgenes of about 12 kbp. Due to the immune responses induced by this vector type they are preferably used for genetic vaccination or tumor treatments.

The deletion of all viral coding sequences resulted in helper-dependent Ad (HD-Ad, also known as high-capacity or "gutless") vectors (Kochanek 1999; Segura et al. 2008; Brunetti-Pierri and Ng 2011). This type of vector can accommodate up to 36 kbp of non-viral DNA. Since there is no viral gene expression from these vectors which are completely devoid of viral genes, their *in vivo* toxicity and immunogenicity are substantially reduced. This enables applications in classic gene therapy approaches that require long-term gene expression for example in the liver (Schiedner et al. 1998; Brunetti-Pierri et al. 2009, 2013), eye (Kreppel et al. 2002), and brain (Thomas et al. 2000) in immunocompetent animal models. Transgene expression from this vector type lasted for up to 7 years in non-human primates after a single vector dose and without noticeable side effects (Brunetti-Pierri et al. 2013). In order to allow for long-term transgene expression mediated by HD-Ad vectors in proliferating cells various approaches have been developed that are either based on replicating episomes (Kreppel and Kochanek 2004; Ehrhardt et al. 2008; Gallaher et al. 2009; Voigtlander et al. 2013) or utilize transposase or other viral mechanisms for stable integration into the host cell genome (Picard-Maureau et al. 2004; Müther et al. 2009; Hausl et al. 2010, 2011; Zhang et al. 2013). HD-Ad vectors are also available based on canine Ad (Bru et al. 2010) and are promising tools for gene transfer into the central nervous system (Junyent and Kremer 2015).

Finally, so-called oncolytic Ad vectors have been developed, which conditionally replicate in tumor cells and may be used as an adjuvant therapy complementing (or even substituting) standard cancer therapies (Nettelbeck 2003; O'Shea 2005; Yamamoto and Curiel 2010; Seymour and Fisher 2011; Bauzon and Hermiston 2012; M. Bauzon and T. Hermiston 2014). The second gene therapy-based commercial drug worldwide Oncorine H101 is based on such a selectively replicating Ad vector (see http://www.sunwaybio.com.cn/en/product.html). Importantly, also chimeric oncolytic Ad vectors have been used in a clinical trial and demonstrated safety (Kim et al. 2012, 2013). Enadenotucirev is a recent development and based on a chimeric replication-competent Ad11p/Ad3 adenovirus (Kuhn et al. 2008). This virus has been engineered by directed evolution (Bauzon and Hermiston 2012) to evade the barriers imposed by human Ad. Therefore, it can be delivered systemically. Since its genome is about 3 kbp smaller compared to Ad11p it can also be equipped with therapeutic genes.

It is important to note that typically the different vector types possess the same capsid as the wild type virus counterpart. Many of the barriers faced by these vectors upon delivery *in vivo* can directly be attributed to interactions of the vector capsids with non-target cellular and non-cellular compartments in the patient. These capsid-mediated interactions hinder efficient transduction of target cells and determine the toxicity of the vector particles to a significant degree. As a consequence, research efforts during the last years were focused on the manipulation of different capsid areas with the aim to decrease unwanted interactions with non-target cellular and non-cellular compartments. A variety of genetic, evolutionary, chemical, and combined approaches have been developed to characterize and evade the barriers for *in vivo* delivery of Ad vectors.

4.3   BARRIERS

4.3.1   INTERACTIONS WITH BLOOD COAGULATION FACTORS

In 1997, CAR was shown to be the primary receptor for Ad5 *in vitro* (Bergelson et al. 1997). However, the ablation of CAR binding by mutating the fiber protein largely failed to modulate the *in vivo* tropism of Ad5, although the mutant virions did not bind CAR *in vitro* (Einfeld et al. 2001; Smith et al. 2003). It took 8 years until Shayakhmetov et al. (2005) published the first evidence that host blood factors might play an important role for the transduction of liver and Ad5-induced hepatotoxicity. The authors demonstrated novel interactions between Ad virions, blood coagulation factor IX, and C4b-binding protein (an inhibitory protein of the complement system), and suggested fiber to be the primary site for these interactions (Shayakhmetov et al. 2005). Shortly after, Parker et al. (2006) showed that in fact multiple vitamin K-dependent coagulation zymogens promoted the transduction of hepatocytes by Ad vectors. *In vitro* the vitamin K-dependent blood factors VII, IX, X, and protein C improved transduction by Ad5, whereas FXI and FXII did not. Finally, 11 years after discovery of the primary receptor for Ad5, three groups independently demonstrated that blood coagulation factor X binds with up to nanomolar affinity to the hexon capsomere of Ad5 and a series of other Ad types/species and mediates transduction of hepatocytes *in vivo* (Kalyuzhniy et al. 2008; Vigant et al. 2008; Waddington et al. 2008). Blood coagulation factors VII, IX, and X comprise a glutamate rich "GLA" domain, that is carboxylated on the γ-C of the glutamate residues in a vitamin K-dependent manner. Inhibition of γ-carboxylation by the vitamin K antagonist warfarin *in vivo* significantly diminished hepatocyte transduction by Ad5 (Parker et al. 2006). Even more, physiologic levels (8 μg/mL) of γ-carboxylated blood coagulation factor X restored hepatocyte transduction by Ad5 (Kalyuzhniy et al. 2008; Waddington et al. 2008). Surface plasmon resonance measurements revealed that FX binding to Ad was dependent on both Ca^{2+} and the presence of the GLA-domain (that binds seven Ca^{2+} ions) (Kalyuzhniy et al. 2008; Waddington et al. 2008). Studies with non-human primates revealed that the binding of FX to Ad5 was responsible for liver transduction not only in mice but also in *Microcebus murinus* (Alba et al. 2012).

Importantly, by mutating the relevant FX-binding sites in the Ad capsid it was possible to reduce or ablate FX-binding and the liver tropism of Ad5 (Alba et al. 2009, 2010). Ads bear seven hypervariable regions (HVRs) on the outer surface of their hexon proteins. These HVRs are targets for neutralizing antibodies (Sumida et al. 2005; Pichla-Gollon et al. 2007) and vary between Ad types (Mizuta et al. 2009). Analysis of the Ad5 hexon suggested that FX interacts with HVR5 and HVR7 of the Ad5 hexon protein (Alba et al. 2009). In particular, a point mutation of E451 in HVR7 which is conserved between FX-binding Ad types significantly reduced binding of FX to the virus (Alba et al. 2009).

While molecular analysis quickly revealed the basis for binding of FX to various Ad types/species it appeared to be more difficult to unravel the mechanism(s) by which Ad–FX complexes transduce hepatocytes and potential other biological functions of Ad–FX complex formation.

It was shown that Ad–FX complexes can bind to heparan sulfates and heparin (Waddington et al. 2008; Jonsson et al. 2009; Bradshaw et al. 2010; Duffy et al. 2011). Thus, the host blood FX appears to retarget Ad particles to a novel receptor by bridging the virus (bound via FX's GLA-domain) to heparan sulfate proteoglycans (HSPGs) (bound via FX's SP domain). However, HSPGs are expressed on many different cell types and extracellular matrix proteins and Bradshaw et al. suggested that the extent of N- or O-sulfation on heparan sulfates in the liver is crucial for the specificity of FX-mediated hepatocyte transduction by Ad. Nevertheless, there appear to be additional ways how Ad can transduce hepatocytes *in vivo*.

The first hint that Ad5 does not necessarily depend on complex formation with FX for *in vivo* transduction of hepatocytes was observed by Prill et al. (2011). The authors attached the inert polymer polyethyleneglycol (PEG) specifically to the hexon in a way that prevented the FX-mediated enhancement of transduction *in vitro*. *In vivo*, however, vectors PEGylated with 5 kDa PEG even showed an improved hepatocyte transduction compared to wild type Ad5–FX complexes. Very importantly, Xu et al. (2013) demonstrated that FX was not necessary for liver transduction in transgenic mice which lacked antibodies, C1q, or C4 complement components. Very recently, Zaiss et al. (2015) demonstrated that Ad5 vectors, after intravenous injection, were able to transduce mouse livers with substantially reduced heparin sulfate content. Taken together this data strongly suggests that non-FX-binding vectors can transduce liver by an unknown pathway(s) and that even FX–Ad5 complexes may be able to use other receptors than HSPGs.

The knowledge that FX is an important determinant for the liver tropism of Ad5 led to the development of novel approaches to deliver Ad intravenously, for example, for the treatment of disseminated tumors. Shashkova et al. combined depletion of Kupffer cells and warfarin pretreatment prior to a single intravenous injection of a replication-competent Ad5 virus. They observed a significantly improved antitumor activity and suggested that detargeting an oncolytic Ad from liver macrophages and hepatocytes was an effective strategy against disseminated tumor sites.

Interestingly, the formation of complexes between Ad5 and FX was shown not only to affect the virus tropism but also modulate its immunogenicity. Doronin et al. (2012) demonstrated that the uptake of Ad–FX complexes into cells could trigger the activation of nuclear factor κB-dependent early-response genes downstream of TLR4/MyD88/TRIF/TRAF6 signaling. The authors discussed that the Ad-mediated "misplacement" of FX into cells (and particularly into macrophages) acts as a danger signal.

Overall, despite significant progress during the past years our comprehension of the mechanisms by which Ad and engineered vectors can transduce the liver remains partial. It will be paramount to fully understand the biology of the virus/vector–host interactions that dictate the tropism of the virus, its toxicity, and its immunogenicity. Of note, interactions between the virions and host blood components appear to be very closely intertwined. In consequence, Ad vectors rationally engineered to overcome one specific barrier (e.g., FX-imposed hepatocyte tropism) may become susceptible to elimination by another barrier in a sometimes surprising and difficult to predict way as outlined in the following paragraphs.

4.3.2 INTERACTIONS WITH NATURAL ANTIBODIES AND COMPLEMENT: FACTOR X AS A NATURAL SHIELD AGAINST ATTACK BY COMPLEMENT

Ad can activate the complement system via the classical and the non-classical pathway (Tian et al. 2009). Germ line-encoded natural IgM antibodies recognize diverse highly repetitive structures like viral capsids and do not depend on prior exposure to specific antigens. Complexes of natural antibodies and virions can activate the classical complement proteins C1, C2, and C4 (Rambach et al. 2008; Xu et al. 2008; Tian et al. 2009). Polyreactive natural IgM antibodies can bind to Ad5 and inhibit liver transduction upon intravenous vector delivery in a concentration-dependent manner (Qiu et al. 2015). This phenomenon is at least in parts responsible for the differences in liver transduction observed in different mouse strains. Both complement and natural antibodies increase the clearance of Ad by liver-resident Kupffer cells (Xu et al. 2008).

Surprisingly, in 2013 Xu et al. linked the ability of Ad5-based vectors to bind FX to a decreased susceptibility for complement-mediated neutralization (Xu et al. 2013). The authors demonstrated that Ad vectors ablated for FX-binding mediated hepatocyte transduction after intravenous delivery into mice lacking natural antibodies to almost the same degree as FX-binding wild type vectors in mice with natural IgM antibodies. As already mentioned above, such FX binding-ablated Ad vectors were able to transduce livers of mice lacking the complement components C1q and C4. Thus, FX may not only bridge Ad to HPSGs but may also act as a shield that protects the virions from attack by complement. Consistently, ablating FX-binding to Ad5 decreases the hepatocyte tropism of the particles mainly by rendering them more susceptible to neutralization by complement and natural antibodies.

Future studies on improved Ad vectors for delivery through the blood stream must take into account the high sensitivity of Ad5 (and presumably other types) against complement. Also, vectors based on other Ad types and chimeric vectors need to be studied to a much greater detail concerning FX-influenced complement-mediated effects. Overall, the complexity of the network of interactions between Ad and host blood components is a paradigm for the importance of a complete understanding of vector–host and in particular vector–blood interactions (see also Baker et al. 2013). Only thorough analysis of the network connections will allow a rational design of improved vectors—largely independent of the Ad type used.

4.3.3 SEQUESTRATION BY KUPFFER CELLS, LIVER SINUSOIDAL ENDOTHELIUM AND PLATELETS

After intravenous delivery, a very large degree of Ad5 virions becomes sequestered by macrophages, in particular by liver-resident Kupffer cells. Kupffer cells are
essential in clearing pathogens and foreign particles from the blood stream. In mice,
Kupffer cells can sequester up to 98% of intravenously delivered Ad5 vector particles
(Alemany et al. 2000). This sequestration of a large majority of vector particles by
immune cells quickly removes the particles from circulation and thus renders them
unable to reach the actual target tissue and cell.

The phagocytic uptake of Ad5 by Kupffer cells induces a very rapid proinflammatory necrotic death that is controlled by interferon-regulatory factor 3 (Manickan
et al. 2006; Di Paolo et al. 2013). Thus, both cells and virions are quickly destroyed
after uptake. Only at high vector doses of Ad5 (1E11 pfu/kg) it appears possible to
transduce a fraction of Kupffer cells, presumably after uptake of the particles via
integrin binding (Wheeler et al. 2001).

The proinflammatory sequestration by Kupffer cells is one of the major determinants for acute toxicity and the very narrow therapeutic index that has been
described for systemically delivered Ad vectors (Tao et al. 2001) and is associated with significant acute toxic and hemodynamic effects (Schiedner et al. 2003a;
Smith et al. 2008, 2011).

Xu et al. (2008) showed that natural antibodies and complement increased the
clearance of Ad5-based vectors by Kupffer cells. In addition, the authors identified
scavenger receptors as a predominant mechanism for the clearance of Ad by Kupffer
cells. Injection of polyinosinic acid immediately prior to injection of Ad5 increased
transgene expression in a number of tissues including liver up to 15-fold (Haisma
et al. 2008). Comparable effects on transgene expression in liver were achieved by
a transient depletion of Kupffer cells by clodronate liposomes (Kuzmin et al. 1997;
Schiedner et al. 2003b). Piccolo et al. (2014) discovered peptidic inhibitors for
scavenger receptor A and SREC-I which allowed to increase liver transduction by
Ad5-based vectors by blocking phagocytic uptake of the virions into Kupffer cells
(Piccolo et al. 2014).

Another cell type that contributes to scavenging of Ad virions from the blood
stream are liver sinusoidal endothelial cells (LSECs). LSECs represent 25% of liver
cells (Jacobs et al. 2010) and line the sinusoids of the liver. They express the scavenger receptors SREC-I and SREC-II on their surface (Plüddemann et al. 2007) and
are a major component of the reticuloendothelial system. While it is clear from many
independent experiments that the vast majority of intravenously injected Ad vector
particles become quickly scavenged and thus eliminated, it is not yet clear what the
quantitative contribution of the LSECs is in that process. A study by Ganesan et al.
(2011) suggested that the LSECs may play a very important role for the scavenging of
Ad vector particles. Like Kupffer cells LSECs are hard to transduce by Ad5 *in vivo*
(Hegenbarth et al. 2000).

Taken together, the sequestration of Ad vector particles in the liver by both Kupffer
cells and endothelial cells is one of the most important barriers for *in vivo* delivery of
this vector type. While it appears that both complement activation and opsonins play

a central role in the interaction of Ad vector particles with Kupffer cells, the precise mechanisms and especially the relevant capsid site have not yet been described to great detail. Attempts to systemically deliver hepatotropic Ad-based vectors for gene transfer to tissues distinct from liver (such as disseminated tumors or skeletal muscle) have to prevent sequestration by the liver, that is, have to prevent transduction of hepatocytes, scavenging by LSECs, and scavenging by Kupffer cells. Although mechanistically not yet fully understood minimally invasive percutaneous injection techniques using balloon catheters have been shown to be advantageous for therapeutic hepatocyte gene transfer with Ad vectors (Brunetti-Pierri et al. 2007, 2009).

Ad virions bind to platelets. This was first described in rabbits (Cichon et al. 1999), and the resulting thrombocytopenia is an important (yet transient) side-effect invariably observed in small and large animal models as well as in clinical trials after systemic Ad vector delivery (Sung et al. 2001; Lozier et al. 2002; Morral et al. 2002; Raper et al. 2002; Wolins et al. 2003). The molecular basis underlying this phenomenon is only poorly understood. Despite controversial results, CAR may be partially involved in Ad binding to human platelets (Shimony et al. 2009) and also integrins may play a role (Gupalo et al. 2011). Furthermore, Ad binding can activate platelets and activated platelets can bind Ad (Shimony et al. 2009). In addition, it is known that the von Willebrand factor plays a crucial role for Ad5-induced thrombocytopenia (Othman et al. 2007). Very recently, using chimeric vectors with fibers from different Ad types Raddi et al. (2016) published evidence that the Ad5 fiber shaft plays a pivotal role for thrombocytopenia after intravenous vector injection and associated that phenomenon with reduced serum cytokine levels, in particular IL-6, induced by the chimeric vectors. Importantly, Raddi et al. ruled out any significant role of spleen, macrophages, and vitamin K-dependent zymogens in Ad-induced thrombocytopenia.

4.3.4 SEQUESTRATION BY ERYTHROCYTES

In 2003, Cichon et al. published that human Ad serotype 5-based virions bind to human erythrocytes (Cichon et al. 2003). In addition, Cichon et al. described that within a very short time 98% of vector (or wild type virus) bind to erythrocytes in human whole blood. However, the molecular basis for this phenomenon remained unclear until 2009 when Carlisle et al. demonstrated that human but not murine erythrocytes carry the primary receptor for all but subgroup B Ades: CAR (Carlisle et al. 2009). Erythrocytes thus sequester the large majority of intravenously injected vector particles. Very importantly, Carlisle et al. showed that in addition to sequestration by CAR, human erythrocytes carry the complement receptor CR1, which binds Ad5 in the presence of plasma (with complement and anti-Ad antibodies). This highly efficient sequestration of Ad via two independent receptors by erythrocytes was shown to severely inhibit liver transduction in non-obese diabetic/SCID mice which were transfused with human erythrocytes (Carlisle et al. 2009). By fractionation of blood components from whole blood after exposure to Ad5, Carlisle et al. demonstrated that 90% of Ad5 particles were associated with erythrocytes whereas only 4% were recovered from plasma. Negligible amounts of virus were found associated with lymphocytes/monocytes and neutrophils, which do not express CAR. Interestingly, the authors did not discover binding of Ad5 to platelets in their assay.

The erythrocyte barrier is obviously one of the most important hurdles that have to be overcome prior to successful systemic delivery of Ad in humans since it appears to be a very effective defense against a blood-borne infection by Ad. It also stresses the importance of identifying and using relevant animal models in preclinical studies.

4.3.5   INNATE IMMUNE ACTIVATION AND ACUTE TOXICITY

As a consequence of the barriers mentioned above, Ad vectors are very potent activators of innate immune pathways. The induction of strong innate immune responses by the vector particles quickly after injection is at least in part a consequence of the vector–host interactions described above and hampers systemic and local vector delivery (Worgall et al. 1997; Muruve 2004; Cotter and Muruve 2005). Upon interaction of the vector particles with antigen-presenting cells like macrophages and dendritic cells in liver, spleen, and blood, proinflammatory cytokines/chemokines are released such as interleukin- 6 (IL-6), tumor necrosis factor-α, interferon-γ, inducible protein-10, and RANTES (Lieber et al. 1997; Muruve et al. 1999; Schnell et al. 2001; Zaiss et al. 2002; Liu et al. 2003). After intravenous delivery of first-generation Ad vectors into mice or non-human primates, a dose-dependent and strong activation of innate immunity was observed (Schnell et al. 2001). Importantly, this innate immune activation is independent of cellular transduction. Structurally intact but transcriptionally inactive vector particles that can be generated by treatment with psoralen and ultraviolet light, exhibited an activation of innate immunity very similar to the one induced by transcriptionally active vector particles (Schnell et al. 2001). In line with this, HD-Ad vectors devoid of viral coding sequences have also been shown to strongly activate innate immunity in mice and non-human primates (Brunetti-Pierri et al. 2004).

The induction of strong innate immune responses is certainly problematic if not fatal for classic gene therapy approaches. However, it is believed to be advantageous for applications based on local vector injection like genetic vaccination or adjuvant tumor therapies. Nevertheless, it is obvious that the safe and successful delivery of Ad vector particles requires precise control of acute inflammatory responses. This may in some cases be achieved by the co-delivery of anti-inflammatory drugs like steroids (De Geest et al. 2005), but a broad application of systemically delivered Ad vectors for various different diseases can only be achieved when vectors are generated that exhibit at least a significantly dampened or completely abrogated acute inflammatory response. Even in the context of genetic vaccination a controlled or even programmable induction of innate immunity may yield superior results.

4.3.6   ACQUIRED ANTI-VECTOR HUMORAL IMMUNITY

Wild type Ad5 exhibits a mild pathogenicity in immunocompetent individuals and a large number of individuals have been infected with Ad type 5 at least once during their lifetime. The resulting widespread anti-Ad immunity in humans imposes an important hurdle for the use of serotype 5-based Ad vectors (Yang et al. 1995; Molnar-Kimber et al. 1998), since preexisting humoral anti-Ad antibodies efficiently blunt transduction of target cells either by opsonization and subsequent complement activation and/or phagocytosis or by interfering with specific mechanisms during the transduction process.

Interestingly, most of human anti-Ad antibodies are directed against the hexon protein (Sumida et al. 2005), although neutralizing antibodies against the capsid proteins fiber and penton base have been described, too (Gahéry-Ségard et al. 1998; Bradley et al. 2012a,b; Hong et al. 2003). Hexon-directed antibodies can interfere with the transport of virions along microtubules after escape from the endosome (Smith et al. 2008). Fiber-directed antibodies can subvert the natural virus tropism and target Ad to phagolysosomes in macrophages (Zaiss et al. 2009). *In vivo* the targeting of opsonized virus to antigen-presenting cells is likely to be the most important pathway. 30%–50% of Americans, 60% of Japanese, and up to 76% of Europeans have preexisting immunity to Ad5 (D'Ambrosio et al. 1982; Holterman et al. 2004). In certain areas of sub-Saharan Africa even up to 100% of the population have anti-Ad5 antibodies (Sumida et al. 2005).

4.3.7 DEMANDS FOR AN IDEAL Ad VECTOR

In view of the barriers outlined above and keeping in mind that probably several Ad vector–host interactions have not yet been discovered, an ideal Ad vector for systemic or local delivery should

- Not bind FX to prevent sequestration by hepatocytes and hepatotoxicity
- Not bind CAR to prevent sequestration by human erythrocytes
- Not be susceptible to complement (to prevent binding to CR1 on erythrocytes and phagocytosis by immune cells)
- Not be recognized by scavenger receptors
- Not interact with platelets
- Should exhibit low toxicity and should be targeted to a specific receptor on the desired target cell

It is the immense complexity of the interactions that needs be understood in order to design truly improved Ad vectors in a rational way.

4.4 STRATEGIES AND TECHNOLOGIES TO OVERCOME BARRIERS

4.4.1 CHEMICAL CAPSID MODIFICATIONS

Compared to genetic modifications, chemical modifications have the advantage that they can attach molecules derived from different substance classes to the capsid surface of the virions. This includes molecules that cannot be encoded genetically (e.g., carbohydrates (Espenlaub et al. 2008)) or those that are fully synthetic. In particular, inert synthetic polymers that act as steric shields have been attached to the surface of Ad vectors.

4.4.1.1 Chemical Capsid Modifications with Synthetic Polymers

Polymer shielding uses (synthetic) polymers, which are hydrophilic and bulky, to create a steric shield around the capsid surface. Typically, the synthetic polymers are covalently attached to the virus surface by chemical reactions that maintain the structural integrity of the capsid. The size and the chemical nature of the polymer

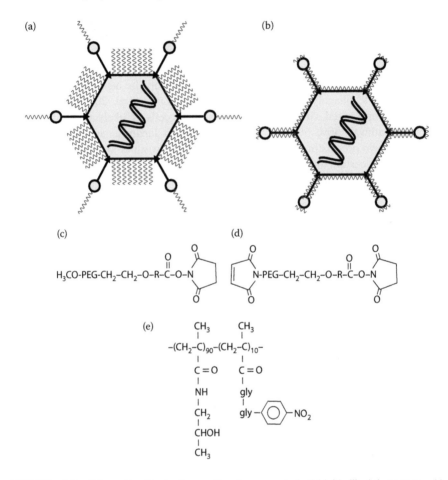

FIGURE 4.2 Schematic illustrations of polymer-coated "shielded" Ad vectors. (a) Illustration of an Ad covalently coated with a synthetic polymer that reacted with one end with the capsid surface ("semitelechelic polymer"). The other end protrudes from the capsid surface and is freely movable. This is typical for PEGylated Ads. (b) Illustration of an Ad covalently coated with a synthetic polymer that reacted with the capsid surface at multiple reactive sites per polymer molecule. This is typical for HPMAylated Ads. The protruding distal ends of semitelechelic PEG molecules and unreacted groups on HPMA molecules can be used to couple ligands for targeting to the polymer shield. (c) Semitelechelic PEG molecule with one reactive end bearing an NHS ester that is reactive towards amine groups which are abundant on the surface of Ad capsids. (d) Heterobifunctional PEG molecule that is suitable for coupling of ligands to the capsid surface. (e) pHPMA molecules for multivalent coupling to the capsid surface.

as well as the mode of covalent attachment to the capsid surface determine to which degree the polymers sterically prevent interactions of the virus surface with the environment. Figure 4.2a,b shows a schematic illustration of polymer-conjugated Ad.

One of the most important advantages of chemical capsid modifications compared to genetic approaches is that the chemical reaction to attach the polymer can be performed after vector production and purification. Thus, there is no interference with vector production and conventional and well-characterized producer cells can be used and stocks with high vector titers can easily be obtained using standard methods. In addition, chemical capsid modification can be performed in a way that simultaneously modifies up to several thousands of amino acid residues on the capsid surface. Such extensive alterations of the capsid surface would likely be impossible to achieve by genetic modifications.

It has to be noted though—in particular in the context of conditionally replication-competent vectors—that the surface modifications performed by chemistry after vector production will not be passed onto progeny virions (in contrast of course to genetic modifications). However, chemical and genetic modifications as well as the use of different Ad types do not exclude each other and a combination of the approaches appears promising (Nguyen et al. 2016).

Two types of synthetic polymers have been used for covalent modification of Ad gene transfer vectors. The first is based on PEG, the second is based on poly-N-hydroxypropylmethacrylamide (pHMPA).

4.4.1.2   Polyethylene Glycol

PEG is a synthetic polymer and consists of the repetitive subunits (-CH_2CH_2O-). It can be synthesized in linear or branched forms of different lengths. The typical molecular weight of PEG for shielding purposes is 200–40,000 Da. PEG is uncharged and characterized by a very high hydrophilicity. This results in extensive hydration of the polymer molecules, which contributes to the shielding of surfaces to which the PEG moieties are coupled. PEG exhibits low immunogenicity and a very low toxicity. It has been approved by the Food and Drug Administration for use in cosmetics, foods, and drugs and several PEGylated therapeutic proteins are in clinical use. These include PEG-adenosine deaminase, PEG-interleukin 2, and PEG-α interferon (Nieforth et al. 1996; Kelleher et al. 1998; Davis 2003; Lipton et al. 2007).

The modification of peptidic or protein-based therapeutics with PEG is an established procedure (Delgado et al. 1992; Haag and Kratz 2006; Kumar et al. 2006; Parveen and Sahoo 2006). The covalent attachment of PEG to protein moieties has been shown to reduce their antigenicity and immunogenicity. In addition, it increases the solubility and helps to maintain the protein bioactivity *in vivo*. The conjugation of PEG can protect proteins from proteolytic degradation and typically increases a protein's serum half-life.

The termini of PEG molecules are comprised of hydroxyl groups (OH), which lack chemical reactivity toward proteins. Therefore, PEG needs to be chemically activated prior to conjugation to proteins/peptides. The N-terminus, ϵ-amine groups of lysine residues, or thiols of cysteine residues typically serve as target sites on the proteins to attach PEG. While one PEG end is chemically activated to react with the target residues on the protein, the other end is often capped with an inert methoxy group. Thus, the molecule is reactive only at one end. This configuration is called "semitelechelic" (see Figure 4.2c). Optionally both PEG ends can be activated with reactive groups (Figure 4.2d). When both functional groups exhibit the same reactivity, the PEG molecules are called homobifunctional. In contrast, PEG molecules with different functional linkers on both ends are called heterobifunctional.

While there is a wide variety of commercially available PEG derivatives with different functional groups at one or both PEG ends, up to date all PEGylation approaches performed with Ad were based on modifications of the ε-amine group of lysine residues or the thiol group of cysteine residues. The amine group is an attractive target for PEGylation because it is the most abundant functional group on the Ad vector capsid surface: on human Ad5 there are approximately 18,000 amine groups (O'Riordan et al. 1999).

The most important activation groups to modify amine groups on Ad gene transfer vectors (and also on non-viral delivery systems) are based on N-hydroxysuccinimide esters (NHS), which are highly reactive towards epsilon-amine groups in lysine residues at pH > 7.4. NHS-esters are electrophilically activated succinimidyl esters and readily react with unprotonated amine groups under formation of physiologically stable amide bonds.

Thiol groups on cysteine side chains can be reacted with PEG moieties, which are activated by maleimide or dithiopyridyl groups. Maleimide-containing linkers form stable thioether bonds with the vector surface whereas dithiopyridyl groups undergo a disufide exchange reaction and form bioreversible (i.e., reducible) disulfide bonds with the vector surface. O'Riordan et al. (1999) and Croyle et al. (2000) were the first to utilize amine-directed PEG conjugation to modify the surface of Ad gene transfer vectors.

It is important to note that by modifying the size of the PEG polymer and/or the density of the PEG shield (i.e., the number of PEG molecules attached per virion), PEGylated Ad vector particles can be generated either under maintenance of their natural tropism (including sequestration by the liver, see for example O'Riordan et al. (1999); Croyle et al. (2000, 2001)) or under ablation of their tropism (including a significant reduction of the liver tropism, see for example Wortmann et al. (2008); Doronin et al. (2009)). As a rule of thumb PEG moieties >10 kDa and conjugated to amine groups on the vector surface at high densities (>8000 molecules per virion) will significantly alter the tropism of the vector particles *in vitro* and can ablate liver tropism *in vivo*. Smaller PEG moieties (<10 kDa) coupled to amine groups and lower densities do not often impact on the vector particle tropism or even increase liver transduction (Doronin et al. 2009). Therefore, studies on amine-PEGylated Ad vectors always have to be carefully evaluated with respect to the size of the PEG and the reaction conditions used.

4.4.1.3 Poly(N-(2-Hydroxypropyl)Methacrylamide)

The second synthetic polymer used to covalently modify the surface of Ad gene transfer vectors is pHPMA (see Figure 4.2e). It is also a hydrophilic and non-toxic polymer with low immunogenicity. It was originally used as plasma expander (Sprincl et al. 1976) and served for drug delivery in the form of HPMA–copolymer–doxorubicin conjugates (Vasey et al. 1999). Compared to free doxorubicin, the

HPMA-modified derivatives demonstrated improved plasma and tumor pharmaco-kinetics, importantly in the absence of polymer-related toxicity.

Fisher et al. (2001) pioneered in using pHPMA to form a covalent shield around Ad type 5 gene transfer vectors. One important difference between the typically used semitelecheclic PEG moieties and the HPMA-based polymers is that the former are often attached to the vector surface with only one end while the other end is protruding from the surface (see Figure 4.2a) whereas the latter are attached to the virus surface at multiple sites per polymer molecule (see Figure 4.2b). The chemistry of both PEG and HPMA molecules is versatile and derivatives, which form bioresponsive bonds or include charges or cleavable backbones have been developed (for reviews see Kopecek and Kopecková (2010); Lammers and Ulbrich (2010)).

HPMA-derivatives are typically coupled to amine groups on the vector capsid surface (as in Fisher et al. (2001)), but can also be coupled to surface thiols (Prill et al. 2014). pHPMA modification of Ad vectors can very efficiently blunt the hepatocyte tropism of the vector particles and modified vectors exhibit a significantly increased plasma circulation when compared to their unmodified counterparts (Green et al. 2004).

4.4.1.4 Polymer Shields Can Dampen the Induction of Innate Immunity, Toxicity, and Adaptive Immune Responses

Green et al. (2004) demonstrated that polymer modification of Ad5 with pHPMA allowed for increased circulation in plasma and reduced the acute vector toxicity after intravenous vector injection in mice. For PEG-modified vectors Croyle et al. obtained evidence for a decreased acute toxicity and decreased innate immune responses after intratracheal (Croyle et al. 2001) and intravenous (Croyle et al. 2002) injection of amine-PEGylated vectors. Importantly, these vectors were modified with small PEG molecules and the vector tropism (including liver) was maintained. Interestingly and importantly, Croyle also observed a significantly reduced induction of adaptive vector-directed immune responses (in particular Th1-type responses: interferon-γ, IL-2) and a significant evasion from nAbs with PEGylated Ad vectors (Croyle et al. 2002). Evasion from nAbs had also been shown by O'Riordan et al. (1999).

A detailed report on innate immune responses induced by PEGylated vectors was published by Mok et al. (2005) using first-generation (ΔE1) and HD-Ad vectors (Mok et al. 2005). Their data revealed that Ad vectors carrying about 15,000 PEG molecules per particle induced lower IL-6 levels 6 hours after intravenous vector injection compared to unPEGylated vector particles or particles with a lower degree of PEGylation. Analysis of the kinetics of serum IL-6 levels over 50 hours demonstrated a marked decrease of the IL-6 peak by up to 70% after injection of a densely PEGylated Ad vector. The direct side-by-side comparison of PEGylated and unPEGylated first-generation and HD-Ad vectors in that study showed that PEGylation dampened the secretion of IL-6 for both vector types. Interestingly, the levels of the liver enzymes alanine aminotransferase and aspartate aminotransferase were not influenced by vector PEGylation of any vector type. Mok et al. also analyzed the interaction of PEGylated Ad vectors with Kupffer cells *in vivo* and with mouse RAW246.7 macrophages *in vitro*. The analyses revealed that PEGylated vector particles were taken up by macrophages to a much lower degree than their unPEGylated counterparts. This observation was in agreement with the decreased IL-6 levels observed after injection of PEGylated Ad vectors.

Subsequent studies by independent groups confirmed that PEGylated Ad vectors (even after modification with different PEG linkers, densities, and in different mouse strains) exhibited a significantly decreased induction of innate responses upon systemic injection in mice (Croyle et al. 2005; De Geest et al. 2005). Studies in human whole blood revealed that Ad vectors after amine-directed PEGylation with large 20 K PEG moieties at high densities induced lower levels of the cytokines IL-8, RANTES and MCP-1 and exhibited reduced vector binding to cells in the absence of neutralizing antibodies (Danielsson et al. 2010).

4.4.1.5 Humoral Anti-Vector Immunity and the Complement System

Fisher et al. (2001) first demonstrated that a modification of Ad vectors with pHPMA polymers allowed to evade from neutralizing antibodies. In addition to evasion from high titer nAb, the authors demonstrated efficient retargeting of the coated particles to the FGF receptor.

Several groups showed that PEGylated Ad vectors induced lower levels of adaptive immune responses to the vector. Even more, readministration of the vector in the presence of anti-vector humoral immunity was feasible to lung and liver (Croyle et al. 2001, 2002). Wortmann et al. (2008) demonstrated that PEGylated, fully detargeted Ad vectors (PEGylated with a dense shield of 20 kDa PEG molecules) were still able to induce cellular and humoral immune responses against their transgene product in the presence of neutralizing anti-Ad antibodies in a setting of genetic vaccination (Wortmann et al. 2008).

These results indicate that a polymer shield helps to protect Ad vectors from the neutralizing antibodies *in vitro* and *in vivo*. However, it has to be noted that all of the studies used different assays to determine the titers of anti-Ad antibodies. Furthermore, human anti-Ad antibodies generated upon a productive infection with the virus may very well differ from mouse anti-Ad antibodies which are generated after exposition to a replication defective Ad-derived vector. Unfortunately, this makes it very difficult to draw conclusions on the ability of polymer-modified vectors to evade from neutralizing antibodies. Side-by-side comparisons, ideally using human material (e.g., in the form of intravenous immunoglobulins, IVIG) and standardized assays are required here.

4.4.1.6 Blood Coagulation Factor Binding and Hepatocyte Tropism

Modification of Ad5 with pHPMA (Green et al. 2004) and with a dense shield of 20 kDa PEG (Wortmann et al. 2008; Doronin et al. 2009) was shown to ablate the liver tropism of intravenously injected Ad vectors. Although physical measurements to determine the binding of blood coagulation factor X to PEGylated adenoviral capsids is still lacking, it may be assumed that dense shields with large PEG moieties at least significantly reduce FX binding and thus directly influence the hepatocyte tropism of the vector particles.

However, it has to be mentioned that a polymer shield which is dense enough to prevent unwanted interactions can render the vector particles non-infectious and inert and thus useless for gene transfer purposes. Strategies have been developed to attach ligands at random positions on top of the vector shields (for HPMA see Fisher et al. (2001), for PEG Romanczuk et al. (1999); Lanciotti et al. (2003), for a review see Kreppel and Kochanek (2008)) for retargeting purposes. However, only very few ligands (and in particular those with relatively broad specificity) were successfully used in such approaches. This may be attributed to the fact that it is very difficult to attach a ligand at a defined position of the polymer shield. Campos and Barry (2006) published data on the effect of ligand positioning on the capsid for targeting purposes . They noticed that while a ligand can be functional when presented by the Ad fiber, the same ligand induced aberrant intracellular trafficking when presented on a hexon or IX. Later similar observations and refinements were made by Corjon et al. (2008). It appears that both the position of the ligand on the capsid and the intracellular fate of the ligand determine the success of targeting approaches to a large degree.

From a scientific point of view it may be noted that Ad vectors shielded by amine-directed attachment of synthetic polymers represent some kind of "black box" and are of limited scientific use to describe and understand biological barriers for Ad vectors. The history of hepatocyte detargeting nicely illustrates this fact. Long circulating, liver-detargeted Ad vector particles had been generated years before the discovery and molecular description of the role of FX for Ad hepatocyte tropism (Green et al. 2004). However, for successful vector development it should be kept in mind that only a comprehensive understanding of biological mechanisms is a robust basis for the development of truly improved Ad gene transfer vectors.

4.4.1.7   Defined Attachment of Shielding Polymers to Selected Capsid Sites

In order to overcome the retargeting difficulties mentioned above and to enable a precise position-specific shielding of only barrier-relevant capsid sites, Kreppel et al. have developed a combined genetic and chemical capsid modification technology which enables targeted PEGylation or HPMAylation (see e.g., Kreppel et al. (2005); Corjon et al. (2008); Espenlaub et al. (2010); Prill et al. (2011, 2014) and Figure 4.3).

The position-specific shielding with only few PEG (or HPMA) molecules is based on the genetic introduction of a specific, new chemical reactivity at defined positions of the viral capsid surface. This chemical reactivity is used, after production of the virus vector, to chemically couple molecules to the capsid surface in a specific, efficient, scalable, and flexible way. To equip the Ad capsid surface with a new chemical reactivity Kreppel et al. genetically introduced cysteine residues at defined sites of the fiber protein, protein IX, or hexon (Kreppel et al. 2005; Corjon et al. 2008; Prill et al. 2011). Cysteine residues, whose thiol groups can efficiently and specifically be modified in physiological buffers were chosen based on the facts that (i) all Ad capsid proteins are devoid of disulfide bridges due to their synthesis in the cytoplasm and thus no misfolding of the modified capsid proteins due to aberrant disulfide bridge formation was expected, and (ii) Ad vector particles do not carry cysteine side-chains on their surface. Both targeting ligands and shielding polymers can be efficiently coupled to various selected sites on the capsid.

By introducing the cysteine in the hypervariable region 5 of the hexon (which is involved in FX-binding) and the subsequent PEGylation, Prill et al. precisely modulated the liver tropism of Ad5-based vectors—into and away from hepatocytes (Prill

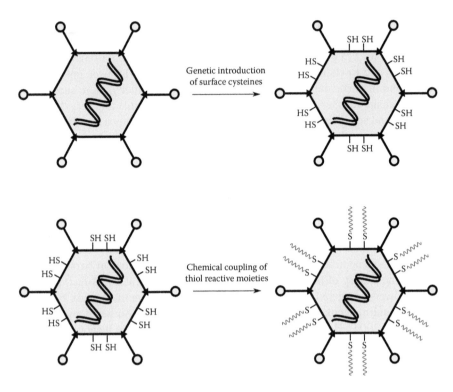

FIGURE 4.3 Position-specific, targeted shielding by combining genetics and chemistry. Cysteine residues are introduced genetically into selected capsid sites. Vectors are produced with conventional producer cells and purified under reducing conditions (upper part). After purification the vector particles are reacted with thiol reactive polymers or ligand molecules.

et al. 2011, 2014). Only 360 small PEG molecules (<2 kDa) coupled to the HVR5 loop of the hexon were sufficient to ablate the hepatocyte tropism of Ad vectors. Furthermore, by increasing the PEG size to 5 kDa Prill et al. demonstrated that a position-specific shielding of HVR5 of the hexon allowed for a significant evasion from Kupffer cell scavenging (Prill et al. 2011). This finding was later corroborated and refined by a work of Khare et al. (2012). By introducing cysteines for site-specific PEGylation of different loops of the hexon capsomere, Khare et al. showed that the hexon HVRs 1, 2, 5, and 7 are likely involved in the recognition and scavenging by Kupffer cells. This work nicely demonstrated that a position-specific targeted PEGylation can generate potent vectors with the ability to evade from specific barriers and at the same time can serve as a valuable tool to describe and understand barriers on a molecular level. By moving the genetically introduced cysteine along specific capsid sites, the capsomeres can be screened for barrier-relevant motifs in suitable assays (see Figure 4.4).

Furthermore, position-specific shielding appears to be a way to keep in balance the natural vector particle dynamics and the need to shield from unwanted interactions. Instead of generating largely inert particles with dense polymer shields, targeted PEGylation allows for minimal shielding of relevant capsid sites only.

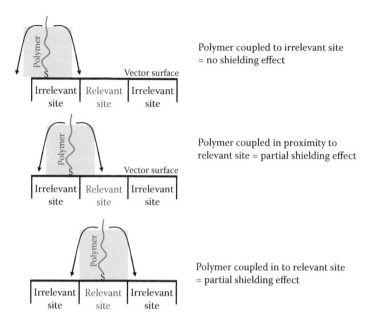

FIGURE 4.4 Position-specific polymer shielding to identify capsid areas that are involved in vector inactivation by biological barriers. Vectors with genetically introduced cysteines at different capsid positions (upper to lower panel) are reacted with shielding polymers and subjected to assays analyzing vector–host interactions, for example binding to complement receptor 1 on erythrocytes. A vector can only be shielded if the polymer was coupled to a capsid site that is relevant for the analyzed vector–host interaction. By "moving" the cysteines along the capsid surface capsid sites involved in vector–host interactions can be identified.

4.4.1.8 Sequestration by Immune Cells, Platelets and Liver Sinusoidal Endothelium

As already outlined above, polymer shielding can significantly reduce the sequestration of Ad vector particles by macrophages. Concerning the interaction with platelets, Croyle et al. (2005) demonstrated that 3 days after intravenous injection of an unmodified HD-Ad vector a sharp drop in platelet counts occurred, which was resolved only 4 days later. In contrast, after injection of PEGylated HD-Ad vectors the platelet counts did not decrease, suggesting that PEGylation had prevented disseminated intravascular coagulation. Hofherr et al. (2007) demonstrated that vector PEGylation blunted the activation of platelets, endothelial cells, and thrombocytopenia to a significant degree.

Sequestration by LSECs has not yet been analyzed in great detail for polymer-shielded Ad vectors. Since a large fraction of Ad vector particles appears to be scavenged by LSECs (Ganesan et al. 2011), it may be assumed that long-circulating Ad vectors, which stably circulate for at least 20 min in blood (Green et al. 2004) should evade from scavenging by LSECs. Nevertheless, a more detailed analysis, in particular at very early time points after injection, is mandatory for polymer-shielded Ad vectors.

4.4.1.9 Vector Sequestration by Erythrocytes

Danielsson et al. (2010) only showed modest effects of PEG shielding on sequestration by erythrocytes. In contrast, Carlisle et al. (2009) and Subr et al. (2009) demonstrated convincingly that Ad vectors shielded by HPMA evade from sequestration by erythrocytes.

Subr et al. (2009) coupled pHPMA derivatives bearing positively charged quaternary amine groups to Ad5-based vectors and analyzed that they did not bind to human erythrocytes. The use of positively charged polymer side chains likely established an electrostatic interaction of the coating polymer with negatively charged loops, in particular with the HVR1 of the hexon. Compared to uncharged HPMA this created a dense wrapping of the vector particles that obviously resulted in better shielding. Importantly, to enable triggered uncoating and reactivation of the densely coated Ad5, the polymer backbone also contained bioresponsive disulfide bonds. Antibody and complement-mediated binding of Ad5 to human erythrocytes was reduced from >95% (without coat) to 25% (with coat). This data demonstrated that a rational molecular design of shielding polymers can be very successful in overcoming otherwise difficult barriers. It may again be noted, that Ad5 is sequestered by human erythrocytes by two independent mechanisms: CAR-binding and CR1-binding. While CAR-binding can be ablated by genetic point mutations or fiber exchange, the antibody/complement binding can probably not be overcome by simple mutations, but requires polymer shielding.

4.5 SYNOPSIS

While the concept of gene therapy to treat a wide variety of different diseases by the introduction of new genetic material into somatic cells has successfully been proven in several *ex vivo* clinical trials, the broad applicability of this concept requires the development of vectors that allow for efficient and safe delivery into patients either locally into nonimmune-privileged tissues or systemically into the blood stream.

However, an efficient (systemic) *in vivo* delivery of Ad vectors is currently hampered by numerous biological barriers imposed by a complex network of vector–host interactions. The development of novel Ad vector systems and a precise molecular description of the barriers are mandatory before this vector type can become clinical routine for a wide variety of patients and diseases.

A wide variety of techniques is available to modify the surface of Ad vectors in order to control unwanted vector–host interactions. These include genetic modifications, the use of alternative Ad types and species, the generation of chimeric vectors and the use of chemical modification technologies. All of these techniques and technologies have specific advantages and disadvantages and it seems likely that combining them will be a clinically successful approach.

The recent history of Ad vector development has shown that the network of vector–host interactions is full of surprises and overcoming one barrier (FX-dictated tropism) can impose another barrier (susceptibility to complement attack). As illustrated by Figure 4.5, such challenges can only be appropriately addressed by a combination of smart technologies—and the utility of chemical modification is certainly one of them.

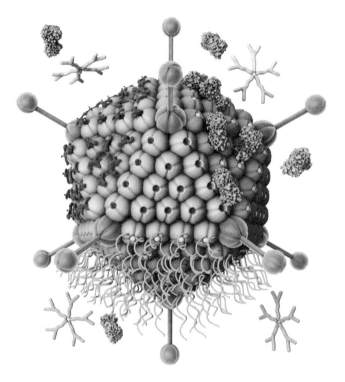

FIGURE 4.5 Adenovirus vector particles undergo multiple interactions with non-cellular and cellular host components. The upper left part shows the natural situation of Ad5 particles, which bind blood coagulation factor X (red) that protects to some degree from natural IgMs (gray) and complement proteins (blue), but mediates liver tropism. After genetic modification of the capsid to ablate factor X binding and the particle's hepatocyte tropism (yellow point mutations, upper right part), FX cannot shield the vector particles and complement and natural IgM efficiently inactivate the vector particles. A combination of genetic ablation of FX binding and chemical shielding by a synthetic polymer (green) can generate vector particles resistant to complement and detargeted from liver (lower part). This figure has been designed by Moritz and Lea Krutzke.

REFERENCES

Aiuti, A., F. Cattaneo, S. Galimberti, U. Benninghoff, B. Cassani, L. Callegaro, S. Scaramuzza et al. 2009. Gene therapy for immunodeficiency due to adenosine deaminase deficiency. *The New England Journal of Medicine* 360 (5): 447–58. doi: 10.1056/NEJMoa0805817.

Aiuti, A., S. Slavin, M. Aker, F. Ficara, S. Deola, A. Mortellaro, S. Morecki et al. 2002. Correction of ADA-SCID by stem cell gene therapy combined with nonmyeloablative conditioning. *Science (New York, N.Y.)* 296 (5577): 2410–13. doi: 10.1126/science.1070104.

Alba, R., A. C. Bradshaw, L. Coughlan, L. Denby, R. A. McDonald, S. N. Waddington, S. M. K. Buckley et al. 2010. Biodistribution and retargeting of FX-binding ablated Ad serotype 5 vectors. *Blood* 116 (15): 2656–64. doi: 10.1182/blood-2009-12-260026.

Alba, R., A. C. Bradshaw, N. Mestre-Francés, J.-M. Verdier, D. Henaff, and A. H. Baker. 2012. Coagulation factor X mediates Ad type 5 liver gene transfer in non-human primates (*Microcebus murinus*). *Gene Therapy* 19 (1): 109–13. doi: 10.1038/gt.2011.87.

Alba, R., A. C. Bradshaw, A. L. Parker, D. Bhella, S. N. Waddington, S. A. Nicklin, N. van Rooijen et al. 2009. Identification of coagulation factor (F)X binding sites on the Ad serotype 5 Hexon: Effect of mutagenesis on FX interactions and gene transfer. *Blood* 114 (5): 965–71. doi: 10.1182/blood-2009-03-208835.

Alemany, R., K. Suzuki, and D. T. Curiel. 2000. Blood clearance rates of Ad type 5 in mice. *The Journal of General Virology* 81 (Pt 11): 2605–9. doi: 10.1099/0022-1317-81-11-2605.

Armentano, D., C. C. Sookdeo, K. M. Hehir, R. J. Gregory, J. A. St George, G. A. Prince, S. C. Wadsworth, and A. E. Smith. 1995. Characterization of an Ad gene transfer vector containing an E4 deletion. *Human Gene Therapy* 6 (10): 1343–53. doi: 10.1089/hum.1995.6.10-1343.

Bai, M., L. Campisi, and P. Freimuth. 1994. Vitronectin receptor antibodies inhibit infection of HeLa and A549 cells by Ad type 12 but not by Ad type 2. *Journal of Virology* 68 (9): 5925–32.

Bainbridge, J. W. B., M. S. Mehat, V. Sundaram, S. J. Robbie, S. E. Barker, C. Ripamonti, A. Georgiadis et al. 2015. Long-term effect of gene therapy on Leber's congenital amaurosis. *The New England Journal of Medicine* 372 (20): 1887–97. doi: 10.1056/NEJMoa1414221.

Bainbridge, J. W. B., A. J. Smith, S. S. Barker, S. Robbie, R. Henderson, K. Balaggan, A. Viswanathan et al. 2008. Effect of gene therapy on visual function in Leber's congenital amaurosis. *The New England Journal of Medicine* 358 (21): 2231–39. doi: 10.1056/NEJMoa0802268.

Baker, A. H., S. A. Nicklin, and D. M. Shayakhmetov. 2013. FX and host defense evasion tactics by Ad. *Molecular Therapy: The Journal of the American Society of Gene Therapy* 21 (6): 1109–11. doi: 10.1038/mt.2013.100.

Bauzon, M., and T. Hermiston. 2014. Armed therapeutic viruses—A disruptive therapy on the horizon of cancer immunotherapy. *Frontiers in Immunology* 5: 74. doi: 10.3389/fimmu.2014.00074.

Bauzon, M., and T. W. Hermiston. 2012. Oncolytic viruses: The power of directed evolution. *Advances in Virology* 2012: 586389. doi: 10.1155/2012/586389.

Bergelson, J. M., J. A. Cunningham, G. Droguett, E. A. Kurt-Jones, A. Krithivas, J. S. Hong, M. S. Horwitz, R. L. Crowell, and R. W. Finberg. 1997. Isolation of a common receptor for coxsackie B viruses and Ades 2 and 5. *Science (New York, N.Y.)* 275 (5304): 1320–23.

Bradley, R. R., D. M. Lynch, M. J. Iampietro, E. N. Borducchi, and D. H. Barouch. 2012a. Adenovirus serotype 5 neutralizing antibodies target both hexon and fiber following vaccination and natural infection. *Journal of Virology* 86 (1): 625–29. doi: 10.1128/JVI.06254-11.

Bradley, R. R., L. F. Maxfield, D. M. Lynch, M. J. Iampietro, E. N. Borducchi, and D. H. Barouch. 2012b. Adenovirus serotype 5-specific neutralizing antibodies target multiple hexon hypervariable regions. *Journal of Virology* 86 (2): 1267–72. doi: 10.1128/JVI.06165-11.

Bradshaw, A. C., A. L. Parker, M. R. Duffy, L. Coughlan, N. van Rooijen, V.-M. Kähäri, S. A. Nicklin, and A. H. Baker. 2010. Requirements for receptor engagement during infection by Ad complexed with blood coagulation factor X. *PLoS Pathogens* 6 (10): e1001142. doi: 10.1371/journal.ppat.1001142.

Bru, T., S. Salinas, and E. J. Kremer. 2010. An update on canine Ad type 2 and its vectors. *Viruses* 2 (9): 2134–53. doi: 10.3390/v2092134.

Brunetti-Pierri, N., and P. Ng. 2011. Helper-dependent adenoviral vectors for liver-directed gene therapy. *Human Molecular Genetics* 20 (R1): R7–13. doi: 10.1093/hmg/ddr143.

Brunetti-Pierri, N., T. Ng, D. Iannitti, W. Cioffi, G. Stapleton, M. Law, J. Breinholt et al. 2013. Transgene expression up to 7 years in nonhuman primates following hepatic transduction with helper-dependent adenoviral vectors. *Human Gene Therapy* 24 (8): 761–65. doi: 10.1089/hum.2013.071.

Brunetti-Pierri, N., D. J. Palmer, A. L. Beaudet, K. Dee Carey, M. Finegold, and P. Ng. 2004. Acute toxicity after high-dose systemic injection of helper-dependent adenoviral vectors into nonhuman primates. *Human Gene Therapy* 15 (1): 35–46. doi: 10.1089/10430340460732445.

Brunetti-Pierri, N., G. E. Stapleton, M. Law, J. Breinholt, D. J. Palmer, Y. Zuo, N. C. Grove et al. 2009. Efficient, long-term hepatic gene transfer using clinically relevant HDAd doses by balloon occlusion catheter delivery in nonhuman primates. *Molecular Therapy: The Journal of the American Society of Gene Therapy* 17 (2): 327–33. doi: 10.1038/mt.2008.257.

Brunetti-Pierri, N., G. E. Stapleton, D. J. Palmer, Y. Zuo, V. P. Mane, M. J. Finegold, A. L. Beaudet, M. M. Leland, C. E. Mullins, and P. Ng. 2007. Pseudo-hydrodynamic delivery of helper-dependent adenoviral vectors into non-human primates for liver-directed gene therapy. *Molecular Therapy: The Journal of the American Society of Gene Therapy* 15 (4): 732–40. doi: 10.1038/sj.mt.6300102.

Campos, S. K., and M. A. Barry. 2006. Comparison of Ad fiber, protein IX, and hexon capsomeres as scaffolds for vector purification and cell targeting. *Virology* 349 (2): 453–62. doi: 10.1016/j.virol.2006.01.032.

Capone, S., A. M. D'Alise, V. Ammendola, S. Colloca, R. Cortese, A. Nicosia, and A. Folgori. 2013. Development of chimpanzee Ades as vaccine vectors: Challenges and successes emerging from clinical trials. *Expert Review of Vaccines* 12 (4): 379–93. doi: 10.1586/erv.13.15.

Carlisle, R. C., Y. Di, A. M. Cerny, A. F.-P. Sonnen, R. B. Sim, N. K. Green, V. Subr et al. 2009. Human erythrocytes bind and inactivate type 5 Ad by presenting coxsackie virus-Ad receptor and complement receptor 1. *Blood* 113 (9): 1909–18. doi: 10.1182/blood-2008-09-178459.

Cartier, N., S. Hacein-Bey-Abina, C. C. Bartholomae, P. Bougnères, M. Schmidt, C. Von Kalle, A. Fischer, M. Cavazzana-Calvo, and P. Aubourg. 2012. Lentiviral hematopoietic cell gene therapy for X-linked adrenoleukodystrophy. *Methods in Enzymology* 507: 187–98. doi: 10.1016/B978-0-12-386509-0.00010-7.

Cartier, N., S. Hacein-Bey-Abina, C. C. Bartholomae, G. Veres, M. Schmidt, I. Kutschera, M. Vidaud et al. 2009. Hematopoietic stem cell gene therapy with a lentiviral vector in X-linked adrenoleukodystrophy. *Science (New York, N.Y.)* 326 (5954): 818–23. doi: 10.1126/science.1171242.

Cassany, A., J. Ragues, T. Guan, D. Bégu, H. Wodrich, M. Kann, G. R. Nemerow, and L. Gerace. 2015. Nuclear import of Ad DNA involves direct interaction of hexon with an N-terminal domain of the nucleoporin Nup214. *Journal of Virology* 89 (3): 1719–30. doi: 10.1128/JVI.02639-14.

Cichon, G., S. Boeckh-Herwig, D. Kuemin, C. Hoffmann, H. H. Schmidt, E. Wehnes, W. Haensch et al. 2003. Titer determination of Ad5 in blood: A cautionary note. *Gene Therapy* 10 (12): 1012–17. doi: 10.1038/sj.gt.3301961.

Cichon, G., H. H. Schmidt, T. Benhidjeb, P. Löser, S. Ziemer, R. Haas, N. Grewe et al. 1999. Intravenous administration of recombinant Ades causes thrombocytopenia, anemia and erythroblastosis in rabbits. *The Journal of Gene Medicine* 1 (5): 360–71. doi: 10.1002/(SICI)1521-2254(199909/10)1:5 <360::AID-JGM54> 3.0.CO;2-Q.

Colloca, S., E. Barnes, A. Folgori, V. Ammendola, S. Capone, A. Cirillo, L. Siani et al. 2012. Vaccine vectors derived from a large collection of simian Ades induce potent cellular immunity across multiple species. *Science Translational Medicine* 4 (115): 115ra2. doi: 10.1126/scitranslmed.3002925.

Corjon, S., A. Wortmann, T. Engler, N. van Rooijen, S. Kochanek, and F. Kreppel. 2008. Targeting of Ad vectors to the LRP receptor family with the high-affinity ligand RAP via combined genetic and chemical modification of the pIX capsomere. *Molecular Therapy: The Journal of the American Society of Gene Therapy* 16 (11): 1813–24. doi: 10.1038/mt.2008.174.

Cotter, M. J., and D. A. Muruve. 2005. The induction of inflammation by Ad vectors used for gene therapy. *Frontiers in Bioscience: A Journal and Virtual Library* 10: 1098–1105.

Croyle, M. A., N. Chirmule, Y. Zhang, and J. M. Wilson. 2001. "Stealth" Ades blunt cell-mediated and humoral immune responses against the virus and allow for significant gene expression upon readministration in the lung. *Journal of Virology* 75 (10): 4792–4801. doi: 10.1128/JVI.75.10.4792-4801.2001.

Croyle, M. A., N. Chirmule, Y. Zhang, and J. M. Wilson. 2002. PEGylation of E1-deleted Ad vectors allows significant gene expression on readministration to liver. *Human Gene Therapy* 13 (15): 1887–1900. doi: 10.1089/104303402760372972.

Croyle, M. A., H. T. Le, K. D. Linse, V. Cerullo, G. Toietta, A. Beaudet, and L. Pastore. 2005. PEGylated helper-dependent adenoviral vectors: Highly efficient vectors with an enhanced safety profile. *Gene Therapy* 12 (7): 579–87. doi: 10.1038/sj.gt.3302441.

Croyle, M. A., Q. C. Yu, and J. M. Wilson. 2000. Development of a rapid method for the PEGylation of Ades with enhanced transduction and improved stability under harsh storage conditions. *Human Gene Therapy* 11 (12): 1713–22. doi: 10.1089/10430340050111368.

D'Ambrosio, E., N. Del Grosso, A. Chicca, and M. Midulla. 1982. Neutralizing antibodies against 33 human Ades in normal children in Rome. *The Journal of Hygiene* 89 (1): 155–61.

Danielsson, A., G. Elgue, B. M. Nilsson, B. Nilsson, J. D. Lambris, T. H. Tötterman, S. Kochanek, F. Kreppel, and M. Essand. 2010. An ex vivo loop system models the toxicity and efficacy of PEGylated and unmodified Ad serotype 5 in whole human blood. *Gene Therapy* 17 (6): 752–62. doi: 10.1038/gt.2010.18.

Davis, F. F. 2003. PEG-adenosine deaminase and PEG-asparaginase. *Advances in Experimental Medicine and Biology* 519: 51–58. doi: 10.1007/0-306-47932-X_3.

De Geest, B., J. Snoeys, S. Van Linthout, J. Lievens, and D. Collen. 2005. Elimination of innate immune responses and liver inflammation by PEGylation of adenoviral vectors and methylprednisolone. *Human Gene Therapy* 16 (12): 1439–51. doi: 10.1089/hum.2005.16.1439.

Delgado, C., G. E. Francis, and D. Fisher. 1992. The uses and properties of PEG-linked proteins. *Critical Reviews in Therapeutic Drug Carrier Systems* 9 (3–4): 249–304.

Di Paolo, N. C., K. Doronin, L. K. Baldwin, T. Papayannopoulou, and D. M. Shayakhmetov. 2013. The transcription factor IRF3 triggers "defensive suicide" necrosis in response to viral and bacterial pathogens. *Cell Reports* 3 (6): 1840–46. doi: 10.1016/j.celrep.2013.05.025.

Doronin, K., J. W. Flatt, N. C. Di Paolo, R. Khare, O. Kalyuzhniy, M. Acchione, J. P. Sumida et al. 2012. Coagulation factor X activates innate immunity to human species C Ad. *Science (New York, N.Y.)* 338 (6108): 795–98. doi: 10.1126/science.1226625.

Doronin, K., E. V. Shashkova, S. M. May, S. E. Hofherr, and M. A. Barry. 2009. Chemical modification with high molecular weight polyethylene glycol reduces transduction of hepatocytes and increases efficacy of intravenously delivered oncolytic Ad. *Human Gene Therapy* 20 (9): 975–88. doi: 10.1089/hum.2009.028.

Duffy, M. R., A. C. Bradshaw, A. L. Parker, J. H. McVey, and A. H. Baker. 2011. A cluster of basic amino acids in the factor X serine protease mediates surface attachment of Ad/FX complexes. *Journal of Virology* 85 (20): 10914–19. doi: 10.1128/JVI.05382-11.

Ehrhardt, A., R. Haase, A. Schepers, M. J. Deutsch, H. J. Lipps, and A. Baiker. 2008. Episomal vectors for gene therapy. *Current Gene Therapy* 8 (3): 147–61.

Einfeld, D. A., R. Schroeder, P. W. Roelvink, A. Lizonova, C. R. King, I. Kovesdi, and T. J. Wickham. 2001. Reducing the native tropism of Ad vectors requires removal of both CAR and integrin interactions. *Journal of Virology* 75 (23): 11284–91. doi: 10.1128/JVI.75.23.11284-11291.2001.

Engelhardt, J. F., X. Ye, B. Doranz, and J. M. Wilson. 1994. Ablation of E2A in recombinant Ades improves transgene persistence and decreases inflammatory response in mouse liver. *Proceedings of the National Academy of Sciences of the United States of America* 91 (13): 6196–6200.

Espenlaub, S., S. Corjon, T. Engler, C. Fella, M. Ogris, E. Wagner, S. Kochanek, and F. Kreppel. 2010. Capsomer-specific fluorescent labeling of adenoviral vector particles allows for detailed analysis of intracellular particle trafficking and the performance of bioresponsive bonds for vector capsid modifications. *Human Gene Therapy* 21 (9): 1155–67. doi: 10.1089/hum.2009.171.

Espenlaub, S., A. Wortmann, T. Engler, S. Corjon, S. Kochanek, and F. Kreppel. 2008. Reductive amination as a strategy to reduce Ad vector promiscuity by chemical capsid modification with large polysaccharides. *The Journal of Gene Medicine* 10 (12): 1303–14. doi: 10.1002/jgm.1262.

Fisher, K. D., Y. Stallwood, N. K. Green, K. Ulbrich, V. Mautner, and L. W. Seymour. 2001. Polymer-coated Ad permits efficient retargeting and evades neutralising antibodies. *Gene Therapy* 8 (5): 341–48. doi: 10.1038/sj.gt.3301389.

Gahéry-Ségard, H., F. Farace, D. Godfrin, J. Gaston, R. Lengagne, T. Tursz, P. Boulanger, and J. G. Guillet. 1998. Immune response to recombinant capsid proteins of Ad in humans: Antifiber and anti-penton base antibodies have a synergistic effect on neutralizing activity. *Journal of Virology* 72 (3): 2388–97.

Gallaher, S. D., J. S. Gil, O. Dorigo, and A. J. Berk. 2009. Robust *in vivo* transduction of a genetically stable Epstein-Barr virus episome to hepatocytes in mice by a hybrid viral vector. *Journal of Virology* 83 (7): 3249–57. doi: 10.1128/JVI.01721-08.

Ganesan, L. P., S. Mohanty, J. Kim, K. Reed Clark, J. M. Robinson, and C. L. Anderson. 2011. Rapid and efficient clearance of blood-borne virus by liver sinusoidal endothelium. *PLoS Pathogens* 7 (9): e1002281. doi: 10.1371/journal.ppat.1002281.

Gao, G. P., Y. Yang, and J. M. Wilson. 1996. Biology of Ad vectors with E1 and E4 deletions for liver-directed gene therapy. *Journal of Virology* 70 (12): 8934–43.

Gaspar, H. B., E. Bjorkegren, K. Parsley, K. C. Gilmour, D. King, J. Sinclair, F. Zhang et al. 2006. Successful reconstitution of immunity in ADA-SCID by stem cell gene therapy following cessation of PEG-ADA and use of mild preconditioning. *Molecular Therapy: The Journal of the American Society of Gene Therapy* 14 (4): 505–13. doi: 10.1016/j.ymthe.2006.06.007.

Gaspar, H. B., S. Cooray, K. C. Gilmour, K. L. Parsley, F. Zhang, S. Adams, E. Bjorkegren et al. 2011. Hematopoietic stem cell gene therapy for adenosine deaminase-deficient severe combined immunodeficiency leads to long-term immunological recovery and metabolic correction. *Science Translational Medicine* 3 (97): 97ra80. doi:10.1126/scitranslmed.3002716.

Gaudet, D., J. de Wal, K. Tremblay, S. Déry, S. van Deventer, A. Freidig, D. Brisson, and J. Méthot. 2010. Review of the clinical development of alipogene tiparvovec gene therapy for lipoprotein lipase deficiency. *Atherosclerosis. Supplements* 11 (1): 55–60. doi: 10.1016/j.atherosclerosissup.2010.03.004.

Gaudet, D., J. Méthot, S. Déry, D. Brisson, C. Essiembre, G. Tremblay, K. Tremblay et al. 2013. Efficacy and long-term safety of alipogene tiparvovec (AAV1-LPLS447X) gene therapy for lipoprotein lipase deficiency: An open-label trial. *Gene Therapy* 20 (4): 361–69. doi: 10.1038/gt.2012.43.

Gaudet, D., J. Méthot, and J. Kastelein. 2012. Gene therapy for lipoprotein lipase deficiency. *Current Opinion in Lipidology* 23 (4): 310–20. doi: 10.1097/MOL.0b013e3283555a7e.

Gazzola, M., C. J. Burckhardt, B. Bayati, M. Engelke, U. F. Greber, and P. Koumoutsakos. 2009. A stochastic model for microtubule motors describes the *in vivo* cytoplasmic transport of human Ad. *PLoS Computational Biology* 5 (12): e1000623. doi: 10.1371/journal.pcbi.1000623.

Greber, U. F. 1998. Virus assembly and disassembly: The Ad cysteine protease as a trigger factor. *Reviews in Medical Virology* 8 (4): 213–22.

Greber, U. F. 2002. Signalling in viral entry. *Cellular and Molecular Life Sciences: CMLS* 59 (4): 608–26.

Greber, U. F., and M. Way. 2006. A superhighway to virus infection. *Cell* 124 (4): 741–54. doi: 10.1016/j.cell.2006.02.018.

Greber, U. F., P. Webster, J. Weber, and A. Helenius. 1996. The role of the Ad protease on virus entry into cells. *The EMBO Journal* 15 (8): 1766–77.

Greber, U. F., M. Willetts, P. Webster, and A. Helenius. 1993. Stepwise dismantling of Ad 2 during entry into cells. *Cell* 75 (3): 477–86.

Green, C. A., E. Scarselli, C. J. Sande, A. J. Thompson, C. M. de Lara, K. S. Taylor, K. Haworth et al. 2015. Chimpanzee Ad- and MVA-vectored respiratory syncytial virus vaccine is safe and immunogenic in adults. *Science Translational Medicine* 7 (300): 300ra126. doi: 10.1126/scitranslmed.aac5745.

Green, N. K., C. W. Herbert, S. J. Hale, A. B. Hale, V. Mautner, R. Harkins, T. Hermiston, K. Ulbrich, K. D. Fisher, and L. W. Seymour. 2004. Extended plasma circulation time and decreased toxicity of polymer-coated Ad. *Gene Therapy* 11 (16): 1256–63. doi: 10.1038/sj.gt.3302295.

Gupalo, E., L. Buriachkovskaia, and M. Othman. 2011. Human platelets express CAR with localization at the sites of intercellular interaction. *Virology Journal* 8: 456. doi: 10.1186/1743-422X-8-456.

Haag, R., and F. Kratz. 2006. Polymer therapeutics: Concepts and applications. *Angewandte Chemie (International Ed. in English)* 45 (8): 1198–1215. doi: 10.1002/anie.200502113.

Haisma, H. J., J. A. A. M. Kamps, G. K. Kamps, J. A. Plantinga, M. G. Rots, and A. R. Bellu. 2008. Polyinosinic acid enhances delivery of Ad vectors *in vivo* by preventing sequestration in liver macrophages. *The Journal of General Virology* 89 (Pt 5): 1097–1105. doi: 10.1099/vir.0.83495-0.

Hausl, M. A., W. Zhang, N. Müther, C. Rauschhuber, H. G. Franck, E. P. Merricks, T. C. Nichols, M. A. Kay, and A. Ehrhardt. 2010. Hyperactive sleeping beauty transposase enables persistent phenotypic correction in mice and a canine model for hemophilia B. *Molecular Therapy: The Journal of the American Society of Gene Therapy* 18 (11): 1896–1906. doi: 10.1038/mt.2010.169.

Hausl, M., W. Zhang, R. Voigtländer, N. Müther, C. Rauschhuber, and A. Ehrhardt. 2011. Development of Ad hybrid vectors for sleeping beauty transposition in large mammals. *Current Gene Therapy* 11 (5): 363–74.

Hauswirth, W. W., T. S. Aleman, S. Kaushal, A. V. Cideciyan, S. B. Schwartz, L. Wang, T. J. Conlon et al. 2008. Treatment of leber congenital amaurosis due to RPE65 mutations by ocular subretinal injection of adeno-associated virus gene vector: Short-term results of a phase I trial. *Human Gene Therapy* 19 (10): 979–90. doi: 10.1089/hum.2008.107.

Hegenbarth, S., R. Gerolami, U. Protzer, P. L. Tran, C. Brechot, G. Gerken, and P. A. Knolle. 2000. Liver sinusoidal endothelial cells are not permissive for Ad type 5. *Human Gene Therapy* 11 (3): 481–86. doi: 10.1089/10430340050015941.

Hofherr, S. E., H. Mok, F. C. Gushiken, J. A. Lopez, and M. A. Barry. 2007. Polyethylene glycol modification of Ad reduces platelet activation, endothelial cell activation, and thrombocytopenia. *Human Gene Therapy* 18 (9): 837–48. doi: 10.1089/hum.2007.0051.

Holterman, L., R. Vogels, R. van der Vlugt, M. Sieuwerts, J. Grimbergen, J. Kaspers, E. Geelen et al. 2004. Novel replication-incompetent vector derived from Ad type 11 (Ad11) for vaccination and gene therapy: Low seroprevalence and non-cross-reactivity with Ad5. *Journal of Virology* 78 (23): 13207–15. doi: 10.1128/JVI.78.23.13207-13215.2004.

Hong, S. S., N. A. Habib, L. Franqueville, S. Jensen, and P. A. Boulanger. 2003. Identification of Ad (ad) penton base neutralizing epitopes by use of sera from patients who had received conditionally replicative Ad (addl1520) for treatment of liver tumors. *Journal of Virology* 77 (19): 10366–75.

Imperiale, M. J. and S. Kochanek. 2004. Adenovirus Vectors: Biology, design, and production. *Current Topics in Microbiology and Immunology* 273: 335–57.

Jacobs, F., E. Wisse, and B. De Geest. 2010. The role of liver sinusoidal cells in hepatocyte-directed gene transfer. *The American Journal of Pathology* 176 (1): 14–21. doi: 10.2353/ajpath.2010.090136.

Jonsson, M. I., A. E. Lenman, L. Frängsmyr, C. Nyberg, M. Abdullahi, and N. Arnberg. 2009. Coagulation factors IX and X enhance binding and infection of Ad types 5 and 31 in human epithelial cells. *Journal of Virology* 83 (8): 3816–25. doi: 10.1128/JVI.02562-08.

Junyent, F. and E. J. Kremer. 2015. CAV-2-why a canine virus is a neurobiologist's best friend. *Current Opinion in Pharmacology* 24 (October): 86–93. doi: 10.1016/j.coph.2015.08.004.

Kalyuzhniy, O., N. C. Di Paolo, M. Silvestry, S. E. Hofherr, M. A. Barry, P. L. Stewart, and D. M. Shayakhmetov. 2008. Adenovirus serotype 5 hexon is critical for virus infection of hepatocytes *in vivo*. *Proceedings of the National Academy of Sciences of the United States of America* 105 (14): 5483–88. doi: 10.1073/pnas.0711757105.

Kang, H. J., C. C. Bartholomae, A. Paruzynski, A. Arens, S. Kim, S. S. Yu, Y. Hong et al. 2011. Retroviral gene therapy for X-linked chronic granulomatous disease: Results from phase I/II trial. *Molecular Therapy: The Journal of the American Society of Gene Therapy* 19 (11): 2092–2101. doi: 10.1038/mt.2011.166.

Kelleher, A. D., M. Roggensack, S. Emery, A. Carr, M. A. French, and D. A. Cooper. 1998. Effects of IL-2 therapy in asymptomatic HIV-infected individuals on proliferative responses to mitogens, recall antigens and HIV-related antigens. *Clinical and Experimental Immunology* 113 (1): 85–91.

Kelly, C., L. Swadling, S. Capone, A. Brown, R. Richardson, J. Halliday, A. von Delft et al. 2015. Chronic hepatitis C virus infection subverts vaccine induced T-cell immunity in humans. *Hepatology (Baltimore, Md.)*, October. doi: 10.1002/hep.28294.

Khare, R., V. S. Reddy, G. R. Nemerow, and M. A. Barry. 2012. Identification of Ad serotype 5 hexon regions that interact with scavenger receptors. *Journal of Virology* 86 (4): 2293–2301. doi: 10.1128/JVI.05760-11.

Kim, K. H., I. Dmitriev, J. P. O'Malley, M. Wang, S. Saddekni, Z. You, M. A. Preuss et al. 2012. A phase I clinical trial of Ad5.SSTR/TK.RGD, a novel infectivity-enhanced bicistronic Ad, in patients with recurrent gynecologic cancer. *Clinical Cancer Research: An Official Journal of the American Association for Cancer Research* 18 (12): 3440–51. doi: 10.1158/1078-0432.CCR-11-2852.

Kim, K. H., I. P. Dmitriev, S. Saddekni, E. A. Kashentseva, R. D. Harris, R. Aurigemma, S. Bae et al. 2013. A phase I clinical trial of Ad5/3-Δ24, a novel serotype-chimeric, infectivity-enhanced, conditionally-replicative Ad (CRAd), in patients with recurrent ovarian cancer. *Gynecologic Oncology* 130 (3): 518–24. doi: 10.1016/j.ygyno.2013.06.003.

Knipe, D. M. and P. M. Howley. 2013. *Fields Virology*. 6th revised edition. Philadelphia, PA: Lippincott Williams&Wilki.

Kochanek, S. 1999. High-capacity adenoviral vectors for gene transfer and somatic gene therapy. *Human Gene Therapy* 10 (15): 2451–59. doi: 10.1089/10430349950016807.

Kopecek, J. and P. Kopecková. 2010. HPMA copolymers: Origins, early developments, present, and future. *Advanced Drug Delivery Reviews* 62 (2): 122–49. doi: 10.1016/j.addr.2009.10.004.

Kreppel, F., J. Gackowski, E. Schmidt, and S. Kochanek. 2005. Combined genetic and chemi-
    cal capsid modifications enable flexible and efficient de- and retargeting of Ad vectors.
    Molecular Therapy: The Journal of the American Society of Gene Therapy 12 (1):
    107–17. doi: 10.1016/j.ymthe.2005.03.006.
Kreppel, F. and S. Kochanek. 2004. Long-term transgene expression in proliferating cells
    mediated by episomally maintained high-capacity Ad vectors. *Journal of Virology* 78
    (1): 9–22.
Kreppel, F. and S. Kochanek. 2008. Modification of Ad gene transfer vectors with synthetic
    polymers: A scientific review and technical guide. *Molecular Therapy: The Journal of
    the American Society of Gene Therapy* 16 (1): 16–29. doi: 10.1038/sj.mt.6300321.
Kreppel, F., T. T. Luther, I. Semkova, U. Schraermeyer, and S. Kochanek. 2002. Long-term
    transgene expression in the RPE after gene transfer with a high-capacity adenoviral
    vector. *Investigative Ophthalmology & Visual Science* 43 (6): 1965–70.
Kuhn, I., P. Harden, M. Bauzon, C. Chartier, J. Nye, S. Thorne, T. Reid et al. 2008. Directed
    evolution generates a novel oncolytic virus for the treatment of colon cancer. *PloS One*
    3 (6): e2409. doi: 10.1371/journal.pone.0002409.
Kumar, T., R. Shantha, K. Soppimath, and S. K. Nachaegari. 2006. Novel delivery technolo-
    gies for protein and peptide therapeutics. *Current Pharmaceutical Biotechnology* 7 (4):
    261–76.
Kuzmin, A. I., M. J. Finegold, and R. C. Eisensmith. 1997. Macrophage depletion increases
    the safety, efficacy and persistence of Ad-mediated gene transfer *in vivo*. *Gene Therapy*
    4 (4): 309–16. doi: 10.1038/sj.gt.3300377.
Lammers, T. and K. Ulbrich. 2010. HPMA copolymers: 30 years of advances. *Advanced
    Drug Delivery Reviews* 62 (2): 119–21. doi: 10.1016/j.addr.2009.12.004.
Lanciotti, J., A. Song, J. Doukas, B. Sosnowski, G. Pierce, R. Gregory, S. Wadsworth, and
    C. O'Riordan. 2003. Targeting adenoviral vectors using heterofunctional polyethylene
    glycol FGF2 conjugates. *Molecular Therapy: The Journal of the American Society of
    Gene Therapy* 8 (1): 99–107.
Lasaro, M. O. and H. C. J. Ertl. 2009. New insights on Ad as vaccine vectors. *Molecular
    Therapy: The Journal of the American Society of Gene Therapy* 17 (8): 1333–39. doi:
    10.1038/mt.2009.130.
Liang, M. 2012. Clinical development of oncolytic viruses in China. *Current Pharmaceutical
    Biotechnology* 13 (9): 1852–57.
Lieber, A., C. Y. He, L. Meuse, D. Schowalter, I. Kirillova, B. Winther, and M. A. Kay. 1997.
    The role of Kupffer cell activation and viral gene expression in early liver toxicity after
    infusion of recombinant Ad vectors. *Journal of Virology* 71 (11): 8798–8807.
Lipton, J. H., N. Khoroshko, A. Golenkov, K. Abdulkadyrov, K. Nair, D. Raghunadharao, T.
    Brummendorf, K. Yoo, B. Bergstrom, and Pegasys CML Study Group. 2007. Phase II,
    randomized, multicenter, comparative study of peginterferon-alpha-2a (40 kD) (pega-
    sys) versus interferon alpha-2a (roferon-A) in patients with treatment-naïve, chronic-
    phase chronic myelogenous leukemia. *Leukemia & Lymphoma* 48 (3): 497–505.
    doi: 10.1080/10428190601175393.
Liu, Q., A. K. Zaiss, P. Colarusso, K. Patel, G. Haljan, T. J. Wickham, and D. A. Muruve.
    2003. The role of capsid-endothelial interactions in the innate immune response to Ad
    vectors. *Human Gene Therapy* 14 (7): 627–43. doi: 10.1089/104303403321618146.
Lozier, J. N., G. Csako, T. H. Mondoro, D. M. Krizek, M. E. Metzger, R. Costello, J. G.
    Vostal, M. E. Rick, R. E. Donahue, and R. A. Morgan. 2002. Toxicity of a first-
    generation adenoviral vector in rhesus macaques. *Human Gene Therapy* 13 (1): 113–24.
    doi: 10.1089/10430340152712665.
Lu, Z.-Z., X.-H. Zou, L.-X. Dong, J.-G. Qu, J.-D. Song, M. Wang, L. Guo, and T. Hung. 2009.
    Novel recombinant Ad type 41 vector and its biological properties. *The Journal of Gene
    Medicine* 11 (2): 128–38. doi: 10.1002/jgm.1284.

Mack, C. A., W. R. Song, H. Carpenter, T. J. Wickham, I. Kovesdi, B. G. Harvey, C. J. Magovern et al. 1997. Circumvention of anti-Ad neutralizing immunity by administration of an adenoviral vector of an alternate serotype. *Human Gene Therapy* 8 (1): 99–109. doi: 10.1089/hum.1997.8.1-99.

Maguire, A. M., F. Simonelli, E. A. Pierce, E. N. Pugh, F. Mingozzi, J. Bennicelli, S. Banfi et al. 2008. Safety and efficacy of gene transfer for Leber's congenital amaurosis. *The New England Journal of Medicine* 358 (21): 2240–48. doi: 10.1056/NEJMoa0802315.

Maier, O., S. A. Marvin, H. Wodrich, E. M. Campbell, and C. M. Wiethoff. 2012. Spatiotemporal dynamics of Ad membrane rupture and endosomal escape. *Journal of Virology* 86 (19): 10821–28. doi: 10.1128/JVI.01428-12.

Manickan, E., J. S. Smith, J. Tian, T. L. Eggerman, J. N. Lozier, J. Muller, and A. P. Byrnes. 2006. Rapid Kupffer cell death after intravenous injection of Ad vectors. *Molecular Therapy: The Journal of the American Society of Gene Therapy* 13 (1): 108–17. doi: 10.1016/j.ymthe.2005.08.007.

Martinez, R., A. M. Burrage, C. M. Wiethoff, and H. Wodrich. 2013. High temporal resolution imaging reveals endosomal membrane penetration and escape of Ades in real time. *Methods in Molecular Biology (Clifton, N.J.)* 1064: 211–26. doi: 10.1007/978-1-62703-601-6_15.

Mei, Y.-F., K. Lindman, and G. Wadell. 2002. Human Ades of subgenera B, C, and E with various tropisms differ in both binding to and replication in the epithelial A549 and 293 cells. *Virology* 295 (1): 30–43. doi: 10.1006/viro.2002.1359.

Meier, O. and U. F. Greber. 2004. Adenovirus endocytosis. *The Journal of Gene Medicine* 6 Suppl 1 (February): S152–63. doi: 10.1002/jgm.553.

Mizuta, K., Y. Matsuzaki, S. Hongo, A. Ohmi, M. Okamoto, H. Nishimura, T. Itagaki et al. 2009. Stability of the seven hexon hypervariable region sequences of Ad types 1–6 isolated in Yamagata, Japan between 1988 and 2007. *Virus Research* 140 (1-2): 32–39. doi: 10.1016/j.virusres.2008.10.014.

Mok, H., D. J. Palmer, P. Ng, and M. A. Barry. 2005. Evaluation of polyethylene glycol modification of first-generation and helper-dependent adenoviral vectors to reduce innate immune responses. *Molecular Therapy: The Journal of the American Society of Gene Therapy* 11 (1): 66–79. doi: 10.1016/j.ymthe.2004.09.015.

Molnar-Kimber, K. L., D. H. Sterman, M. Chang, E. H. Kang, M. ElBash, M. Lanuti, A. Elshami et al. 1998. Impact of preexisting and induced humoral and cellular immune responses in an Ad-based gene therapy phase I clinical trial for localized mesothelioma. *Human Gene Therapy* 9 (14): 2121–33. doi: 10.1089/hum.1998.9.14-2121.

Morral, N., W. K. O'Neal, K. Rice, M. Michelle Leland, P. A. Piedra, E. Aguilar-Córdova, K. Dee Carey, A. L. Beaudet, and C. Langston. 2002. Lethal toxicity, severe endothelial injury, and a threshold effect with high doses of an adenoviral vector in baboons. *Human Gene Therapy* 13 (1): 143–54. doi: 10.1089/10430340152712692.

Muruve, D. A. 2004. The innate immune response to Ad vectors. *Human Gene Therapy* 15 (12): 1157–66. doi: 10.1089/hum.2004.15.1157.

Muruve, D. A., M. J. Barnes, I. E. Stillman, and T. A. Libermann. 1999. Adenoviral gene therapy leads to rapid induction of multiple chemokines and acute neutrophil-dependent hepatic injury *in vivo*. *Human Gene Therapy* 10 (6): 965–76. doi: 10.1089/10430349950018364.

Müther, N., N. Noske, and A. Ehrhardt. 2009. Viral hybrid vectors for somatic integration—Are they the better solution? *Viruses* 1 (3): 1295–1324. doi: 10.3390/v1031295.

Nakano, M. Y., K. Boucke, M. Suomalainen, R. P. Stidwill, and U. F. Greber. 2000. The first step of Ad type 2 disassembly occurs at the cell surface, independently of endocytosis and escape to the cytosol. *Journal of Virology* 74 (15): 7085–95.

Nettelbeck, D. M. 2003. Virotherapeutics: Conditionally replicative Ades for viral oncolysis. *Anti-Cancer Drugs* 14 (8): 577–84.

Nguyen, T. V., G. J. Heller, M. E. Barry, C. M. Crosby, M. A. Turner, and M. A. Barry. 2016. Evaluation of polymer shielding for Ad serotype 6 (Ad6) for systemic virotherapy against human prostate cancers. *Molecular Therapy Oncolytics* 3.

Nieforth, K. A., R. Nadeau, I. H. Patel, and D. Mould. 1996. Use of an indirect pharmacodynamic stimulation model of MX protein induction to compare *in vivo* activity of interferon alfa-2a and a polyethylene glycol-modified derivative in healthy subjects. *Clinical Pharmacology and Therapeutics* 59 (6): 636–46. doi: 10.1016/S0009-9236(96)90003-X.

Nilsson, E. C., R. J. Storm, J. Bauer, S. M. C. Johansson, A. Lookene, J. Ångström, M. Hedenström et al. 2011. The GD1a glycan is a cellular receptor for Ades causing epidemic keratoconjunctivitis. *Nature Medicine* 17 (1): 105–9. doi: 10.1038/nm.2267.

O'Riordan, C. R., A. Lachapelle, C. Delgado, V. Parkes, S. C. Wadsworth, A. E. Smith, and G. E. Francis. 1999. PEGylation of Ad with retention of infectivity and protection from neutralizing antibody in vitro and in vivo. *Human Gene Therapy* 10 (8): 1349–58. doi: 10.1089/10430349950018021.

O'Shea, C. C. 2005. Viruses—Seeking and destroying the tumor program. *Oncogene* 24 (52): 7640–55. doi: 10.1038/sj.onc.1209047.

Othman, M., A. Labelle, I. Mazzetti, H. S. Elbatarny, and D. Lillicrap. 2007. Adenovirus-induced thrombocytopenia: The role of von Willebrand factor and P-selectin in mediating accelerated platelet clearance. *Blood* 109 (7): 2832–39. doi: 10.1182/blood-2006-06-032524.

Ott, M. G., M. Schmidt, K. Schwarzwaelder, S. Stein, U. Siler, U. Koehl, H. Glimm et al. 2006. Correction of X-linked chronic granulomatous disease by gene therapy, augmented by insertional activation of MDS1-EVI1, PRDM16 or SETBP1. *Nature Medicine* 12 (4): 401–9. doi: 10.1038/nm1393.

Ott, M. G., R. Seger, S. Stein, U. Siler, D. Hoelzer, and M. Grez. 2007. Advances in the treatment of chronic granulomatous disease by gene therapy. *Current Gene Therapy* 7 (3): 155–61.

Parker, A. L., S. N. Waddington, C. G. Nicol, D. M. Shayakhmetov, S. M. Buckley, L. Denby, G. Kemball-Cook et al. 2006. Multiple vitamin K-dependent coagulation zymogens promote Ad-mediated gene delivery to hepatocytes. *Blood* 108 (8): 2554–61. doi: 10.1182/blood-2006-04-008532.

Parveen, S. and S. K. Sahoo. 2006. Nanomedicine: Clinical applications of polyethylene glycol conjugated proteins and drugs. *Clinical Pharmacokinetics* 45 (10): 965–88. doi: 10.2165/00003088-200645100-00002.

Peng, Z. 2005. Current status of gendicine in China: Recombinant human Ad-p53 agent for treatment of cancers. *Human Gene Therapy* 16 (9): 1016–27. doi: 10.1089/hum.2005.16.1016.

Peruzzi, D., S. Dharmapuri, A. Cirillo, B. E. Bruni, A. Nicosia, R. Cortese, S. Colloca, G. Ciliberto, N. La Monica, and L. Aurisicchio. 2009. A novel chimpanzee serotype-based adenoviral vector as delivery tool for cancer vaccines. *Vaccine* 27 (9): 1293–1300. doi: 10.1016/j.vaccine.2008.12.051.

Picard-Maureau, M., F. Kreppel, D. Lindemann, T. Juretzek, O. Herchenröder, A. Rethwilm, S. Kochanek, and M. Heinkelein. 2004. Foamy virus–Ad hybrid vectors. *Gene Therapy* 11 (8): 722–28. doi: 10.1038/sj.gt.3302216.

Piccolo, P., P. Annunziata, P. Mithbaokar, and N. Brunetti-Pierri. 2014. SR-A and SREC-I binding peptides increase HDAd-mediated liver transduction. *Gene Therapy* 21 (11): 950–57. doi: 10.1038/gt.2014.71.

Pichla-Gollon, S. L., M. Drinker, X. Zhou, F. Xue, J. J. Rux, G.-P. Gao, J. M. Wilson, H. C. J. Ertl, R. M. Burnett, and J. M. Bergelson. 2007. Structure-based identification of a major neutralizing site in an Ad hexon. *Journal of Virology* 81 (4): 1680–89. doi: 10.1128/JVI.02023-06.

Plüddemann, A., C. Neyen, and S. Gordon. 2007. Macrophage scavenger receptors and host-derived ligands. *Methods (San Diego, Calif.)* 43 (3): 207–17. doi: 10.1016/j.ymeth.2007.06.004.

Prill, J.-M., S. Espenlaub, U. Samen, T. Engler, E. Schmidt, F. Vetrini, A. Rosewell et al. 2011. Modifications of Ad hexon allow for either hepatocyte detargeting or targeting with potential evasion from Kupffer cells. *Molecular Therapy: The Journal of the American Society of Gene Therapy* 19 (1): 83–92. doi: 10.1038/mt.2010.229.

Prill, J.-M., V. Subr, N. Pasquarelli, T. Engler, A. Hoffmeister, S. Kochanek, K. Ulbrich, and F. Kreppel. 2014. Traceless bioresponsive shielding of Ad hexon with HPMA copolymers maintains transduction capacity *in vitro* and *in vivo*. *PloS One* 9 (1): e82716. doi: 10.1371/journal.pone.0082716.

Qiu, Q., Z. Xu, J. Tian, R. Moitra, S. Gunti, A. L. Notkins, and A. P. Byrnes. 2015. Impact of natural IgM concentration on gene therapy with Ad type 5 vectors. *Journal of Virology* 89 (6): 3412–16. doi: 10.1128/JVI.03217-14.

Raddi, N., F. Vigant, O. Wagner-Ballon, S. Giraudier, J. Custers, S. Hemmi, and K. Benihoud. 2016. Pseudotyping serotype 5 Ad with the fiber from other serotypes uncovers a key role of the fiber protein in Ad 5-induced thrombocytopenia. *Human Gene Therapy* 27 (2): 193–201. doi: 10.1089/hum.2015.154.

Rambach, G., R. Würzner, and C. Speth. 2008. Complement: An efficient sword of innate immunity. *Contributions to Microbiology* 15: 78–100. doi: 10.1159/000136316.

Raper, S. E., N. Chirmule, F. S. Lee, N. A. Wivel, A. Bagg, G.-P. Gao, J. M. Wilson, and M. L. Batshaw. 2003. Fatal systemic inflammatory response syndrome in a ornithine transcarbamylase deficient patient following adenoviral gene transfer. *Molecular Genetics and Metabolism* 80 (1–2): 148–58.

Raper, S. E., M. Yudkoff, N. Chirmule, G.-P. Gao, F. Nunes, Z. J. Haskal, E. E. Furth et al. 2002. A pilot study of *in vivo* liver-directed gene transfer with an adenoviral vector in partial ornithine transcarbamylase deficiency. *Human Gene Therapy* 13 (1): 163–75. doi: 10.1089/10430340152712719.

Räty, J. K., J. T. Pikkarainen, T. Wirth, and S. Ylä-Herttuala. 2008. Gene therapy: The first approved gene-based medicines, molecular mechanisms and clinical indications. *Current Molecular Pharmacology* 1 (1): 13–23.

Reddy, V. S., S. K. Natchiar, L. Gritton, T.-M. Mullen, P. L. Stewart, and G. R. Nemerow. 2010b. Crystallization and preliminary X-ray diffraction analysis of human Ad. *Virology* 402 (1): 209–14. doi: 10.1016/j.virol.2010.03.028.

Reddy, V. S., S. K. Natchiar, P. L. Stewart, and G. R. Nemerow. 2010a. Crystal structure of human Ad at 3.5 a resolution. *Science (New York, N.Y.)* 329 (5995): 1071–75. doi: 10.1126/science.1187292.

Roelvink, P. W., A. Lizonova, J. G. Lee, Y. Li, J. M. Bergelson, R. W. Finberg, D. E. Brough, I. Kovesdi, and T. J. Wickham. 1998. The coxsackievirus-Ad receptor protein can function as a cellular attachment protein for Ad serotypes from subgroups A, C, D, E, and F. *Journal of Virology* 72 (10): 7909–15.

Romanczuk, H., C. E. Galer, J. Zabner, G. Barsomian, S. C. Wadsworth, and C. R. O'Riordan. 1999. Modification of an adenoviral vector with biologically selected peptides: A novel strategy for gene delivery to cells of choice. *Human Gene Therapy* 10 (16): 2615–26. doi: 10.1089/10430349950016654.

Rosen, L. 1960. A hemagglutination-inhibition technique for typing Ades. *American Journal of Hygiene* 71 (January): 120–28.

Rowe, W. P., R. J. Huebner, L. K. Gilmore, R. H. Parrott, and T. G. Ward. 1953. Isolation of a cytopathogenic agent from human adenoids undergoing spontaneous degeneration in tissue culture. *Proceedings of the Society for Experimental Biology and Medicine. Society for Experimental Biology and Medicine (New York, N.Y.)* 84 (3): 570–73.

Roy, S., G. P. Kobinger, J. Lin, J. Figueredo, R. Calcedo, D. Kobasa, and J. M. Wilson. 2007. Partial protection against H5N1 influenza in mice with a single dose of a chimpanzee Ad vector expressing nucleoprotein. *Vaccine* 25 (39–40): 6845–51. doi: 10.1016/j.vaccine.2007.07.035.

Roy, S., L. H. Vandenberghe, S. Kryazhimskiy, R. Grant, R. Calcedo, X. Yuan, M. Keough et al. 2009. Isolation and characterization of Ades persistently shed from the gastrointestinal tract of non-human primates. *PLoS Pathogens* 5 (7): e1000503. doi: 10.1371/journal.ppat.1000503.

Roy, S., Y. Zhi, G. P. Kobinger, J. Figueredo, R. Calcedo, J. R. Miller, H. Feldmann, and J. M. Wilson. 2006. Generation of an adenoviral vaccine vector based on simian Ad 21. *The Journal of General Virology* 87 (Pt 9): 2477–85. doi: 10.1099/vir.0.81989-0.

Schiedner, G., W. Bloch, S. Hertel, M. Johnston, A. Molojavyi, V. Dries, G. Varga, N. Van Rooijen, and S. Kochanek. 2003a. A Hemodynamic response to intravenous Ad vector particles is caused by systemic Kupffer cell-mediated activation of endothelial cells. *Human Gene Therapy* 14 (17): 1631–41. doi: 10.1089/104303403322542275.

Schiedner, G., S. Hertel, M. Johnston, V. Dries, N. van Rooijen, and S. Kochanek. 2003b. Selective depletion or blockade of Kupffer cells leads to enhanced and prolonged hepatic transgene expression using high-capacity adenoviral vectors. *Molecular Therapy: The Journal of the American Society of Gene Therapy* 7 (1): 35–43.

Schiedner, G., N. Morral, R. J. Parks, Y. Wu, S. C. Koopmans, C. Langston, F. L. Graham, A. L. Beaudet, and S. Kochanek. 1998. Genomic DNA transfer with a high-capacity Ad vector results in improved in vivo gene expression and decreased toxicity. *Nature Genetics* 18 (2): 180–83. doi: 10.1038/ng0298-180.

Schnell, M. A., Y. Zhang, J. Tazelaar, G. P. Gao, Q. C. Yu, R. Qian, S. J. Chen et al. 2001. Activation of innate immunity in nonhuman primates following intraportal administration of adenoviral vectors. *Molecular Therapy: The Journal of the American Society of Gene Therapy* 3 (5 Pt 1): 708–22. doi: 10.1006/mthe.2001.0330.

Segura, M. M., R. Alba, A. Bosch, and M. Chillón. 2008. Advances in helper-dependent adenoviral vector research. *Current Gene Therapy* 8 (4): 222–35.

Seymour, L. W., and K. D. Fisher. 2011. Adenovirus: Teaching an old dog new tricks. *Human Gene Therapy* 22 (9): 1041–42. doi: 10.1089/hum.2011.2517.

Shayakhmetov, D. M., A. Gaggar, S. Ni, Z.-Y. Li, and A. Lieber. 2005. Adenovirus binding to blood factors results in liver cell infection and hepatotoxicity. *Journal of Virology* 79 (12): 7478–91. doi: 10.1128/JVI.79.12.7478-7491.2005.

Shayakhmetov, D. M., T. Papayannopoulou, G. Stamatoyannopoulos, and A. Lieber. 2000. Efficient gene transfer into human CD34(+) cells by a retargeted Ad vector. *Journal of Virology* 74 (6): 2567–83.

Shimony, N., G. Elkin, D. Kolodkin-Gal, L. Krasny, S. Urieli-Shoval, and Y. S. Haviv. 2009. Analysis of adenoviral attachment to human platelets. *Virology Journal* 6: 25. doi: 10.1186/1743-422X-6-25.

Small, J. C. and H. C. J. Ertl. 2011. Viruses—From pathogens to vaccine carriers. *Current Opinion in Virology* 1 (4): 241–45. doi: 10.1016/j.coviro.2011.07.009.

Smith, J. G., A. Cassany, L. Gerace, R. Ralston, and G. R. Nemerow. 2008. Neutralizing antibody blocks Ad infection by arresting microtubule-dependent cytoplasmic transport. *Journal of Virology* 82 (13): 6492–6500. doi: 10.1128/JVI.00557-08.

Smith, J. S., Z. Xu, J. Tian, D. J. Palmer, P. Ng, and A. P. Byrnes. 2011. The role of endosomal escape and mitogen-activated protein kinases in adenoviral activation of the innate immune response. *PLoS One* 6 (10): e26755. doi: 10.1371/journal.pone.0026755.

Smith, J. S., Z. Xu, J. Tian, S. C. Stevenson, and A. P. Byrnes. 2008. Interaction of systemically delivered Ad vectors with Kupffer cells in mouse liver. *Human Gene Therapy* 19 (5): 547–54. doi: 10.1089/hum.2008.004.

Smith, T. A. G., N. Idamakanti, J. Marshall-Neff, M. L. Rollence, P. Wright, M. Kaloss, L. King et al. 2003. Receptor interactions involved in adenoviral-mediated gene delivery after systemic administration in non-human primates. *Human Gene Therapy* 14 (17): 1595–1604. doi: 10.1089/104303403322542248.

Sprincl, L., J. Exner, O. Stěrba, and J. Kopecek. 1976. New types of synthetic infusion solutions. III. Elimination and retention of poly-[N-(2-hydroxypropyl)methacrylamide] in a test organism. *Journal of Biomedical Materials Research* 10 (6): 953–63. doi: 10.1002/jbm.820100612.

Subr, V., L. Kostka, T. Selby-Milic, K. Fisher, K. Ulbrich, L. W. Seymour, and R. C. Carlisle. 2009. Coating of Ad type 5 with polymers containing quaternary amines prevents binding to blood components. *Journal of Controlled Release: Official Journal of the Controlled Release Society* 135 (2): 152–58. doi: 10.1016/j.jconrel.2008.12.009.

Sumida, S. M., D. M. Truitt, A. A. C. Lemckert, R. Vogels, J. H. H. V. Custers, M. M. Addo, S. Lockman et al. 2005. Neutralizing antibodies to Ad serotype 5 vaccine vectors are directed primarily against the Ad hexon protein. *Journal of Immunology (Baltimore, Md.: 1950)* 174 (11): 7179–85.

Sung, M. W., H. C. Yeh, S. N. Thung, M. E. Schwartz, J. P. Mandeli, S. H. Chen, and S. L. Woo. 2001. Intratumoral Ad-mediated suicide gene transfer for hepatic metastases from colorectal adenocarcinoma: Results of a phase I clinical trial. *Molecular Therapy: The Journal of the American Society of Gene Therapy* 4 (3): 182–91. doi: 10.1006/mthe.2001.0444.

Suomalainen, M., M. Y. Nakano, K. Boucke, S. Keller, and U. F. Greber. 2001. Adenovirus-activated PKA and p38/MAPK pathways boost microtubule-mediated nuclear targeting of virus. *The EMBO Journal* 20 (6): 1310–19. doi: 10.1093/emboj/20.6.1310.

Suomalainen, M., M. Y. Nakano, S. Keller, K. Boucke, R. P. Stidwill, and U. F. Greber. 1999. Microtubule-dependent plus- and minus end-directed motilities are competing processes for nuclear targeting of Ad. *The Journal of Cell Biology* 144 (4): 657–72.

Tamanoi, F. and B. W. Stillman. 1982. Function of Ad terminal protein in the initiation of DNA replication. *Proceedings of the National Academy of Sciences of the United States of America* 79 (7): 2221–25.

Tao, N., G. P. Gao, M. Parr, J. Johnston, T. Baradet, J. M. Wilson, J. Barsoum, and S. E. Fawell. 2001. Sequestration of adenoviral vector by Kupffer cells leads to a nonlinear dose response of transduction in liver. *Molecular Therapy: The Journal of the American Society of Gene Therapy* 3 (1): 28–35. doi: 10.1006/mthe.2000.0227.

Tatsis, N., and H. C. J. Ertl. 2004. Adenoviruses as vaccine vectors. *Molecular Therapy: The Journal of the American Society of Gene Therapy* 10 (4): 616–29. doi: 10.1016/j.ymthe.2004.07.013.

Testa, F., A. M. Maguire, S. Rossi, E. A. Pierce, P. Melillo, K. Marshall, S. Banfi et al. 2013. Three-year follow-up after unilateral subretinal delivery of adeno-associated virus in patients with leber congenital amaurosis type 2. *Ophthalmology* 120 (6): 1283–91. doi: 10.1016/j.ophtha.2012.11.048.

Thomas, C. E., G. Schiedner, S. Kochanek, M. G. Castro, and P. R. Löwenstein. 2000. Peripheral infection with Ad causes unexpected long-term brain inflammation in animals injected intracranially with first-generation, but not with high-capacity, Ad vectors: Toward realistic long-term neurological gene therapy for chronic diseases. *Proceedings of the National Academy of Sciences of the United States of America* 97 (13): 7482–87. doi: 10.1073/pnas.120474397.

Tian, J., Z. Xu, J. S. Smith, S. E. Hofherr, M. A. Barry, and A. P. Byrnes. 2009. Adenovirus activates complement by distinctly different mechanisms *in vitro* and *in vivo*: Indirect complement activation by virions *in vivo*. *Journal of Virology* 83 (11): 5648–58. doi: 10.1128/JVI.00082-09.

Tomko, R. P., R. Xu, and L. Philipson. 1997. HCAR and MCAR: The human and mouse cellular receptors for subgroup C Ades and group B coxsackieviruses. *Proceedings of the National Academy of Sciences of the United States of America* 94 (7): 3352–56.

Trotman, L. C., N. Mosberger, M. Fornerod, R. P. Stidwill, and U. F. Greber. 2001. Import of Ad DNA involves the nuclear pore complex receptor CAN/Nup214 and histone H1. *Nature Cell Biology* 3 (12): 1092–1100. doi: 10.1038/ncb1201-1092.

Vasey, P. A., S. B. Kaye, R. Morrison, C. Twelves, P. Wilson, R. Duncan, A. H. Thomson et al. 1999. Phase I clinical and pharmacokinetic study of PK1 [N-(2-hydroxypropyl) methacrylamide copolymer doxorubicin]: First member of a new class of chemotherapeutic agents-drug-polymer conjugates. Cancer research campaign phase I/II committee. *Clinical Cancer Research: An Official Journal of the American Association for Cancer Research* 5 (1): 83–94.

Vigant, F., D. Descamps, B. Jullienne, S. Esselin, E. Connault, P. Opolon, T. Tordjmann, E. Vigne, M. Perricaudet, and K. Benihoud. 2008. Substitution of hexon hypervariable region 5 of Ad serotype 5 abrogates blood factor binding and limits gene transfer to liver. *Molecular Therapy: The Journal of the American Society of Gene Therapy* 16 (8): 1474–80. doi: 10.1038/mt.2008.132.

Voigtlander, R., R. Haase, M. Mück-Hausl, W. Zhang, P. Boehme, H.-J. Lipps, E. Schulz, A. Baiker, and A. Ehrhardt. 2013. A novel adenoviral hybrid-vector system carrying a plasmid replicon for safe and efficient cell and gene therapeutic applications. *Molecular Therapy. Nucleic Acids* 2: e83. doi: 10.1038/mtna.2013.11.

Volpers, C., and S. Kochanek. 2004. Adenoviral vectors for gene transfer and therapy. *The Journal of Gene Medicine* 6 Suppl 1 (February): S164–71. doi: 10.1002/jgm.496.

Waddington, S. N., J. H. McVey, D. Bhella, A. L. Parker, K. Barker, H. Atoda, R. Pink et al. 2008. Adenovirus serotype 5 hexon mediates liver gene transfer. *Cell* 132 (3): 397–409. doi: 10.1016/j.cell.2008.01.016.

Wang, H., Z.-Y. Li, Y. Liu, J. Persson, I. Beyer, T. Möller, D. Koyuncu et al. 2011. Desmoglein 2 is a receptor for Ad serotypes 3, 7, 11 and 14. *Nature Medicine* 17 (1): 96–104. doi: 10.1038/nm.2270.

Wang, Q., G. Greenburg, D. Bunch, D. Farson, and M. H. Finer. 1997. Persistent transgene expression in mouse liver following *in vivo* gene transfer with a delta E1/delta E4 Ad vector. *Gene Therapy* 4 (5): 393–400. doi: 10.1038/sj.gt.3300404.

Wheeler, M. D., S. Yamashina, M. Froh, I. Rusyn, and R. G. Thurman. 2001. Adenoviral gene delivery can inactivate Kupffer cells: Role of oxidants in NF-kappaB activation and cytokine production. *Journal of Leukocyte Biology* 69 (4): 622–30.

Wickham, T. J., P. Mathias, D. A. Cheresh, and G. R. Nemerow. 1993. Integrins alpha v beta 3 and alpha v beta 5 promote Ad internalization but not virus attachment. *Cell* 73 (2): 309–19.

Wickham, T. J., E. Tzeng, L. L. Shears, P. W. Roelvink, Y. Li, G. M. Lee, D. E. Brough, A. Lizonova, and I. Kovesdi. 1997. Increased *in vitro* and *in vivo* gene transfer by Ad vectors containing chimeric fiber proteins. *Journal of Virology* 71 (11): 8221–29.

Wiethoff, C. M., H. Wodrich, L. Gerace, and G. R. Nemerow. 2005. Adenovirus protein VI mediates membrane disruption following capsid disassembly. *Journal of Virology* 79 (4): 1992–2000. doi: 10.1128/JVI.79.4.1992-2000.2005.

Wilson, J. M. 2005. Gendicine: The first commercial gene therapy product. *Human Gene Therapy* 16 (9): 1014–15. doi: 10.1089/hum.2005.16.1014.

Wolfrum, N. and U. F. Greber. 2013. Adenovirus signalling in entry. *Cellular Microbiology* 15 (1): 53–62. doi: 10.1111/cmi.12053.

Wolins, N., J. Lozier, T. L. Eggerman, E. Jones, E. Aguilar-Córdova, and J. G. Vostal. 2003. Intravenous administration of replication-incompetent Ad to rhesus monkeys induces thrombocytopenia by increasing *in vivo* platelet clearance. *British Journal of Haematology* 123 (5): 903–5.

Worgall, S., G. Wolff, E. Falck-Pedersen, and R. G. Crystal. 1997. Innate immune mechanisms dominate elimination of adenoviral vectors following *in vivo* administration. *Human Gene Therapy* 8 (1): 37–44. doi: 10.1089/hum.1997.8.1-37.

Wortmann, A., S. Vöhringer, T. Engler, S. Corjon, R. Schirmbeck, J. Reimann, S. Kochanek, and F. Kreppel. 2008. Fully detargeted polyethylene glycol-coated Ad vectors are potent genetic vaccines and escape from pre-existing anti-Ad antibodies. *Molecular Therapy: The Journal of the American Society of Gene Therapy* 16 (1): 154–62. doi: 10.1038/sj.mt.6300306.

Wu, E. and G. R. Nemerow. 2004. Virus yoga: The role of flexibility in virus host cell recognition. *Trends in Microbiology* 12 (4): 162–69. doi: 10.1016/j.tim.2004.02.005.

Xu, Z., Q. Qiu, J. Tian, J. S. Smith, G. M. Conenello, T. Morita, and A. P. Byrnes. 2013. Coagulation factor X shields Ad type 5 from attack by natural antibodies and complement. *Nature Medicine* 19 (4): 452–57. doi: 10.1038/nm.3107.

Xu, Z., J. Tian, J. S. Smith, and A. P. Byrnes. 2008. Clearance of Ad by Kupffer cells is mediated by scavenger receptors, natural antibodies, and complement. *Journal of Virology* 82 (23): 11705–13. doi: 10.1128/JVI.01320-08.

Yamamoto, M., and D. T. Curiel. 2010. Current issues and future directions of oncolytic Ades. *Molecular Therapy: The Journal of the American Society of Gene Therapy* 18 (2): 243–50. doi: 10.1038/mt.2009.266.

Yang, Y., Q. Li, H. C. Ertl, and J. M. Wilson. 1995. Cellular and humoral immune responses to viral antigens create barriers to lung-directed gene therapy with recombinant Ades. *Journal of Virology* 69 (4): 2004–15.

Yang, Y., F. A. Nunes, K. Berencsi, E. E. Furth, E. Gönczöl, and J. M. Wilson. 1994. Cellular immunity to viral antigens limits E1-deleted Ades for gene therapy. *Proceedings of the National Academy of Sciences of the United States of America* 91 (10): 4407–11.

Zaiss, A. K., E. M. Foley, R. Lawrence, L. S. Schneider, H. Hoveida, P. Secrest, A. B. Catapang et al. 2015. Hepatocyte heparan sulfate is required for adeno-associated virus 2 but dispensable for Ad 5 liver transduction *in vivo*. *Journal of Virology* 90 (1): 412–20. doi: 10.1128/JVI.01939-15.

Zaiss, A.-K., Q. Liu, G. P. Bowen, N. C. W. Wong, J. S. Bartlett, and D. A. Muruve. 2002. Differential activation of innate immune responses by Ad and adeno-associated virus vectors. *Journal of Virology* 76 (9): 4580–90.

Zaiss, A. K., A. Vilaysane, M. J. Cotter, S. A. Clark, H. Christopher Meijndert, P. Colarusso, R. M. Yates, V. Petrilli, J. Tschopp, and D. A. Muruve. 2009. Antiviral antibodies target Ad to phagolysosomes and amplify the innate immune response. *Journal of Immunology (Baltimore, Md.: 1950)* 182 (11): 7058–68. doi: 10.4049/jimmunol.0804269.

Zhang, W., M. Solanki, N. Müther, M. Ebel, J. Wang, C. Sun, Z. Izsvak, and A. Ehrhardt. 2013. Hybrid adeno-associated viral vectors utilizing transposase-mediated somatic integration for stable transgene expression in human cells. *PloS One* 8 (10): e76771. doi: 10.1371/journal.pone.0076771.

5 Design and Applications of Adenovirus-Based Hybrid Vectors

Anja Ehrhardt and Wenli Zhang

CONTENTS

5.1 INTRODUCTION: DEFINITION OF ADENOVIRUS-BASED HYBRID VECTORS

Gene therapy represents an emerging field and recent successful clinical trials using viral vectors have gained great attention. Lessons learned from clinical trials stimulated the field of vector development and led to the continued improvement of vector efficacy and safety.

The most commonly used viral vectors for gene transfer such as adeno-associated virus (AAV), retroviral, and adenoviral vectors can be divided into two basic groups with respect to their genome status in transduced cells. Either the viral vector genome remains episomally and therefore extrachromosomal within the transduced target cell or the genetic cargo is integrated into the host chromosomes. Both strategies of infecting eukaryotic cells harbor advantages and disadvantages. For instance

125

on the one side somatic integration leads to stable expression of the therapeutic DNA whereas on the other side integration may cause genotoxicity. Episomal persistence of the genetic cargo may provide the safer option with respect to genotoxicity but if a retention mechanism is lacking, the therapeutic DNA is lost after cell division. The choice of the optimal viral vector is driven by the application and the specific requirements related to phenotypic correction of the target disease.

Recombinant adenoviruses (Ads) efficiently transfer foreign DNA to a broad variety of dividing and nondividing cells and they can be produced at high titers. Therefore, Ad-based vectors (AdVs) were widely investigated as a tool in basic research, biotechnology, preclinical, and clinical applications. For clinical applications oncolytic and therefore replication-competent or replication-deficient adenoviral vectors have been studied in depth. Ad-based hybrid vectors which are discussed here are predominantly based on replication-deficient Ads which are deleted for one or more adenoviral genes. This includes first-generation AdVs (FGAd) deleted for the early adenoviral gene E1 or E1 and E3, second-generation AdVs (SGAd) which are in contrast to FGAd deleted for an additional early adenoviral gene, and high-capacity adenoviral (HDAd) vectors lacking all viral-coding sequences.

After transduction of eukaryotic cells with replication-deficient adenoviral Ad vectors and subsequent uptake of the adenoviral DNA molecule into the nucleus, the viral DNA is established as an extrachromosomal and linear DNA fragment (Jager and Ehrhardt 2009). This fragment is protected at its ends by the Ad-derived terminal proteins (TP) which are covalently bound to the adenoviral inverted terminal repeats (ITR). During cell division the extrachromosomal viral DNA is diluted and lost due to the lack of a sufficient nuclear retention mechanism and the inability to replicate the viral genome. Therefore, no viral genomes are segregated into the daughter cells after cell division is completed (Figure 5.1). As a consequence the number of adenoviral genomes rapidly declines during cell cycling resulting in loss of vector-derived transgene expression. Thus, in dividing cells such as stem cells or regenerating tissues, stable long-term effects after adenoviral gene transfer is not expected. Hence, to achieve long-term effects after Ad transduction Ad hybrid vectors were generated.

Adenoviral hybrid vectors combine the best features of two worlds for achieving efficient gene transfer and stabilized transgene expression and long-term effects. For efficient transduction of desired target cells, these hybrid vectors take advantage of efficient Ad-based gene transfer into a broad variety of different cell types and use this feature in concert with genetic elements for maintenance of the therapeutic DNA and prolonged phenotypic effects in transduced cells.

The principle of Ad-based vectors is schematically shown in Figure 5.1. Attachment to the target cell, endosomal uptake, nuclear import, and release of the adenoviral DNA molecule of Ad-based hybrid vectors are identical to commonly used AdVs. The AdV genome ends up in the nucleus as a linear DNA molecule (Figure 5.1(i)). Along with the adenoviral genome, genetic elements contained in the Ad-based hybrid vector genome are also transported into the nucleus where they can initiate their function. In principle, three different types of genetic elements can be explored. Ad-based hybrid vectors were used to deliver genetic elements of different origin for somatic integration into the host chromosomes (Figure 5.1(ii)).

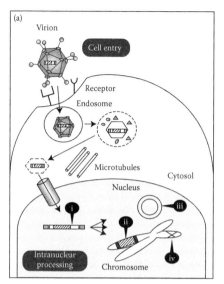
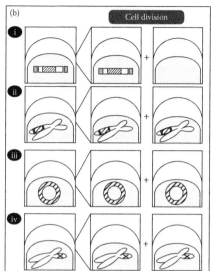

FIGURE 5.1 Principle of Ad-based hybrid vectors. Adenoviral hybrid vectors combine efficient infection rates of Ads with genetic elements for stable transduction and maintenance of the therapeutic effect. The virion is taken up by the cell (a) and after trafficking through the cytoplasm and nuclear entry the adenoviral DNA of the replication-deficient adenoviral vector (AdV) enters the nucleus (i). Without genetic elements for stable maintenance of the therapeutic DNA the adenoviral DNA molecule is lost during cell division (b). Three basic principles for achieving stabilized transgene expression relying on (ii) somatic integration into host chromosomes, extrachromosomal maintenance mediated by DNA replicons (iii), and gene editing were explored (iv). For the later three options the DNA can be maintained during cell division (b).

Meanwhile DNA elements based on viral or nonviral DNA replicons were investigated. These DNA replicons are maintained and replicated extrachromosomally (Figure 5.1(iii)). The newest generations of Ad-based hybrid vectors take advantage of recently developed designer nuclease (DN)-mediated DNA editing which can be used for sequence-specific modification of target DNA (Figure 5.1(iv)). Critical features of Ad-based hybrid vectors which need to be considered when designing or applying any kind of Ad-based hybrid vector are listed in Table 5.1.

5.2 MOLECULAR DESIGN AND APPLICATIONS OF Ad-BASED HYBRID VECTORS

The molecular design of Ad-based hybrid vectors needs to be carefully considered. Efficient transduction of desired target cells, functionality and efficacy of the delivered genetic element needs to be evaluated. Furthermore, safety and toxicity profiles need to be kept at the minimum level and efficient production of the Ad-hybrid vectors at high viral titers are mandatory for applications *in vivo*. Toward this end, continuous improvement and optimization of these vectors are a prerequisite to translate this vector type into preclinical and clinical studies. The following paragraphs describe

TABLE 5.1

Critical Features for the Design of Ad-Based Hybrid Vectors

- Production of Ad-based hybrid vectors at high titers should be possible
- Vectors should display an optimized safety and toxicity profile
- Genotoxicity should be avoided or minimized
- Vectors should allow targeted delivery to the desired target cells
- Long-term effects should be possible
- Regulation of transgene expression may be desirable
- Type of adenoviral vector (early generation of HDAd vectors)

the basic molecular design, functions, and efficacy of these Ad-based hybrid vectors which can be divided into four basic types.

5.2.1 Hybrid Vectors for Somatic Integration Into Host Chromosomes

AdVs display relatively low somatic integration efficiencies of approximately 10^{-4}–10^{-6} integration events per infectious genome that entered the target cells. As a consequence viral genomes are predominantly maintained episomally in the nuclei of infected cells (Stephen et al. 2008, Jager and Ehrhardt 2009). Therefore, transgene expression will be progressively diluted due to the non-replicative nature of the viral genome during cell division, which makes stable transgene expression in rapidly dividing cells challenging. To overcome these limitations imposed by transient expression of cargo genes, researchers aimed at combining the broad tropism and efficient transduction efficiencies of AdVs with integration machineries for somatic integration into host chromosomes. With respect to the molecular design in most cases a two-vector system was used of which the first vector delivers the transgene, while the second vector offers all necessary genetic elements to facilitate somatic integration. Available Ad-based hybrid vectors for somatic integration into host chromosomes are summarized in Table 5.2.

5.2.1.1 The Ad–Transposase Hybrid Vector System

The Sleeping Beauty (SB) transposon system was first molecularly reconstructed from ancient inactive copies of Tc1/mariner-like elements which were found in several fish genomes (Ivics et al. 1997). The wild type transposon consists of the SB transposase and inverted repeats (IRs) flanking the transposase-encoding sequence. For transposition, SB transposase recognizes the IRs and mediates the insertion into TA-dinucleotide sites present in the host genome by a "cut-and-paste" mechanism (Figure 5.2a). In gene therapy approaches, SB components are split into an artificial transposon containing the therapeutic cargo sequence flanked by the IRs and the transposase provided *in trans* for mobilization of the transposon (Izsvak and Ivics 2004). Over the past decade the intrinsic activity of the SB system has been enhanced through a variety of measures including the creation of a series of hyperactive SB transposases and engineering of transposons with optimized 3′ and 5′ IRs. For instance, the hyperactive versions of SB transposase such as SB11, HSB5, and SB100X were developed

TABLE 5.2

Described Ad-Based Hybrid Vector Systems for Somatic Integration

AdV/PM[a]	Milestone	Reference
HDAd[b]/SB[c]	First prototype of an Ad–SB transposase hybrid vector showing efficacy *in vitro* and *in vivo*	Yant et al. (2002)
HDAd/HSB5	Detailed vector dose setting; first evaluation in larger animal model	Hausl et al. (2010)
HDAd/HSB5	Detailed integration sites analysis; vector safety issue, therapeutic window exploration	Zhang et al. (2013a)
FGAd[d]/PB[e]	Initial characterization of the hybrid adenoviral–PB gene delivery system for stable transgene integration in cultured cells and *in vivo*	Smith et al. (2015)
HDAd/L1[f]	First prototype of an Ad–L1 transposon hybrid vector resulted in retrotransposition and stable integration of the transgene	Soifer et al. (2001)
HDAd/L1	The Ad–L1 retrotransposon hybrid virus efficiently retrotranspose in nondividing cells arrested in the G1/S phase of the cell cycle and in differentiated primary human somatic cells	Kubo et al. (2006)
HDAd/PhiC31[g]	The adenoviral–phiC31 integrase hybrid vector system for stable transgene integration in cultured cells and *in vivo*	Ehrhardt et al. (2007)
HDAd/PhiC31[g]	Showed circularization of adenoviral genome for PhiC31-mediated integration was unnecessary and even reduced the efficacy	Robert et al. (2012)
FGAd/ retro-LTR-Luc	Combines the high titer and versatility of adenoviral vectors with the long-term gene expression and integration of retroviral vectors	Zheng et al. (2000)
FGAd/PFV[h]	The development of an Ad/PFV hybrid vector for transgene integration and stable expression	Russell et al. (2004)
HDAd/ AAV[i]-Rep[j]	The first HDAd/AAV-Rep hybrid vector which overcomes the poor tolerance of Rep contained in earlier generation Ad production	Recchia et al. (1999)
HDAd/ AAV-Rep	Single hybrid Ad/AAV vector carrying a double reporter gene integration cassette flanked by AAV ITRs and tightly regulated, drug-inducible Rep expression cassettes	Recchia et al. (2004)
HDAd/AAV	Capsid-modified HDAd (F35) hybrid with AAV to integrate large transgenes up to 27 kb	Wang and Lieber (2006)
HDAd/AAV	These retargeted vectors efficiently complemented the genetic defect of dystrophin-defective myoblasts and myotubes	Gonçalves et al. (2008)
FGAd/ AAV-Rep	Reengineered Rep which was tolerated during FGAd replication; identification of Rep inhibitory sequence	Sitaraman et al. (2011)
FGAd/AAV	Generation of mini-Ad/AAV vectors as by-products of FGAd/AAV vector amplification	Lieber et al. (1999)

(Continued)

TABLE 5.2 (*Continued*)
Described Ad-Based Hybrid Vector Systems for Somatic Integration

AdV/PM[a]	Milestone	Reference
FGAd/AAV	Hybrid Ad–AAV vector with a chimeric adenoviral capsid from Ad35 for efficient gene transfer into human hematopoietic stem cells	Shayakhmetov et al. (2002)
FGAd/AAV	High-capacity AAV/Ad hybrid vectors with extended delivery DNA fragments of at least 27 kb; accomplishes stable long-term transgene expression in rapidly proliferating cells	Goncalves et al. (2004)

[a] PM: Persisting machineries.
[b] HDAd: High-capacity adenoviral vector.
[c] SB: Sleeping Beauty transposase.
[d] FGAd: First-generation adenoviral vector.
[e] PB: PiggyBac transposon.
[f] L1: The human long interspersed element-1.
[g] PhiC31: Bacteriophage-derived integrase PhiC31.
[h] PFV: Prototype foamy virus.
[i] AAV: Adeno-associated virus.
[j] AAV-Rep: AAV-derived Rep coding sequence responsible for site-specific integration into AAVS1.

showing increased enzymatic activities *in vitro* (Geurts et al. 2003, Ivics et al. 2007, Mates et al. 2009). These novel versions, especially the latest (SB100X) displaying a 100-fold higher activity compared to the original version, were shown to increase integration efficiencies into the host genome. Due to the high transpositional activity in vertebrates, SB transposase is an attractive tool for human gene therapy. It has been shown to provide long-term transgene expression *in vitro* with a random integration profile in the context of nonviral gene transfer based on plasmids (Yant et al. 2000, 2005). Besides *in vitro* studies for transposition, plasmid-based transposons carrying therapeutic transgenes were delivered into mice and primary hematopoietic stem cells were genetically modified for *in vivo* transfer (Yant et al. 2000, Mikkelsen et al. 2003, Mates et al. 2009). However, for a plasmid-based SB system delivery is a major hurdle because hydrodynamic delivery methods (Bell et al. 2007) are not suitable for humans. Therefore, much effort was spent on designing virus/SB hybrid vectors to deliver the SB transposon system efficiently. Until now, the SB transposase machinery was combined with several different viral vectors such as AAV, AdV, herpes simplex virus (HSV), lentiviral, and retroviral vectors which showed efficient delivery and long-term transgene expression (Yant et al. 2002, Peterson et al. 2007, Vink et al. 2009, De Silva et al. 2010, Heinz et al. 2011, Zhang et al. 2013b).

To combine efficient delivery and high transduction efficiency of AdVs with the long-term expression of a transgene expression cassette after SB-mediated somatic integration, the first AdV/SB hybrid vector system was generated in 2002 (Yant et al. 2002). This hybrid vector system consisted of two independent HDAds, in which one vector contained the integration machinery and the other vector the transposon. This specific molecular design was called the two-vector system. Herein, one adenoviral vector contained SB transposase and Flp recombinase-encoding sequences, and the second vector contained the transposon which encompasses the transgene expression cassette flanked by IRs for SB recognition and FRT sites for Flp recombinase recognition. Therefore, after co-transduction of a single cell with both HDAds, Flp recombinase interacts with the FRT sites and excises the transposon from the HDAd genome generating a circular intermediate. Subsequently, the SB transposase recognizes the IRs in the excised circular DNA and integrates the transposon by a "cut-and-paste" mechanism into the genomic DNA (Figure 5.2a). In the first study stable human coagulation factor IX (hFIX) expression levels were obtained in murine liver even after inducing rapid cell cycling of hepatocytes (Yant et al. 2002) indicating that the transgene was stably integrated into the host genome. Later, another study showed that phenotypic correction utilizing the same design of an adenoviral–SB transposase hybrid vector can be achieved for up to 960 days in a hemophilia B dog model. This was also the first study showing transposon-mediated integration in a large mammal (Hausl et al. 2011). Integration site analysis after *in vivo* application of the AdV/SB hybrid vector system revealed a close-to-random integration pattern with respect to integration into gene and non-gene areas. Another study also analyzed the correlation between the vector dose and the efficiency of the hybrid vector system in terms of long-term transgene expression and found a narrow therapeutic index (Zhang et al. 2013a).

Besides the SB transposon system, other transposon-based-vector systems have emerged as promising tools for achieving persistent gene delivery. Note that they all share a similar "cut-and-paste" mechanism and a fairly random integration profile (Skipper et al. 2013). The piggyBac (PB) transposon system for instance derived from the cabbage looper integrates into TTAA sites of the mammalian genome and was shown to mobilize foreign DNA from episomal DNA (Ding et al. 2005). Similar to the SB transposase Ad-based hybrid vector system also the PB transposase machinery was delivered by AdV. In contrast to the SB transposase hybrid vector system the excision step by Flp recombinase can be omitted enabling a more simple process of stable transduction without the first circulation step (Smith et al. 2015). Stable expression of a reporter gene delivered by an Ad-PB transposase hybrid vector was observed in 20%–40% of hepatocytes for up to 20 weeks following standard tail vein injection (Smith et al. 2015). Moreover, due to the fact that PB has the ability to efficiently mobilize transposons larger than 100 kilobases (kb), the authors hypothesized that, if PB transposase is used in concert with HDAd, the transposition cargo can potentially be enlarged to HDAd's delivery limitation of up to 35 kb, allowing transposition of large transgenes.

Another type of transposon which was also combined with AdV is based on retrotransposition and relies on the human long interspersed element-1 (L1). L1 sequences are retrotransposons without long-terminal repeat (LTR) which account for 17% of the human genome. The integration mechanism is initiated by expression of the L1-RNA which is transported into the cytoplasm where the ribonucleoprotein particle (RNP) is formed. This RNP moves into the nucleus via an unknown mechanism followed by an L1-endonuclease mediated cleavage of a consensus target DNA and subsequent genomic integration. The prototype of this hybrid vector system

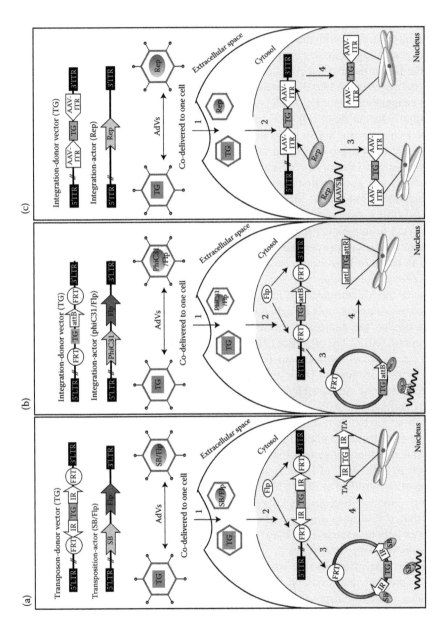

FIGURE 5.2

based on HDAd was introduced by Soifer et al. (2001) resulting in retrotransposition and stable integration of the transgene in cultured human cells. Another group developed an advanced Ad–L1 retrotransposon hybrid virus which led to efficient retrotransposition in nondividing and differentiated primary human somatic cells (Kubo et al. 2006).

5.2.1.2 The Ad–Integrase Hybrid Vector System

The integrase derived from the *Streptomyces* phage PhiC31 was first introduced in 1991 (Kuhstoss and Rao 1991). In contrast to SB transposase, PhiC31 was shown to mediate a unidirectional site-specific recombination between two ~40 bp DNA recognition sequences, the phage attachment site (attP) and the bacterial attachment site (attB) (Thorpe and Smith 1998).

In the mammalian cell setting, PhiC31 integrase mediates integration of extra-chromosomal plasmids bearing an attB site into a limited number of "pseudo-attP" sites present in native mammalian genomes (Groth et al. 2000). A distinguished feature of the PhiC31 integrase for gene therapy is the fact that integration is limited

---

FIGURE 5.2 (Continued) Schematic outline of Ad-based hybrid vectors for somatic integration. Schematically shown is a "two-vector strategy" based on two HDAd vectors co-infecting one cell. The first HDAd vector acts as donor vector and encodes the therapeutic gene (TG) flanked by essential recognition sites of respective persisting machineries (PM), while the second vector supplies the necessary enzymes to promote the integration of the transgene into host chromosomes. All systems share two steps including (1) cellular uptake via different receptors and co-receptors (attachment factors, bridging factors, and entry receptors) and (2) nuclear entry mediated by the nuclear pore complexes followed by nuclear import of the double-stranded linear DNA (dsDNA). (a) Principle of the Ad/transposase hybrid vector system. Exemplarily shown is the HDAd/SB transposase hybrid vector system (Yant et al. 2002). It utilizes the high transduction efficiency of HDAd and the genomic TA-dinucleotide directed integration of the SB transposon. The first HDAd acts as transposon-donor vector and encodes the transgene flanked by transposon-derived IR and FRT sites for Flp recombinase-mediated excision. The second HDAd provides SB transposase and Flp recombinase and plays a role as a "transposition actor." Circularization is initiated by Flp-FRT recombination (3) and subsequently SB-mediated transgene insertion into TA sites of the host genome by the "cut-and-paste" mechanism (4). (b) Principle of the Ad/integrase hybrid vector systems. Exemplarily shown is the HDAd/phiC31 integrase hybrid vector system (Ehrhardt et al. 2007). The first adenoviral vector harbors the PhiC31 recognition site attB and the therapeutic transgene. When PhiC31 protein is provided *in trans* by a second vector, it mediates integration into a limited number of pseudo-attP sites in the mammalian genome by unidirectional recombination (4). The circular form of the PhiC31 substrate is released from the adenoviral vector genome by Flp recombinase (3). (c) Principle of the Ad/AAV-Rep hybrid vector system. The first type (3) is based on HDAd/AAV hybrid vectors that integrate into the host genome site-specifically via the interaction of AAV inverted terminal repeats (ITRs) and AAV Rep protein (Recchia et al. 1999). Rep expressed *in trans* provided by a second vector mediates site-specific integration at the AAV integration site S1 (AAVS1), located on the human chromosome 19 (3). The second type (4) of an AdV/AAV hybrid vector (Shayakhmetov et al. 2002) enters the nucleus and with the help of Ad helper functions the transgene flanked by AAV ITRs recapitulates the lytic phase of the wild type virus by being rescued from the adenoviral genome and integrates into random sites of the host genome.

to "hot spot" sites with up to 15% specificity in the mammalian genome decreasing the risk of insertional mutagenesis. Another advantage of the phiC31 integrase machinery is represented by the fact that it can mediate single-copy integration with unlimited cargo capacity, making it a suitable tool for stable cell line and transgenic animal model generation (Thyagarajan et al. 2001, Olivares et al. 2002, Ehrhardt et al. 2006).

For generation of viral hybrid vectors, PhiC31 integrase was combined with non-integrative lentiviral vectors (NILVs) and HDAd (Ehrhardt et al. 2007, Grandchamp et al. 2014). In the initial study, Ehrhardt et al (2007) utilized HDAd as an efficient delivery vehicle. The strategy and the design of this vector system are schematically outlined in Figure 5.2b. Similar to the Ad/SB transposase hybrid, a two-vector system was designed, in which the transgene expression cassette is circularized from the HDAd genome by Flp-mediated recombination, followed by PhiC31-mediated somatic integration. The integration efficiencies of this vector system was first evaluated *in vitro* via colony-forming assay and then *in vivo* in mice. The Ad-PhiC31 hybrid vector showed stabilized hFIX expression levels and integration into the previously described hot spot mpsL1 present in murine liver (Ehrhardt et al. 2007). Note that in contrast to the SB transposase integration machinery, linear DNA substrates can serve as substrates for PhiC31-mediated integration, although with relatively low efficiency. It is predicted that a molecular design of Ad-PhiC31 hybrid vector maybe further simplified without Flp as pursued already in a second study (Robert et al. 2012). However, in 2006 a study demonstrated that chromosomal rearrangements can be induced after PhiC31 treatment which decreased the interest of using this system for gene therapeutic applications (Ehrhardt et al. 2006).

Besides nonviral integrase, viral integrase elements derived from retroviruses were incorporated into AdV. One representative Ad/RV hybrid vector for somatic integration is AdLTR–luc, in which the adenoviral vector contains a luciferase reporter gene flanked by the retroviral LTRs from the murine leukemia virus (MLV). Gene expression was observed in cultured cells *in vitro* and *in vivo* for up to 3 months post-vector delivery and the transgene integrated randomly into the host genome of both dividing and nondividing cells (Zheng et al. 2000). In 2002 another Ad/RV hybrid vector was described and in sharp contrast to the earlier study somatic integration was mediated by retroviral integrase (Murphy et al. 2002). Similar to the strategy used for the transposon and the PhiC31 system, a provirus circle containing the retroviral LTR was excised from the adenoviral DNA molecule. In the presence of retroviral Gag and Pol proteins, the excised circle was integrated randomly into the host genome by an integrase-directed mechanism. Two years later an Ad/foamy virus (Ad/FV) hybrid vector was developed (Russell et al. 2004). FV is a member of the retrovirus family and it is believed that there is no pathogenicity associated with FV infection rendering FV an attractive tool for stable gene transfer. The Ad/FV hybrid vector resulted in 70% of stably transduced cells.

5.2.1.3 Ad/AAV-Rep Hybrid Vectors

AAV retains the unique capacity for site-specific integration into a transcriptionally silent region of the human genome on human chromosome 19q13.42 (called AAVS1 region). The Rep68/78 polypeptide in conjunction with AAV terminal repeat integrating elements is responsible for this unique ability. Rep68/78 bind to the Rep-binding site located in the AAV-ITR and to a restricted 33-bp cellular sequence within AAVS1. Subsequently, a nonhomologous deletion–insertion recombination results in the integration of the AAV-ITR flanked cassette into targeting genome (Kotin et al. 1990, Samulski et al. 1991). In the absence of Rep proteins, AAV genomes show a bias toward integration into actively transcribed genes (Nakai et al. 2003).

The incorporation of AAV integrating elements into other viruses represents a logical strategy for site-specific genetic replacement therapies and the molecular design is schematically shown in Figure 5.2c. In the past both large DNA viruses like the herpes virus (Heister et al. 2002, Cortes et al. 2008) and Ad were combined with AAV integration. The earliest studies unified the advantageous gene delivery features of an FGAd with the AAV genome simply for high-level production of recombinant AAV (AAV) vectors (Fisher et al. 1996). However, due to inhibition of the FGAd replication by the AAV Rep proteins, this approach was not pursued for a while. The first successful Ad/AAV hybrid vector utilizing Rep-mediated site-specific integration at AAVS1 was developed by Recchia et al. (1999). This system was based on a two-vector HDAd system. One vector carried the Rep78 gene under control of either the T7 or the alpha-1-antitrypsin liver-specific promoter and the other vector contained an AAV-ITR-flanked transgene. Up to 35% of integration events into AAVS1 in HepG2 cells were detected (Recchia et al. 1999). To avoid Rep protein-mediated inhibition of AdV replication inducible systems for Rep expression were explored such as the Cre/loxP-expression-switching system (Ueno et al. 2000) and a tetracycline (Tc)-regulated rep expression system (Recchia 2004). A more recent publication by Sitaraman et al. (2011) reported that a computationally designed AAV Rep78 protein is efficiently maintained within an early generation AdV. Here the AAV Rep gene was reengineered by modifying synonymous codon pairs to phenotypically affect the replicative properties of Ad-expressing Rep78, while preserving the endonuclease function of AAV Rep78. This genetically recoded AAV Rep protein will dramatically simplify the molecular design of AdV/AAV-Rep hybrid virus systems for both Rep-mediated site-specific integration and vector production.

To broaden the tropism of Ad/AAV hybrid vectors, capsid-modified AdVs were used. For instance, Wang et al. packaged HDAd genomes into capsids containing Ad B-group type 35 fiber knob domains (HC-Ad5/35) that can efficiently transduce hematopoietic stem cells as well as human leukemia cells and primary CD34+ cells. They found that 55% of all analyzed integration sites were either within the AAVS1 or globin locus control region (LCR), which demonstrates that a high frequency of targeted integration of a large transgene cassette can be achieved (Wang and Lieber 2006). Besides Ad type 35, Gonçalves et al. described the generation of dual Ad/AAV hybrid vectors with Ad type 50 fiber-modified adenoviral capsids and with all the AAV cis- and trans-acting elements needed for locus-specific insertion of exogenous DNA. In contrast to their isogenic counterparts with conventional Ad type 5 fibers, these retargeted vectors efficiently complemented the genetic defect of dystrophin-defective myoblasts and myotubes. To avoid the well-described interference of the large AAV Rep proteins (i.e., Rep78 and Rep68) on Ad DNA replication, they devised an flp recombinase-dependent gene switch module to repress and activate

Rep68 expression in producer and target cells, respectively (Gonçalves et al. 2006, Gonçalves et al. 2008).

It remains to be mentioned that also Rep-independent AdV/AAV hybrid vector systems for sustained transgene expression were described. The prototype of this hybrid vector type was described by Lieber et al. (1999). Here a mini-Ad genome lacking viral E1, E2, E3, and late genes was used to deliver a recombinant AAV genome encoding a reporter gene. However, the stability of such mini-Ad vectors was short-lived. Therefore, development of AdV/AAV hybrid vectors that have similar genome size as wild type Ad with increased genome stability was necessary. In combination with a capsid-modified vector containing a short-shafted Ad serotype 35 fiber this improved Ad/AAV hybrid vector showed increased insert capacity and tropism for hematopoietic stem cells (Shayakhmetov et al. 2002). Importantly, random integration of single vector copies with intact transgene cassettes was mediated by AAV-ITRs (Shayakhmetov et al. 2002). Another strategy to combine high transduction efficiencies of Ad with AAV's sustained transgene expression ability was developed by Gonçalves et al. (2001). Those AdV/AAV hybrid vectors are based on coupling the AAV DNA replication mechanism to the Ad encapsidation process through packaging of AAV-dependent replicative intermediates provided with Ad packaging elements. By doing this, the cloning capacity limitation of AAV was also overcome yielding lengths of up to 27 kb. Moreover, this AdV/AAV hybrid vector showed also cellular genomic integration of large fragments of foreign DNA and accomplishes stable long-term transgene expression in rapidly proliferating cells (Goncalves et al. 2002, Gonçalves et al. 2004). Analyses of stably transduced cells confirmed that most of them contained a single copy of the full-length hybrid vector genomes with AAV ITR sequences at both ends. However, although AdV/AAV hybrid vectors could be efficiently generated and amplified, co-transfection of two additional plasmids supplying AAV Rep and Ad E2a expression was mandatory.

5.2.2   HYBRID VECTORS FOR EPISOMAL PERSISTENCE OF FOREIGN DNA

Besides somatic integration as outlined above, another promising persisting machinery (PM) to stabilize the Ad delivered TG is to take advantage of the episomal persistence of autonomously replicating genetic elements. These genetic elements include autonomous DNA (plasmid replicon pEPI and Epstein–Barr virus-based replicons, EBV replicon) and RNA replicons (Semliki Forest virus replicons, SFV replicon). Compared to the potential risk of insertional mutagenesis from integration machineries, autonomous replicons are maintained extrachromosomally leading to enhanced safety profiles. A summary of pilot studies describing hybrid vectors for episomal transgene persistence is presented in Table 5.3.

5.2.2.1   Ad/pEPI Hybrid Vectors

The plasmid replicon pEPI was created by replacing the simian virus large T-antigen of a commercial expression vector by a scaffold/matrix attachment region (S/MAR) derived from the human interferon (IFN)-β gene cluster (Piechaczek et al. 1999). The resulting vector pEPI is maintained episomally in the transfected cell as a stochastic event even in the absence of antibiotic selection over hundreds of generations at copy

TABLE 5.3
Described Viral Hybrid Vector Systems for Delivery of Autonomous Replicons

Adv/PM[a]	Milestone	Reference
HDAd/pEPI[b]	Plasmid replicon pEPI with S/MAR sequences[h] can be delivered by HDAd *in vitro* and *in vivo*	Voigtlander et al. (2013)
SGAd[c]/EBV[d] episome	Using SGAd to delivery and maintain an EBV episome in xenotransplantation HeLa cells	Leblois et al. (2000)
SGAd/EBV episome	Using a single vector to carry the inducible EBV episome-excising cassette	Recchia et al. (2004)
HDAd[e]/EBV episome	HDAd for stable transduction of mammalian cells *in vitro*; successful delivery of EBV episomes *in vivo* to the liver of transgenic mice expressing Cre	Dorigo et al. (2004)
HDAd/EBV episome	A binary HDAd system allows for significantly prolonged episomal maintenance of HDAd genomes in proliferating cells	Kreppel and Kochanek (2004)
HDAd/EBV episome	First evidence that an EBV episome can be delivered into murine liver of immunodeficient mice using HDAd	Gallaher et al. (2009)
HDAd/EBV episome	Long-term transgene expression in immunocompetent mice after *in vivo* delivery of an HDAd–EBV hybrid vector based on a one-vector system	Gil et al. (2010)
HDAd/SFV[f] replicon	First Ad hybrid vector system for delivery of SFV RNA replicons	Guan et al. (2006)
SGAd/SFV replicon	SGAd/SFV hybrid vector could be optimized to infect hematopoietic cells with high efficiency	Yang et al. (2013)
FGAd[g]/SFV replicon	The AdV/SFV hybrid vector-based vaccine with complete protection against CSFV	Sun et al. (2013)

[a] PM: Persisting machineries.
[b] pEPI: Plasmid replicon pEPI.
[c] SGAd: Second-generation adenoviral vector.
[d] EBV: Epstein–Barr virus.
[e] HDAd: High-capacity adenoviral vector.
[f] SFV: Semliki Forest virus.
[g] S/MAR: The scaffold/matrix attachment region derived from the 5′-region of the human β-interferon gene.
[h] FGAd: First-generation adenoviral vector.

numbers of 5–10 vector genomes per cell (Stehle et al. 2007). The vector replicates once per cell cycle in early S-phase and components of the pre-replication complex (pre-RC) assemble and initiate replication at various sites of the vector (Schaarschmidt et al. 2004). In the interphase nuclei pEPI is associated with the nuclear matrix via an interaction with the scaffold attachment factor SAF-A (Jenke et al. 2002). In the past years, pEPI vectors have been well established for long-term transgene expression and gene therapeutic approaches *in vitro*. Stable transgene expression and episomal maintenance under nonselective conditions was demonstrated in several established cell lines and in human primary fibroblasts and human CD34+ cells (Papapetrou

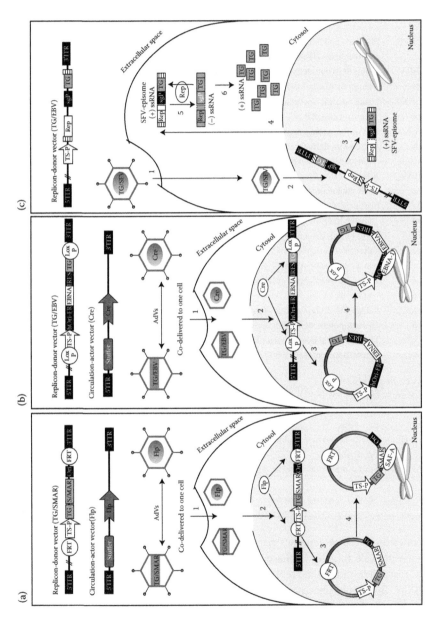

FIGURE 5.3

et al. 2006). In contrast to the episomal maintenance *in vitro*, the situation *in vivo* seems to be challenging. One major problem for this nonviral vector system to overcome was the inefficient delivery of the naked DNA molecule *in vivo*. To overcome this limitation, researchers took advantage of various virus-based delivery vehicles including AdV, HSV, and integration-deficient lentiviral vector (IDLV) (Lufino et al. 2007, Voigtlander et al. 2013, Verghese et al. 2014).

The hybrid vector system described by Voigtlander and colleagues was based on a two-vector system which combined HDAd for efficient delivery with the improved pEPI-derived plasmid replicon pEPito (Figure 5.3a). The plasmid replicon could be excised and circularized from the adenoviral DNA molecule by Flp recombination and was maintained in mammalian cells for up to 6 weeks evidenced by colony-forming assays detecting antibiotic expression and flow cytometry to detect eGFP expression. Afterwards, the extrachromosomal plasmid replicon could be detected by PCR and rescued. Importantly, this Ad/pEPI hybrid vector system also showed *in vivo* stability during cell division (Voigtlander et al. 2013).

5.2.2.2 Ad/EBV Hybrid Vector Systems

The EBV is member of the herpesviridae family. During lytic and latent cell cycle two different origins of replication are involved. For DNA replication in the lytic

---

FIGURE 5.3 (Continued) Schematic outline of Ad-based hybrid vectors for episomal persistence of foreign DNA. Exemplarily shown are principles of three replicons delivered by HDAd. All systems share two steps including (1) cellular uptake via different receptors and co-receptors (attachment factors, bridging factors, and entry receptors) and (2) nuclear entry mediated by the nuclear pore complexes followed by nuclear import of dsDNA. (a) HDAd-based delivery of the plasmid replicon pEPI (Voigtlander et al. 2013). Upon nuclear entry and DNA replicon release, the pEPI-episome resembling a minicircle without bacterial backbone sequences is generated by Flp recombinase. Flp expressed from a second vector recognizes the FRT sites flanking the plasmid replicon pEPI. The recombination product is a circularized pEPI episome (3). The matrix attachment region (S/MAR) containing plasmid replicon pEPI is then replicated in a cell cycle-dependent manner. Among other factors it is maintained by binding via the putative nuclear matrix protein SAF-A to metaphase chromosomes (4). The transgene can be expressed by a tissue-specific promoter (TS-P). (b) Mechanism of EBV episome stability in the context of the HDAd–EBV episome hybrid vector system (Dorigo et al. 2004). In the nucleus the EBV episome is excised by Cre recombinase-mediated recombination (3) between the two loxP sites releasing the circular DNA replicon containing the transgene (TG). In this example replication of the EBV episome is mediated by a 19 kb fragment from human chromosome 10 that functions as an origin of replication (hOri). For retention, the FRs within the episome bind via EBNA-1 to host cell chromosomes (4), thereby promoting efficient segregation to both daughter cells during mitosis. The transgene can be expressed by a TS-P. (c) The SFV replicon mRNA is directly transcribed from the HDAd genome under the control of a TS-P (Guan et al. 2006). Transcription generates single-stranded RNA of positive orientation [(+) ssRNA] (3), which is then exported from the nucleus into the cytoplasm (4). Subsequently, the replicase protein (Rep) will be translated which then converts the SFV mRNA into (–) ssRNA. Meanwhile Rep can also recognize the subgenomic promoter (sgP) in the (–) ssRNA (5), generating a smaller subgenomic mRNA that can then be translated for heterologous protein production (6).

cycle the origin binding protein (OBP) binds to the lytic origin and initiates a rolling circle replication. In the latent cycle the EBV genome is maintained as an episome and the oriP in combination with the EBV nuclear antigen 1 (EBNA-1) is utilized for DNA replication (Pfuller and Hammerschmidt 1996) (Figure 5.3b). Due to the episomal persistence feature, the EBV-based vector has been used for gene therapy to express reporter genes in human embryonic stem cell lines (hESC), or for tumor therapy to express HSV type 1 thymidine kinase (HSV1-TK) (Maruyama-Tabata et al. 2000, Ren et al. 2006, Thyagarajan et al. 2009). Later on, an EBV vector encoding for granulocyte macrophage colony-stimulating factor (GM-CSF) was constructed to successfully transduce different B-cell lines (Hellebrand et al. 2006).

AdV/EBV hybrid vector systems represent the main Ad/autonomous replicon hybrid vector type and the first representative prototype was established by Tan et al. (1999). Here the authors used an early generation AdV containing the EBV replicon DNA flanked by the Cre recombinase recognition sites loxP. Functional DNA sequences included in the EBV replicon were the latent EBV origin of replication (oriP) and DNA sequences encoding EBNA-1. After co-transduction of canine cells with the EBV replicon containing vector and a second vector expressing Cre recombinase, the EBV replicon can be released from the adenoviral genome and established as a stably maintained DNA replicon in the nucleus in up to 37% of transduced cells. However, to reach these high transduction efficiencies high vector doses were needed and they were associated with cellular toxicity which clearly is a major drawback of this system. In the following year another study utilized an SGAd to deliver an EBV replicon (Leblois et al. 2000). Here the authors demonstrated for the first time that the DNA replicon can be delivered and released *in vivo* in a HeLa-cell-based xenotransplantation model by Cre-mediated recombination. The latter systems were based on co-delivery of one adenoviral vector containing the episome and a second adenoviral vector encoding Cre recombinase to release the replicon. To develop a one-vector system another group combined replicon and a transgene encoding Cre recombinase in a single vector by expressing Cre recombinase under the control of a Tc-inducible promoter (Krougliak et al. 2001).

The next generation Ad–EBV episome hybrid vectors were based on HDAd devoid of all adenoviral coding sequences. The first study by Dorigo et al. (2004) introduced a hybrid vector containing an EBNA-1 expression cassette, a family of repeats (FRs) for attachment of the DNA replicon to the host chromosomes, and the EBV-derived dyad symmetry (DS) was replaced by a 19 kb human origin of replication derived from chromosome 10. They demonstrated that this replicon can be excised from the adenoviral vector DNA molecule and maintained for 20 weeks *in vitro*. The second study published by Kreppel and Kochanek (2004) evaluated another Ad–EBV episome hybrid vector which contained the putative human origin of replication from the lamin B2 locus replacing the EBV-derived origin of replication. Interestingly, this study demonstrated that a significant number of HDAd genomes containing the EBV replication and retention machineries were circularized without excision. To test the *in vivo* performance of HDAd delivered EBV episomes, Gallaher et al. (2009) systemically injected immune-deficient mice at a dose reaching transduction rates of 100% of all hepatocytes. They found that reporter gene expression can be maintained for up to 30 weeks. In 2010, another pioneering study demonstrated that an EBV-based episome can be maintained *in vivo* in immunocompetent mice based on adenoviral delivery (Gil et al. 2010). The authors utilized a single HDAd to deliver the replicon and the recombinase for release of the replicon. Notably, expression of the recombinase was driven by a liver-specific promoter which leads to hepatocyte-specific release of the episome.

5.2.2.3 Ad/SFV Hybrid Vectors

Aside from autonomous DNA replicons, RNA replicons derived from SFV were developed. SFV belongs to the group of alphaviruses, which contain a negative-sense RNA genome which is single stranded (Strauss and Strauss 1994). The SFV replicon (Figure 5.3c) is based on the use of self-replicating RNA molecules derived from alphavirus genomes. The SFV replicon only contains the 5′ and 3′ sequences necessary for replication (retention signals) and the replicase gene (Rep), while other viral-coding sequences were replaced by a transgene. Upon cell entry, Rep will first be translated, then it will generate a negative-sense RNA genome from which new copies of RNA will be produced at high efficiency of up to 200,000 copies per cell (Figure 5.3c) (Schlesinger 2001). This robust production of viral RNA also interferes with the host cell protein synthesis machinery and further induces apoptosis, which makes it a suitable candidate for vaccine and antitumor therapy. Note that a phase I clinical trial based on this vector was initiated in patients with advanced kidney carcinoma and melanoma (Murphy et al. 2000, Colmenero et al. 2002). It remains to be mentioned that SFV vectors pose a strong cytotoxic effect in infected cells and the maximum tolerable dose was relatively low. Therefore, the SFV RNA replicons were incorporated into Ad- (Guan et al. 2006, Yang et al. 2013) and baculovirus-based vectors (Pan et al. 2009, Sun et al. 2011, Sun et al. 2013, Wu et al. 2013) for efficient delivery of the replicon.

The first Ad/SFV hybrid vector system was developed in 2006 (Guan et al. 2006). Here the SFV replicon was inserted into a recombinant HDAd genome. The expression of the RNA replicon was controlled by the alpha-fetoprotein promoter allowing tumor-specific expression, while expression of the gene of interest (GOI) was driven under the control of the SFV subgenomic promoter contained in the same expression unit. Based on a hepatocellular carcinoma animal model the authors demonstrated that these vectors induced apoptosis and antitumoral activity when using murine IL-12 as a transgene. Another study aimed at utilizing an Ad-alphavirus hybrid vector for transduction of malignant hematopoietic cells (Yang et al. 2013) which may also pave the way toward using these vectors for production of cell-based vaccines for leukemia patients. In 2011 the first prototype of Ad-alphavirus hybrid vectors was developed which represented the first SFV replicon-based vaccine (Sun et al. 2011). Here the Ad expressed the early E2 gene of the classical swine fever virus (CSFV) which induced robust immune responses against CSFV in mice, rabbits, and pigs. Building upon this study, this Ad/SFV-replicon vector-based vaccine was studied in more detail regarding safety issues, efficiency, and specificity (Sun et al. 2013). In summary, Ad/SFV hybrid vectors are promising tools for applications in which transient transgene expression is required such as vaccinations and immune-based therapies.

5.2.3 HYBRID VECTORS FOR SITE-SPECIFIC GENOME EDITING

The possibility of performing genome editing in mammalian cells revolutionized the field of genome engineering applied in biotechnology and therapy. These technologies are based on the delivery of DNs which can bind to and cut DNA in a sequence-specific manner. Tools available for genome engineering are the zinc-finger nuclease (ZFN), the transcription activator-like effector nuclease (TALEN), and the clustered regularly interspaced short palindromic repeats (CRISPR)/Cas9 technologies. The molecular design of ZFNs and TALENs is based on a fusion protein comprising of a specific DNA-binding domain (DBD) and a DNA cleavage domain predominantly represented by the restriction endonuclease *Fok*I. For both ZFNs and TALENs, a pair of fusion proteins binds to the flanking DNA sequences of the target site, which at the same time need to dimerize via the *Fok*I nuclease domain.

The DBD of ZFNs comprise 3–6 individual zinc finger repeats of which each one can recognize between 9 and 18 base pairs (Smith et al. 2000, Porteus and Baltimore 2003, Miller et al. 2007, Sander et al. 2011, Wood et al. 2011). A major disadvantage of this technology is the complex screening process required to identify functional ZFN pairs with zinc finger modules which requires screening of complete libraries. TALENs can be considered as second-generation DNs which were first introduced as tools for sequence-specific genome editing in mammalian cells in 2011 (Cermak et al. 2011, Hockemeyer et al. 2011, Mussolino et al. 2011). The specificity of the DBD of TALENs is determined by two hypervariable amino acids that are known as repeat-variable di-residues (RVDs) (Mak et al. 2012). For generation of a functional DBD, modular TAL-effector (TALE) repeats are linked together to recognize DNA in a sequence-specific manner. However, in contrast to ZFNs, this technology allows construction of long arrays of TALEs without performing a sophisticated screening process. In 2012 the CRISPR/Cas9 technology was introduced which is an RNA-guided method to specifically cut DNA in mammalian cells (Jinek et al. 2012, Cong et al. 2013). Here the Cas9 protein recognizes the guide RNA (gRNA) which contains homologous sequences to the desired target DNA in the host genome resulting in a DNA double-strand break at the target sequence. The advantage of the CRISPR/Cas9 machinery compared to ZFNs and TALENs is the easy to handle construction. In principle this technology only requires the design of a small gRNA which can then be co-delivered along with the Cas9 transgene into mammalian cells.

ZFN, TALEN, and CRISPR/Cas9 machineries have in common that they cut DNA in a sequence-specific manner and generate a DNA double-strand break at the respective target site. This DNA cut in the target DNA is repaired by a host cell pathway which is known as nonhomologous end joining (NHEJ) (Moore and Haber 1996). NHEJ can be error prone resulting in the introduction of small mutations such as insertions or deletions (in/dels) at the target site. Therefore, the application of DNs can be used for the destruction of a given target site such as disruption of genes present in the host genome (Figure 5.4a) or in virus genomes. However, if in addition to the DN machinery a donor DNA with homology arms to the cut target site is present, homology-directed DNA repair (HDR) can occur. This pathway can then be used for direct gene correction in target cells (Figure 5.4b) or for gene addition at safe harbors in the host genome without genotoxicity (Figure 5.4c).

The delivery of all components required for efficient cutting and subsequent homologous recombination was broadly investigated over the past decade and various viral vector systems were exploited for delivery of DNs and respective donor DNA if desired. This included the most commonly used viral vectors such as lentiviral vectors, AAV vectors, and AdV. The molecular design of adenoviral vectors included early generation vectors deleted for at least one early adenoviral gene but also HDAds were studied. The overall strategy to deliver DNs delivered by adenoviral vectors is schematically shown in Figure 5.4.

Historically, ZFNs were discovered first and therefore, the first AdVs which were deployed for delivery of DNs were AdVs containing expression cassettes encoding ZFNs. In 2008, a study showed that the gene of the co-receptor CCR5 responsible for infection with the human immunodeficiency virus (HIV) can be disrupted by applying the ZFN technology (Perez et al. 2008). The AdV applied in this study was based on a chimeric first-generation human Ad type 5 vector carrying a fiber from human Ad type 35. Translation of this approach into the clinic was shown later by demonstrating that the HIV co-receptor CCR5 in T-lymphocytes can be efficiently disrupted by Ad-based delivery of a CCR5 specific ZFN (Maier et al. 2013, Li et al. 2015). In 2013 it was shown that Ad-based ZFN delivery targeting the AAVS1 on human chromosome 19 as a "safe harbor" can initiate HDR in the presence of a respective donor DNA co-delivered by a non-integrating lentiviral vector (Coluccio et al. 2013). Gene correction utilizing a single adenoviral vector carrying an inducible ZFN-encoding transgene and the donor was shown 1 year later in 2014 (Zhang et al. 2014).

However, after the TALEN technology was introduced which was available to a broader research community due to its less complex and labor-intensive design, the next studies also focused on the Ad-based delivery of TALENs. Initial studies delivered TALEN genes by FGAd using a two-vector strategy. Each fusion protein of a TALEN pair required for efficient gene disruption was expressed from a separate AdV (Holkers et al. 2013). Later there was one study delivering a complete TALEN pair in a single HDAd (Saydaminova et al. 2015) and efficacy of this hybrid vector was shown in hematopoietic stem cells. Note that adenoviral delivery of a complete TALEN pair using a single vector is challenging because expression and activity of a functional TALEN pair during vector amplification can prevent from sufficient virus amplification, potentially due to disruption of the target locus in the vector producing cell line. Therefore, the study by Saydaminova and colleagues utilized microRNA regulated expression (Brown et al. 2006) of TALEN-coding sequences. Another study utilized the HDAd system for delivery of donor DNA into human-disease-specific induced pluripotent stem cell clones. In combination with TALEN-encoding sequences delivered by nonviral approaches this strategy can be explored to perform gene correction in mammalian cells (Suzuki et al. 2014). In another pilot study it was shown that FGAds can also be explored for HDR in mammalian cells (Holkers et al. 2014).

Over the recent years mainly the CRISPR/Cas9 technology was studied in the context of Ad-based gene transfer. Initial work was published in 2014 in which the principle of applying the AdV technology as a tool for efficient delivery of RNA-guided CRISPR/Cas9 nucleases for disruption of various loci in human cells was

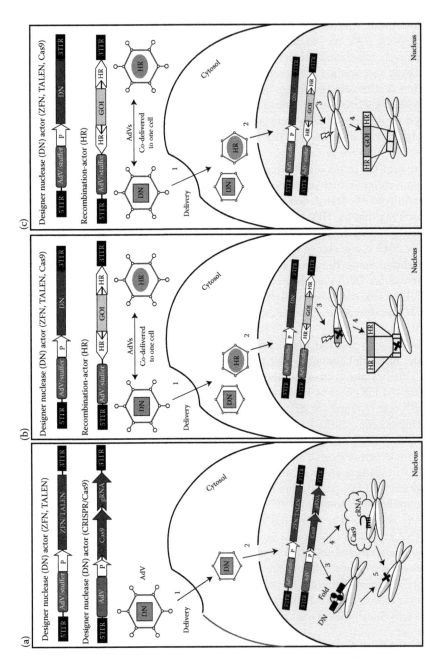

FIGURE 5.4

demonstrated (Maggio et al. 2014). This experimental setup can also be applied for genome editing in small animals which was demonstrated in the same year by delivery of the CRISPR/Cas9 machinery into adult mouse liver (Cheng et al. 2014). Using a similar approach, Wang et al. (2015) employed CRISPR/Cas9-Ad vectors for gene disruption in the murine genome to establish a model for human liver diseases in mice. That gene disruption based on adenoviral delivery of all components for the CRISPR/Cas9 machinery can also be used in an antiviral approach was explored in 2014 (Bi et al. 2014). Here the group applied the CRISPR/Cas9 technology for the interruption of genes contained in the genome of large pathogenic DNA viruses such as Ad and herpes virus (Bi et al. 2014). One year later and in concordance with the work performed with ZFNs (Maier et al. 2013, Li et al. 2015), another study demonstrated the disruption of the HIV co-receptor by using recombinant Ads inserting respective CRISPR/Cas and gRNAs into T-cells (Li et al. 2015).

Regarding CRISPR/Cas9-based genome editing in combination with HDR, two studies were performed. The first study combines adenoviral delivery of components for CRISPR/Cas9-mediated genome editing with Ad donor vectors and found that this combination is sufficient to induce accurate genome editing in mammalian cells (Holkers et al. 2014). Interestingly, a study published in 2015 demonstrated that adenoviral early genes increase the efficiencies of homology-directed repair after introduction of a DNA double-strand break into the genome of mammalian cells (Chu et al. 2015). The CRISPR/Cas9-Ad hybrid vector system was also explored *in vivo* to restore dystrophin function in a mouse model for Duchenne muscular dystrophy (DMD) (Xu et al. 2016). A subsequent study performed *in vitro* was performed by another group also showing gene repair after delivery of an adenoviral hybrid vector to rescue dystrophin synthesis in affected DMD muscle cells (Maggio et al. 2016). Note that a study exploring the CRISPR/Cas9 technology in concert with HDAd is still lacking. A summary of adenoviral hybrid vectors for genome editing is provided in Table 5.4.

FIGURE 5.4 (Continued)   Outline of Ad-based hybrid vectors utilizing DNs for genome editing. Ad-based hybrid vectors based on early generation (AdV) or HDAd (with stuffer DNA) can be utilized to efficiently deliver transgenes encoding DNs (designer nuclease actor = DN) into desired target cells using Ad-specific entry (1) and nuclear import (2). DNs include ZFN, TALENs, or CRISPR/Cas9 technologies. (a) Delivery of DNs can be used to introduce DNA double-strand breaks into the target sequence localized in the host chromosomal DNA or episomal viral DNA (latter not shown). ZFNs or TALENs bind as dimers to the target sequence (3). For CRISPR/Cas9 the gRNA binds to the target sequence (4) and in concert with the Cas9 nuclease this leads to a DNA double-strand breaks. Cellular DNA repair can then lead to introduction of mutations (5) due to error prone nonhomologous end joining. (b) The same strategy as shown in (b) can be used for gene correction. After introduction of a DNA double-strand break at a genomic locus carrying a mutated gene (3), this locus can be repaired by insertion of the functional DNA fragment using HR (4). The functional DNA fragment is introduced into the target cell by second recombination-actor (HR) vector. (c) After cutting at a specific genomic locus using DN encoding Ads (3) and in the presence of a second adenoviral vector (recombination actor = HR), transgenes for instance with a specific GOI can be inserted at specific genomic locus (4). The mechanism is based on HR.

TABLE 5.4
Summary of Relevant Studies Related to Adenoviral Vectors Delivering Designer Nucleases

AdV/DN[a]	Milestone	Reference
AdV[b]/ZFN[c]	First prototype of an adenoviral vector for ZFN delivery and disruption of the HIV co-receptor CCR5[h]	Perez et al. (2008)
AdV/ZFN AdV/HR	Ad-based ZFN delivery to target and to induce homologous recombination at the adeno-associated virus integration site 1 (AAVS1) as a safe harbor in human epithelial stem cells	Coluccio et al. (2013)
AdV/ZFN/HR	Gene correction utilizing a single adenoviral vector carrying the ZFN-encoding vector and the adenoviral donor for homologous recombination	Zhang et al. (2014)
AdV/TALEN[d]	First study showing delivery of a TALEN pair utilizing two early generation adenoviral vectors	Holkers et al. (2013)
AdV/TALEN AdV/HR	First proof of principle study demonstrating induction of homologous recombination in the host genome by co-delivery of early generation AdVs	Holkers et al. (2014)
HDAd[e]/TALEN	Delivery of a complete TALEN pair using a single HDAd	Saydaminova et al. (2015)
HDAd/HR[f]	Delivery of donor DNA for homologous recombination (recombination actor) using HDAd	Suzuki et al. (2014)
AdV/CRISPR[g]	Delivery of RNA-guided CRISPR/Cas9 nucleases for disruption of various loci in human cells	Maggio et al. (2014)
AdV/CRISPR	AdV based in vivo delivery of CRISPR/Cas9 for genome editing in murine liver	Cheng et al. (2014)
AdV/CRISPR	First study showing gene disruption in DNA viruses	Bi et al. (2014)
AdV/CRISPR AdV/HR	This study demonstrates induction of homologous recombination in host chromosomes by co-delivery of two AdVs	Holkers et al. (2014)

[a] DN: Designer nuclease.
[b] AdV: Early generation adenoviral vector deleted for the early genes E1 and E3.
[c] ZFN: Zinc-finger nuclease.
[d] TALEN: Transcription activator-like effector nucleases.
[e] HDAd: High-capacity adenoviral vector.
[f] HR: HDAd functioning as integration actor.
[g] CRISPR/Cas9: Clustered regularly interspaced short palindromic repeats (CRISPR)/Cas9.
[h] CCR5: HIV co-receptor.

5.2.4 HYBRID ADENOVIRAL VECTORS FOR VIRAL VECTOR PRODUCTION

With the advantages to accommodate large inserts, provide high titers, and infect nondividing as well as dividing cells with high efficiency, AdVs have been promising delivery tools for direct gene therapy settings. However, the AdV is also attractive as a carrier to support the production of other viral vectors. In these Ad-hybrid vector systems the AdV can be utilized to deliver all components into a producer cell line to assemble and produce other viral vectors such as retrovirus and lentivirus (Table 5.5).

TABLE 5.5
Ad-Based Hybrid Vector Systems for Viral Vector Production

AdV/vV[a]	Milestone	Reference
FGAd[b]/RV[c]	First AdV/retroviral hybrid vector system for generation of retroviral particles *in vitro* and *in vivo*	Bilbao et al. (1997)
FGAd/MuLV[d]	The construction of second-generation retroviral vector packaging cells that produced RV at comparable titer to traditional way	Lin (1998), Ramsey et al. (1998)
FGAd/MuLV	Hybrid adeno–retroviral vector system to generate a retroviral vector *in situ*; efficacy shown by intratumoral administration	Caplen et al. (1999)
FGAd/MuLV or GALV[e]	Using adenoviral vectors to express a retrovirus protein, a marker gene or a TG[i] in retrovirus producer cell lines; efficient release of helper-free retroviral vectors into the supernatants	Duisit et al. (1999)
HDAd[f]/MuLV-RCR[g]	Establishment of the two-stage transduction mechanism of the hybrid vector, with highly enhanced tumor transduction and antitumor efficacy *in vivo*	Soifer et al. (2002), Kubo et al. (2011)
HDAd/LV[h]	The first hybrid AdV system for lentiviral vector production; all-in-one vector under Tc-inducible promoter control	Kubo and Mitani (2003)
HDAd/LV	Convenient alternative approach for large-scale lentiviral vector production; comparable vector titer was achieved as conventional multiple-plasmid transfection methods	Kuate et al. (2004)
HDAd/LV	Using three separate HCAds for LV production, achieving up to 30-fold higher titer	Hu et al. (2015)

[a] vV: viral Vector.
[b] FGAd: First-generation adenoviral vector.
[c] RV: Retroviral vector.
[d] MuLV: Murine leukemia virus.
[e] GALV: Gibbon ape leukemia virus.
[f] HDAd: High-capacity adenoviral vector.
[g] RCR: Replication-competent retrovirus.
[h] LV: Lentiviral vector.
[i] TG: Transgene.

5.2.4.1 Hybrid AdV/Retrovirus System for Retroviral Vector Production

Retroviral vectors (RV), which irreversibly integrate into the genome, remain a popular tool to permanently modify the host DNA. Clinical success has also been achieved especially in the treatment of diseases of the immune system. RVs are usually produced by co-transfection of multiple plasmids into respective producer cells.

The first adenoviral/retroviral hybrid vector system for RV production (Bilbao et al. 1997) exploits favorable aspects of both adenoviral and RV. In detail, two separate early generation AdVs encoding an RV and packaging functions were generated. After co-infection of HEK293 cells with both AdVs retroviruses are produced and therefore this can be considered as a transient retroviral producer system. The progeny RV particles can then effectively achieve stable transduction of neighboring cells. Next, this hybrid adenoviral/RV system was used to construct second-generation packaging cells that deliver marker genes to target cells. Notably, a feasible gene transfer scheme was developed based on an Ad delivery system to deliver MoMLV structural genes (gag, pol, and env) to cultured cells which led to production of MoMLV with comparable titer to the co-transfection plasmid-based protocol (Lin 1998, Ramsey et al. 1998). In the following year (Duisit et al. 1999) reported the construction of adenoviral/retroviral chimeric vectors, using them to screen through a panel of primate cell lines to evaluate these cell types for their ability to generate RV. Meanwhile, the AdV/retroviral chimeric-vector system was further developed as an *in vivo* transducing agent with the ability to generate a retroviral vector *in vivo* after administration into subcutaneous 9L glioma tumors in rats and human A375 melanoma xenografts in nude mice (Caplen et al. 1999). Soifer et al. used HDAd that offers the ability to accommodate large or multiple inserts as a carrier to deliver a fully functional RV in a single vector (Figure 5.5a). They demonstrated that this HDAd/retrovirus hybrid vector mediates highly efficient gene delivery and permanent integration of transgenes through a two-stage mechanism relying on the following principle: the human Ad type 5-based HDAd can infect both human and mouse cells but it is unable to replicate efficiently in murine cells. Conversely, an ecotropic replicating-competent retroviral (RCR) vector is incapable of infecting human cells, but shows robust replication in murine cells. This motivated researchers to combine these two features of two different virus systems by inserting retroviral DNA sequences into an AdV for RV production in cells which are usually not susceptible to retroviral infection (Soifer et al. 2002). Recently, the same working group further developed this vector system for delivery and *in situ* production of RCR capable of infecting human cancer cells. Here the RCR produced from AdV/RCR-transduced cells revealed significantly higher titers, compared to cells only transduced with RCR at a similar multiplicity of infections (MOIs). A similar trend was also observed in human xenograft tumors *in vivo*, in which enhanced antitumor efficacy was achieved (Kubo et al. 2011).

5.2.4.2 Hybrid AdV Systems for Lentiviral Vector Production

Lentiviral vectors derived from HIV type 1 or HIV have advanced to clinical trials, with successes in treating genetic disease. However, optimization of the production procedure of these vectors should be further improved for achieving higher titers in a less labor-intensive manner. Lentiviral vectors are usually prepared by multi-plasmid DNA co-transfection into producer cells and lentiviral particles can be harvested from the culture supernatant a few days later.

Kubo and Mitani (2003) developed the first hybrid vector using HDAd as a carrier of the lentiviral production machinery. This new lentivirus/Ad encodes sequences required for production of an HIV-based lentiviral vector in a single HDAd backbone. Importantly, here the envelope cassette and the transcriptional activator cassette were controlled under a Tc-inducible promoter. Therefore, the two-stage transduction

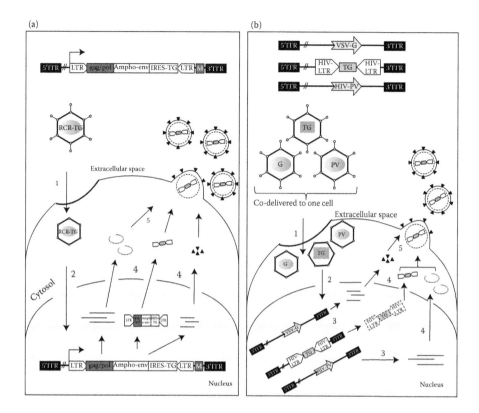

FIGURE 5.5 Schematic outline of hybrid adenoviral vectors for production of other viral vectors. All systems share two steps including (1) cellular uptake via respective receptors and co-receptors and (2) nuclear entry mediated by the nuclear pore complexes followed by nuclear import of the dsDNA. (a) Principle of the Ad/retroviral hybrid vector system. Exemplarily shown is a RCR production system that is based on HDAd (Kubo et al. 2011). All necessary vector components are inserted into a single HDAd. Upon nucleus entry, viral proteins are expressed, the RCR genome flanked by LTR is released from the HDAd genomes (3), which are subsequently exported from the nucleus into the cytoplasm (4), followed by the assembly of new RVs (5). (b) Principle of the Ad/lentiviral hybrid vector system. Exemplarily shown is a lentiviral vector (LV) production system that utilizes HDAd vectors (Hu et al. 2015). All necessary vector components are inserted into three separate HDAds. The first vector encodes VSV-G envelope (G), the second contains the transgene (TG) flanked by HIV long terminal repeats (LTRs), while the third one (PV) encodes HIV *gag*, *pol*, *tat*, and *rev* genes. After co-delivery of these vectors into one cell, viral proteins are expressed, the LV genome represented by LTR flanked TG is rescued from the HDAd genome, which is subsequently exported from the nucleus into the cytoplasm (4), followed by the assembly of new LVs (5).

protocol was achieved by initially infecting producer cells with AdV and production of the lentiviral vector *in situ* in the presence of doxycycline. Using this experimental setup, comparable vector titers were obtained as shown for conventional multiple-plasmid transfection methods. Moreover, efficient spread of the virus and persistent expression of the transgene in different cell lines under long-term culture conditions were observed. A similar strategy was also evaluated by another group, but the vector titers obtained were still not improved compared to the transient co-transfection experiments (Kuate et al. 2004). However, recently an improved protocol relying on HDAd transduction yielded up to 30-fold higher lentiviral titers compared to conventional protocols (Hu et al. 2015). In this latter study three separate HDAds were constructed, each with a single HIV vector component (HIV-PV, FG12, and VSVG3) (Figure 5.5b). This allows mixing and matching the single vectors and the transgene can be easily exchanged by modification of a single vector. Moreover, the authors also showed that the system is scalable and vector production can be accomplished by using adherent cells grown on stacked large plates (cell factory) or suspension cells cultured in spinner flasks. With more cost-effectiveness and more reproducibility than standard calcium phosphate co-transfection, the authors suggest using this optimized protocol for GMP conditions and later on for applications in clinical trials.

5.3   HOW TO CHOOSE THE RIGHT Ad-BASED HYBRID VECTOR?

Each of the Ad-based hybrid vector systems presented in this chapter display advantages and disadvantages and the optimal vector depends on the specific needs of the desired application. As summarized in Table 5.1 several means with respect to the optimal vector design need to be carefully evaluated.

For instance for all integrating hybrid vector systems or vector systems relying on DNs, side effects with respect to genotoxicity need to be carefully evaluated. Ad-based hybrid vectors delivering DNA replicons are potentially favorable with respect to genotoxicity but epigenetic silencing of the transgene or low level integration efficiencies should also be considered.

Moreover, the vector type plays an important role. For instance early generation AdVs such as FGAd and SGAd retain early and late adenoviral genes which can cause cellular toxicity due to the induction of anti-adenoviral immune responses. Especially when performing *in vivo* studies HDAd devoid of all adenoviral coding sequences may be advantageous. However, the production procedure for HDAd is still complex and not available to every laboratory. Besides the production procedure the molecular HDAd vector design also needs to be carefully reviewed. For instance expression of any kind of recombinase or integrase (e.g., Flp recombinase, Cre recombinase, SB transposase, PhiC31 integrase, designer nuclease, and retroviral integrase) can cause side effects if constantly expressed. Therefore, future vectors should consider transient systems such as inducible promoters to drive expression of these proteins. Also Ad capsid components of the Ad-based hybrid vector should be optimized with respect to targeted transduction of desired target cells. This can be achieved by genetic engineering or chemical capsid modifications. In summary, the Ad-based hybrid vector should be designed in a customized fashion and adapted to the specific application. If further pursued, the development of Ad-based hybrid vectors may present a valuable and alternative treatment option presenting an important field in precision medicine.

REFERENCES

Bell, J. B., K. M. Podetz-Pedersen, E. L. Aronovich, L. R. Belur, R. S. McIvor, and P. B. Hackett. 2007. Preferential delivery of the Sleeping Beauty transposon system to livers of mice by hydrodynamic injection. *Nat Protoc* 2 (12):3153–65. doi: 10.1038/nprot.2007.471.

Bi, Y., L. Sun, D. Gao et al. 2014. High-efficiency targeted editing of large viral genomes by RNA-guided nucleases. *PLoS Pathog* 10 (5):e1004090. doi: 10.1371/journal.ppat.1004090.

Bilbao, G., M. Feng, C. Rancourt, W. H. Jackson, Jr., and D. T. Curiel. 1997. Adenoviral/retroviral vector chimeras: A novel strategy to achieve high-efficiency stable transduction *in vivo*. *FASEB J* 11 (8):624–34.

Brown, B. D., M. A. Venneri, A. Zingale, L. Sergi Sergi, and L. Naldini. 2006. Endogenous microRNA regulation suppresses transgene expression in hematopoietic lineages and enables stable gene transfer. *Nat Med* 12 (5):585–91. doi: 10.1038/nm1398.

Caplen, N. J., J. N. Higginbotham, J. R. Scheel et al. 1999. Adeno-retroviral chimeric viruses as *in vivo* transducing agents. *Gene Ther* 6 (3):454–9. doi: 10.1038/sj.gt.3300835.

Cermak, T., E. L. Doyle, M. Christian et al. 2011. Efficient design and assembly of custom TALEN and other TAL effector-based constructs for DNA targeting. *Nucl Acids Res* 39 (12):e82. doi: 10.1093/nar/gkr218.

Cheng, R., J. Peng, Y. Yan et al. 2014. Efficient gene editing in adult mouse livers via adenoviral delivery of CRISPR/Cas9. *FEBS Lett* 588 (21):3954–8. doi: 10.1016/j.febslet.2014.09.008.

Chu, V. T., T. Weber, B. Wefers et al. 2015. Increasing the efficiency of homology-directed repair for CRISPR-Cas9-induced precise gene editing in mammalian cells. *Nat Biotechnol* 33 (5):543–8. doi: 10.1038/nbt.3198.

Colmenero, P., M. Chen, E. Castanos-Velez, P. Liljestrom, and M. Jondal. 2002. Immunotherapy with recombinant SFV-replicons expressing the P815A tumor antigen or IL-12 induces tumor regression. *Int J Cancer* 98 (4):554–60.

Coluccio, A., F. Miselli, A. Lombardo et al. 2013. Targeted gene addition in human epithelial stem cells by zinc-finger nuclease-mediated homologous recombination. *Mol Ther* 21 (9):1695–704. doi: 10.1038/mt.2013.143.

Cong, L., F. A. Ran, D. Cox et al. 2013. Multiplex genome engineering using CRISPR/Cas systems. *Science* 339 (6121):819–23. doi: 10.1126/science.1231143.

Cortes, M. L., A. Oehmig, O. Saydam et al. 2008. Targeted integration of functional human ATM cDNA into genome mediated by HSV/AAV hybrid amplicon vector. *Mol Ther* 16 (1):81–8. doi: 10.1038/sj.mt.6300338.

de Silva, S., M. A. Mastrangelo, L. T. Lotta, Jr., C. A. Burris, H. J. Federoff, and W. J. Bowers. 2010. Extending the transposable payload limit of Sleeping Beauty (SB) using the Herpes Simplex Virus (HSV)/SB amplicon-vector platform. *Gene Ther* 17 (3):424–31. doi: 10.1038/gt.2009.144.

Ding, S., X. Wu, G. Li, M. Han, Y. Zhuang, and T. Xu. 2005. Efficient transposition of the piggyBac (PB) transposon in mammalian cells and mice. *Cell* 122 (3):473–83. doi: 10.1016/j.cell.2005.07.013.

Dorigo, O., J. S. Gil, S. D. Gallaher et al. 2004. Development of a novel helper-dependent adenovirus-Epstein–Barr virus hybrid system for the stable transformation of mammalian cells. *J Virol* 78 (12):6556–66. doi: 10.1128/JVI.78.12.6556-6566.2004.

Duisit, G., A. Salvetti, P. Moullier, and F. L. Cosset. 1999. Functional characterization of adenoviral/retroviral chimeric vectors and their use for efficient screening of retroviral producer cell lines. *Hum Gene Ther* 10 (2):189–200. doi: 10.1089/10430349950018986.

Ehrhardt, A., J. A. Engler, H. Xu, A. M. Cherry, and M. A. Kay. 2006. Molecular analysis of chromosomal rearrangements in mammalian cells after phiC31-mediated integration. *Hum Gene Ther* 17 (11):1077–94. doi: 10.1089/hum.2006.17.1077.

Ehrhardt, A., S. R. Yant, J. C. Giering, H. Xu, J. A. Engler, and M. A. Kay. 2007. Somatic integration from an adenoviral hybrid vector into a hot spot in mouse liver results in persistent transgene expression levels *in vivo*. *Mol Ther* 15 (1):146–56. doi: 10.1038/sj.mt.6300011.

Fisher, K. J., W. M. Kelley, J. F. Burda, and J. M. Wilson. 1996. A novel adenovirus-adeno-associated virus hybrid vector that displays efficient rescue and delivery of the AAV genome. *Hum Gene Ther* 7 (17):2079–87. doi: 10.1089/hum.1996.7.17-2079.

Gallaher, S. D., J. S. Gil, O. Dorigo, and A. J. Berk. 2009. Robust *in vivo* transduction of a genetically stable Epstein–Barr virus episome to hepatocytes in mice by a hybrid viral vector. *J Virol* 83 (7):3249–57. doi: 10.1128/JVI.01721-08.

Geurts, A. M., Y. Yang, K. J. Clark et al. 2003. Gene transfer into genomes of human cells by the sleeping beauty transposon system. *Mol Ther* 8 (1):108–17.

Gil, J. S., S. D. Gallaher, and A. J. Berk. 2010. Delivery of an EBV episome by a self-circularizing helper-dependent adenovirus: Long-term transgene expression in immunocompetent mice. *Gene Ther* 17 (10):1288–93. doi: 10.1038/gt.2010.75.

Goncalves, M. A., M. Holkers, C. Cudre-Mauroux et al. 2006. Transduction of myogenic cells by retargeted dual high-capacity hybrid viral vectors: Robust dystrophin synthesis in duchenne muscular dystrophy muscle cells. *Mol Ther* 13 (5):976–86. doi: 10.1016/j.ymthe.2005.11.018.

Goncalves, M. A., M. Holkers, G. P. van Nierop, R. Wieringa, M. G. Pau, and A. A. de Vries. 2008. Targeted chromosomal insertion of large DNA into the human genome by a fiber-modified high-capacity adenovirus-based vector system. *PLoS One* 3 (8):e3084. doi: 10.1371/journal.pone.0003084.

Goncalves, M. A., M. G. Pau, A. A. de Vries, and D. Valerio. 2001. Generation of a high-capacity hybrid vector: Packaging of recombinant adenoassociated virus replicative intermediates in adenovirus capsids overcomes the limited cloning capacity of adeno-associated virus vectors. *Virology* 288 (2):236–46. doi: 10.1006/viro.2001.1073.

Goncalves, M. A., I. van der Velde, J. M. Janssen et al. 2002. Efficient generation and amplification of high-capacity adeno-associated virus/adenovirus hybrid vectors. *J Virol* 76 (21):10734–44.

Goncalves, M. A., I. van der Velde, S. Knaan-Shanzer, D. Valerio, and A. A. de Vries. 2004. Stable transduction of large DNA by high-capacity adeno-associated virus/adenovirus hybrid vectors. *Virology* 321 (2):287–96. doi: 10.1016/j.virol.2004.01.007.

Grandchamp, N., D. Altemir, S. Philippe et al. 2014. Hybrid lentivirus-phiC31-int-NLS vector allows site-specific recombination in murine and human cells but induces DNA damage. *PLoS One* 9 (6):e99649. doi: 10.1371/journal.pone.0099649.

Groth, A. C., E. C. Olivares, B. Thyagarajan, and M. P. Calos. 2000. A phage integrase directs efficient site-specific integration in human cells. *Proc Natl Acad Sci U S A* 97 (11):5995–6000. doi: 10.1073/pnas.090527097.

Guan, M., J. R. Rodriguez-Madoz, P. Alzuguren et al. 2006. Increased efficacy and safety in the treatment of experimental liver cancer with a novel adenovirus-alphavirus hybrid vector. *Cancer Res* 66 (3):1620–9. doi: 10.1158/0008-5472.CAN-05-0877.

Hausl, M. A., W. Zhang, N. Muther et al. 2010. Hyperactive sleeping beauty transposase enables persistent phenotypic correction in mice and a canine model for hemophilia B. *Mol Ther* 18 (11):1896–906. doi: 10.1038/mt.2010.169.

Hausl, M., W. Zhang, R. Voigtlander, N. Muther, C. Rauschhuber, and A. Ehrhardt. 2011. Development of adenovirus hybrid vectors for Sleeping Beauty transposition in large mammals. *Curr Gene Ther* 11 (5):363–74.

Heinz, N., A. Schambach, M. Galla et al. 2011. Retroviral and transposon-based tet-regulated all-in-one vectors with reduced background expression and improved dynamic range. *Hum Gene Ther* 22 (2):166–76. doi: 10.1089/hum.2010.099.

Heister, T., I. Heid, M. Ackermann, and C. Fraefel. 2002. Herpes simplex virus type 1/adeno-associated virus hybrid vectors mediate site-specific integration at the adeno-associated virus preintegration site, AAVS1, on human chromosome 19. *J Virol* 76 (14):7163–73.

Hellebrand, E., J. Mautner, G. Reisbach et al. 2006. Epstein–Barr virus vector-mediated gene transfer into human B cells: Potential for antitumor vaccination. *Gene Ther* 13 (2):150–62. doi: 10.1038/sj.gt.3302602.

Hockemeyer, D., H. Wang, S. Kiani et al. 2011. Genetic engineering of human pluripotent cells using TALE nucleases. *Nat Biotechnol* 29 (8):731–4. doi: 10.1038/nbt.1927.

Holkers, M., I. Maggio, S. F. Henriques, J. M. Janssen, T. Cathomen, and M. A. Goncalves. 2014. Adenoviral vector DNA for accurate genome editing with engineered nucleases. *Nat Methods* 11 (10):1051–7. doi: 10.1038/nmeth.3075.

Holkers, M., I. Maggio, J. Liu et al. 2013. Differential integrity of TALE nuclease genes following adenoviral and lentiviral vector gene transfer into human cells. *Nucl Acids Res* 41 (5):e63. doi: 10.1093/nar/gks1446.

Hu, Y., K. O'Boyle, D. Palmer, P. Ng, and R. E. Sutton. 2015. High-level production of replication-defective human immunodeficiency type 1 virus vector particles using helper-dependent adenovirus vectors. *Mol Ther Methods Clin Dev* 2:15004. doi: 10.1038/mtm.2015.4.

Ivics, Z., P. B. Hackett, R. H. Plasterk, and Z. Izsvak. 1997. Molecular reconstruction of Sleeping Beauty, a Tc1-like transposon from fish, and its transposition in human cells. *Cell* 91 (4):501–10.

Ivics, Z., A. Katzer, E. E. Stuwe, D. Fiedler, S. Knespel, and Z. Izsvak. 2007. Targeted Sleeping Beauty transposition in human cells. *Mol Ther* 15 (6):1137–44. doi: 10.1038/sj.mt.6300169.

Izsvak, Z. and Z. Ivics. 2004. Sleeping beauty transposition: Biology and applications for molecular therapy. *Mol Ther* 9 (2):147–56. doi: 10.1016/j.ymthe.2003.11.009.

Jager, L. and A. Ehrhardt. 2009. Persistence of high-capacity adenoviral vectors as replication-defective monomeric genomes *in vitro* and in murine liver. *Hum Gene Ther* 20 (8):883–96. doi: 10.1089/hum.2009.020.

Jenke, B. H., C. P. Fetzer, I. M. Stehle et al. 2002. An episomally replicating vector binds to the nuclear matrix protein SAF-A *in vivo*. *EMBO Rep* 3 (4):349–54. doi: 10.1093/embo-reports/kvf070.

Jinek, M., K. Chylinski, I. Fonfara, M. Hauer, J. A. Doudna, and E. Charpentier. 2012. A programmable dual-RNA-guided DNA endonuclease in adaptive bacterial immunity. *Science* 337 (6096):816–21. doi: 10.1126/science.1225829.

Kotin, R. M., M. Siniscalco, R. J. Samulski et al. 1990. Site-specific integration by adeno-associated virus. *Proc Natl Acad Sci U S A* 87 (6):2211–5.

Kreppel, F. and S. Kochanek. 2004. Long-term transgene expression in proliferating cells mediated by episomally maintained high-capacity adenovirus vectors. *J Virol* 78 (1):9–22.

Krougliak, V. A., N. Krougliak, and R. C. Eisensmith. 2001. Stabilization of transgenes delivered by recombinant adenovirus vectors through extrachromosomal replication. *J Gene Med* 3 (1):51–8. doi: 10.1002/1521-2254(2000)9999:9999 <::AID-JGM150> 3.0.CO;2-#.

Kuate, S., D. Stefanou, D. Hoffmann, O. Wildner, and K. Uberla. 2004. Production of lentiviral vectors by transient expression of minimal packaging genes from recombinant adenoviruses. *J Gene Med* 6 (11):1197–205. doi: 10.1002/jgm.623.

Kubo, S., K. Haga, A. Tamamoto et al. 2011. Adenovirus-retrovirus hybrid vectors achieve highly enhanced tumor transduction and antitumor efficacy *in vivo*. *Mol Ther* 19 (1):76–82. doi: 10.1038/mt.2010.182.

Kubo, S. and K. Mitani. 2003. A new hybrid system capable of efficient lentiviral vector production and stable gene transfer mediated by a single helper-dependent adenoviral vector. *J Virol* 77 (5):2964–71.

Kubo, S., M. C. Seleme, H. S. Soifer et al. 2006. L1 retrotransposition in nondividing and primary human somatic cells. *Proc Natl Acad Sci U S A* 103 (21):8036–41. doi: 10.1073/pnas.0601954103.

Kuhstoss, S. and R. N. Rao. 1991. Analysis of the integration function of the streptomycete bacteriophage phi C31. *J Mol Biol* 222 (4):897–908.

Leblois, H., C. Roche, N. Di Falco, C. Orsini, P. Yeh, and M. Perricaudet. 2000. Stable transduction of actively dividing cells via a novel adenoviral/episomal vector. *Mol Ther* 1 (4):314–22. doi: 10.1006/mthe.2000.0042.

Li, C., X. Guan, T. Du et al. 2015. Inhibition of HIV-1 infection of primary CD4+ T-cells by gene editing of CCR5 using adenovirus-delivered CRISPR/Cas9. *J Gen Virol* 96 (8):2381–93. doi: 10.1099/vir.0.000139.

Lieber, A., D. S. Steinwaerder, C. A. Carlson, and M. A. Kay. 1999. Integrating adenovirus-adeno-associated virus hybrid vectors devoid of all viral genes. *J Virol* 73 (11):9314–24.

Lin, X. 1998. Construction of new retroviral producer cells from adenoviral and retroviral vectors. *Gene Ther* 5 (9):1251–8. doi: 10.1038/sj.gt.3300720.

Lufino, M. M., R. Manservigi, and R. Wade-Martins. 2007. An S/MAR-based infectious episomal genomic DNA expression vector provides long-term regulated functional complementation of LDLR deficiency. *Nucl Acids Res* 35 (15):e98. doi: 10.1093/nar/gkm570.

Maggio, I., M. Holkers, J. Liu, J. M. Janssen, X. Chen, and M. A. Goncalves. 2014. Adenoviral vector delivery of RNA-guided CRISPR/Cas9 nuclease complexes induces targeted mutagenesis in a diverse array of human cells. *Sci Rep* 4:5105. doi: 10.1038/srep05105.

Maggio, I., L. Stefanucci, J. M. Janssen et al. 2016. Selection-free gene repair after adenoviral vector transduction of designer nucleases: Rescue of dystrophin synthesis in DMD muscle cell populations. *Nucl Acids Res* doi: 10.1093/nar/gkv1540.

Maier, D. A., A. L. Brennan, S. Jiang et al. 2013. Efficient clinical scale gene modification via zinc finger nuclease-targeted disruption of the HIV co-receptor CCR5. *Hum Gene Ther* 24 (3):245–58. doi: 10.1089/hum.2012.172.

Mak, A. N., P. Bradley, R. A. Cernadas, A. J. Bogdanove, and B. L. Stoddard. 2012. The crystal structure of TAL effector PthXo1 bound to its DNA target. *Science* 335 (6069):716–9. doi: 10.1126/science.1216211.

Maruyama-Tabata, H., Y. Harada, T. Matsumura et al. 2000. Effective suicide gene therapy *in vivo* by EBV-based plasmid vector coupled with polyamidoamine dendrimer. *Gene Ther* 7 (1):53–60. doi: 10.1038/sj.gt.3301044.

Mates, L., M. K. Chuah, E. Belay et al. 2009. Molecular evolution of a novel hyperactive Sleeping Beauty transposase enables robust stable gene transfer in vertebrates. *Nat Genet* 41 (6):753–61. doi: 10.1038/ng.343.

Mikkelsen, J. G., S. R. Yant, L. Meuse, Z. Huang, H. Xu, and M. A. Kay. 2003. Helper-independent Sleeping Beauty transposon-transposase vectors for efficient nonviral gene delivery and persistent gene expression *in vivo*. *Mol Ther* 8 (4):654–65.

Miller, J. C., M. C. Holmes, J. Wang et al. 2007. An improved zinc-finger nuclease architecture for highly specific genome editing. *Nat Biotechnol* 25 (7):778–85. doi: 10.1038/nbt1319.

Moore, J. K. and J. E. Haber. 1996. Cell cycle and genetic requirements of two pathways of nonhomologous end-joining repair of double-strand breaks in *Saccharomyces cerevisiae*. *Mol Cell Biol* 16 (5):2164–73.

Murphy, A. M., M. M. Morris-Downes, B. J. Sheahan, and G. J. Atkins. 2000. Inhibition of human lung carcinoma cell growth by apoptosis induction using Semliki Forest virus recombinant particles. *Gene Ther* 7 (17):1477–82. doi: 10.1038/sj.gt.3301263.

Murphy, S. J., H. Chong, S. Bell, R. M. Diaz, and R. G. Vile. 2002. Novel integrating adenoviral/retroviral hybrid vector for gene therapy. *Hum Gene Ther* 13 (6):745–60. doi: 10.1089/104303402317322302.

Mussolino, C., R. Morbitzer, F. Lutge, N. Dannemann, T. Lahaye, and T. Cathomen. 2011. A novel TALE nuclease scaffold enables high genome editing activity in combination with low toxicity. *Nucl Acids Res* 39 (21):9283–93. doi: 10.1093/nar/gkr597.

Nakai, H., E. Montini, S. Fuess, T. A. Storm, M. Grompe, and M. A. Kay. 2003. AAV serotype 2 vectors preferentially integrate into active genes in mice. *Nat Genet* 34 (3):297–302. doi: 10.1038/ng1179.

Olivares, E. C., R. P. Hollis, T. W. Chalberg, L. Meuse, M. A. Kay, and M. P. Calos. 2002. Site-specific genomic integration produces therapeutic Factor IX levels in mice. *Nat Biotechnol* 20 (11):1124–8. doi: 10.1038/nbt753.

Pan, Y., Q. Zhao, L. Fang, R. Luo, H. Chen, and S. Xiao. 2009. Efficient gene delivery into mammalian cells by recombinant baculovirus containing a hybrid cytomegalovirus promoter/Semliki Forest virus replicon. *J Gene Med* 11 (11):1030–8. doi: 10.1002/jgm.1390.

Papapetrou, E. P., P. G. Ziros, I. D. Micheva, N. C. Zoumbos, and A. Athanassiadou. 2006. Gene transfer into human hematopoietic progenitor cells with an episomal vector carrying an S/MAR element. *Gene Ther* 13 (1):40–51. doi: 10.1038/sj.gt.3302593.

Perez, E. E., J. Wang, J. C. Miller et al. 2008. Establishment of HIV-1 resistance in CD4+ T cells by genome editing using zinc-finger nucleases. *Nat Biotechnol* 26 (7):808–16. doi: 10.1038/nbt1410.

Peterson, E. B., M. A. Mastrangelo, H. J. Federoff, and W. J. Bowers. 2007. Neuronal specificity of HSV/sleeping beauty amplicon transduction in utero is driven primarily by tropism and cell type composition. *Mol Ther* 15 (10):1848–55. doi: 10.1038/sj.mt.6300267.

Pfuller, R. and W. Hammerschmidt. 1996. Plasmid-like replicative intermediates of the Epstein–Barr virus lytic origin of DNA replication. *J Virol* 70 (6):3423–31.

Piechaczek, C., C. Fetzer, A. Baiker, J. Bode, and H. J. Lipps. 1999. A vector based on the SV40 origin of replication and chromosomal S/MARs replicates episomally in CHO cells. *Nucleic Acids Res* 27 (2):426–8.

Porteus, M. H. and D. Baltimore. 2003. Chimeric nucleases stimulate gene targeting in human cells. *Science* 300 (5620):763. doi: 10.1126/science.1078395.

Ramsey, W. J., N. J. Caplen, Q. Li, J. N. Higginbotham, M. Shah, and R. M. Blaese. 1998. Adenovirus vectors as transcomplementing templates for the production of replication defective retroviral vectors. *Biochem Biophys Res Commun* 246 (3):912–9. doi: 10.1006/bbrc.1998.8726.

Recchia, A. 2004. The Journal of Gene Medicine European Society of Gene Therapy Young Investigator Award 2004. *J Gene Med* 6 (10):1170. doi: 10.1002/jgm.672.

Recchia, A., R. J. Parks, S. Lamartina et al. 1999. Site-specific integration mediated by a hybrid adenovirus/adeno-associated virus vector. *Proc Natl Acad Sci U S A* 96 (6):2615–20.

Recchia, A., L. Perani, D. Sartori, C. Olgiati, and F. Mavilio. 2004. Site-specific integration of functional transgenes into the human genome by adeno/AAV hybrid vectors. *Mol Ther* 10 (4):660–70. doi: 10.1016/j.ymthe.2004.07.003.

Ren, C., M. Zhao, X. Yang et al. 2006. Establishment and applications of Epstein–Barr virus-based episomal vectors in human embryonic stem cells. *Stem Cells* 24 (5):1338–47. doi: 10.1634/stemcells.2005-0338.

Robert, M. A., Y. Zeng, B. Raymond et al. 2012. Efficacy and site-specificity of adenoviral vector integration mediated by the phage phiC31 integrase. *Hum Gene Ther Methods* 23 (6):393–407. doi: 10.1089/hgtb.2012.122.

Russell, R. A., G. Vassaux, P. Martin-Duque, and M. O. McClure. 2004. Transient foamy virus vector production by adenovirus vectors. *Gene Ther* 11 (3):310–6. doi: 10.1038/sj.gt.3302177.

Samulski, R. J., X. Zhu, X. Xiao et al. 1991. Targeted integration of adeno-associated virus (AAV) into human chromosome 19. *EMBO J* 10 (12):3941–50.

Sander, J. D., E. J. Dahlborg, M. J. Goodwin et al. 2011. Selection-free zinc-finger-nuclease engineering by context-dependent assembly (CoDA). *Nat Methods* 8 (1):67–9. doi: 10.1038/nmeth.1542.

Saydaminova, K., X. Ye, H. Wang et al. 2015. Efficient genome editing in hematopoietic stem cells with helper-dependent Ad5/35 vectors expressing site-specific endonucleases under microRNA regulation. *Mol Ther Methods Clin Dev* 1:14057. doi: 10.1038/mtm.2014.57.

Schaarschmidt, D., J. Baltin, I. M. Stehle, H. J. Lipps, and R. Knippers. 2004. An episomal mammalian replicon: Sequence-independent binding of the origin recognition complex. *EMBO J* 23 (1):191–201. doi: 10.1038/sj.emboj.7600029.

Schlesinger, S. 2001. Alphavirus vectors: Development and potential therapeutic applications. *Expert Opin Biol Ther* 1 (2):177–91. doi: 10.1517/14712598.1.2.177.

Shayakhmetov, D. M., C. A. Carlson, H. Stecher, Q. Li, G. Stamatoyannopoulos, and A. Lieber. 2002. A high-capacity, capsid-modified hybrid adenovirus/adeno-associated virus vector for stable transduction of human hematopoietic cells. *J Virol* 76 (3):1135–43.

Sitaraman, V., P. Hearing, C. B. Ward et al. 2011. Computationally designed adeno-associated virus (AAV) Rep 78 is efficiently maintained within an adenovirus vector. *Proc Natl Acad Sci U S A* 108 (34):14294–9. doi: 10.1073/pnas.1102883108.

Skipper, K. A., P. R. Andersen, N. Sharma, and J. G. Mikkelsen. 2013. DNA transposon-based gene vehicles—scenes from an evolutionary drive. *J Biomed Sci* 20:92. doi: 10.1186/1423-0127-20-92.

Smith, J., M. Bibikova, F. G. Whitby, A. R. Reddy, S. Chandrasegaran, and D. Carroll. 2000. Requirements for double-strand cleavage by chimeric restriction enzymes with zinc finger DNA-recognition domains. *Nucl Acids Res* 28 (17):3361–9.

Smith, R. P., J. D. Riordan, C. R. Feddersen, and A. J. Dupuy. 2015. A hybrid adenoviral vector system achieves efficient long-term gene expression in the liver via piggyBac transposition. *Hum Gene Ther* 26 (6):377–85. doi: 10.1089/hum.2014.123.

Soifer, H., C. Higo, H. H. Kazazian, Jr., J. V. Moran, K. Mitani, and N. Kasahara. 2001. Stable integration of transgenes delivered by a retrotransposon-adenovirus hybrid vector. *Hum Gene Ther* 12 (11):1417–28. doi: 10.1089/104303401750298571.

Soifer, H., C. Higo, C. R. Logg et al. 2002. A novel, helper-dependent, adenovirus-retrovirus hybrid vector: Stable transduction by a two-stage mechanism. *Mol Ther* 5 (5 Pt 1):599–608. doi: 10.1006/mthe.2002.0586.

Stehle, I. M., J. Postberg, S. Rupprecht, T. Cremer, D. A. Jackson, and H. J. Lipps. 2007. Establishment and mitotic stability of an extra-chromosomal mammalian replicon. *BMC Cell Biol* 8:33. doi: 10.1186/1471-2121-8-33.

Stephen, S. L., V. G. Sivanandam, and S. Kochanek. 2008. Homologous and heterologous recombination between adenovirus vector DNA and chromosomal DNA. *J Gene Med* 10 (11):1176–89. doi: 10.1002/jgm.1246.

Strauss, J. H. and E. G. Strauss. 1994. The alphaviruses: Gene expression, replication, and evolution. *Microbiol Rev* 58 (3):491–562.

Sun, Y., H. Y. Li, D. Y. Tian et al. 2011. A novel alphavirus replicon-vectored vaccine delivered by adenovirus induces sterile immunity against classical swine fever. *Vaccine* 29 (46):8364–72. doi: 10.1016/j.vaccine.2011.08.085.

Sun, Y., D. Y. Tian, S. Li et al. 2013. Comprehensive evaluation of the adenovirus/alphavirus-replicon chimeric vector-based vaccine rAdV-SFV-E2 against classical swine fever. *Vaccine* 31 (3):538–44. doi: 10.1016/j.vaccine.2012.11.013.

Suzuki, K., C. Yu, J. Qu et al. 2014. Targeted gene correction minimally impacts whole-genome mutational load in human-disease-specific induced pluripotent stem cell clones. *Cell Stem Cell* 15 (1):31–6. doi: 10.1016/j.stem.2014.06.016.

Tan, B. T., L. Wu, and A. J. Berk. 1999. An adenovirus-Epstein–Barr virus hybrid vector that stably transforms cultured cells with high efficiency. *J Virol* 73 (9):7582–9.

Thorpe, H. M. and M. C. Smith. 1998. *In vitro* site-specific integration of bacteriophage DNA catalyzed by a recombinase of the resolvase/invertase family. *Proc Natl Acad Sci U S A* 95 (10):5505–10.

Thyagarajan, B., E. C. Olivares, R. P. Hollis, D. S. Ginsburg, and M. P. Calos. 2001. Site-specific genomic integration in mammalian cells mediated by phage phiC31 integrase. *Mol Cell Biol* 21 (12):3926–34. doi: 10.1128/MCB.21.12.3926-3934.2001.

Thyagarajan, B., K. Scheyhing, H. Xue et al. 2009. A single EBV-based vector for stable episomal maintenance and expression of GFP in human embryonic stem cells. *Regen Med* 4 (2):239–50. doi: 10.2217/17460751.4.2.239.

Ueno, T., H. Matsumura, K. Tanaka et al. 2000. Site-specific integration of a transgene mediated by a hybrid adenovirus/adeno-associated virus vector using the Cre/loxP-expres-
sion-switching system. *Biochem Biophys Res Commun* 273 (2):473–8. doi: 10.1006/
bbrc.2000.2972.

Verghese, S. C., N. A. Goloviznina, A. M. Skinner, H. J. Lipps, and P. Kurre. 2014. S/MAR
sequence confers long-term mitotic stability on non-integrating lentiviral vector episomes without selection. *Nucl Acids Res* 42 (7):e53. doi: 10.1093/nar/gku082.

Vink, C. A., H. B. Gaspar, R. Gabriel et al. 2009. Sleeping beauty transposition from nonin-
tegrating lentivirus. *Mol Ther* 17 (7):1197–204. doi: 10.1038/mt.2009.94.

Voigtlander, R., R. Haase, M. Muck-Hausl et al. 2013. A novel adenoviral hybrid-vector system carrying a plasmid replicon for safe and efficient cell and gene therapeutic applications. *Mol Ther Nucl Acids* 2:e83. doi: 10.1038/mtna.2013.11.

Wang, D., H. Mou, S. Li et al. 2015. Adenovirus-mediated somatic genome editing of Pten by
CRISPR/Cas9 in mouse liver in spite of Cas9-specific immune responses. *Hum Gene
Ther* 26 (7):432–42. doi: 10.1089/hum.2015.087.

Wang, H. and A. Lieber. 2006. A helper-dependent capsid-modified adenovirus vector
expressing adeno-associated virus rep78 mediates site-specific integration of a 27-kilo-
base transgene cassette. *J Virol* 80 (23):11699–709. doi: 10.1128/JVI.00779-06.

Wood, A. J., T. W. Lo, B. Zeitler et al. 2011. Targeted genome editing across species using
ZFNs and TALENs. *Science* 333 (6040):307. doi: 10.1126/science.1207773.

Wu, Q., F. Xu, L. Fang et al. 2013. Enhanced immunogenicity induced by an alphavirus replicon-based pseudotyped baculovirus vaccine against porcine reproductive and respiratory syndrome virus. *J Virol Methods* 187 (2):251–8. doi: 10.1016/j.jviromet.2012.11.018.

Xu, L., K. H. Park, L. Zhao et al. 2016. CRISPR-mediated genome editing restores dystrophin
expression and function in mdx mice. *Mol Ther* 24: 564–9. doi: 10.1038/mt.2015.192.

Yang, Y., F. Xiao, Z. Lu et al. 2013. Development of a novel adenovirus-alphavirus hybrid
vector with RNA replicon features for malignant hematopoietic cell transduction.
Cancer Gene Ther 20 (8):429–36. doi: 10.1038/cgt.2013.37.

Yant, S. R., A. Ehrhardt, J. G. Mikkelsen, L. Meuse, T. Pham, and M. A. Kay. 2002.
Transposition from a gutless adeno-transposon vector stabilizes transgene expression
in vivo. Nat Biotechnol 20 (10):999–1005. doi: 10.1038/nbt738.

Yant, S. R., L. Meuse, W. Chiu, Z. Ivics, Z. Izsvak, and M. A. Kay. 2000. Somatic integration and long-term transgene expression in normal and haemophilic mice using a DNA
transposon system. *Nat Genet* 25 (1):35–41. doi: 10.1038/75568.

Yant, S. R., X. Wu, Y. Huang, B. Garrison, S. M. Burgess, and M. A. Kay. 2005. High-
resolution genome-wide mapping of transposon integration in mammals. *Mol Cell Biol*
25 (6):2085–94. doi: 10.1128/MCB.25.6.2085-2094.2005.

Zhang, W., M. Muck-Hausl, J. Wang et al. 2013a. Integration profile and safety of an adenovirus hybrid-vector utilizing hyperactive sleeping beauty transposase for somatic integration. *PLoS One* 8 (10):e75344. doi: 10.1371/journal.pone.0075344.

Zhang, W., M. Solanki, N. Muther et al. 2013b. Hybrid adeno-associated viral vectors utilizing transposase-mediated somatic integration for stable transgene expression in human
cells. *PLoS One* 8 (10):e76771. doi: 10.1371/journal.pone.0076771.

Zhang, W., D. Wang, S. Liu et al. 2014. Multiple copies of a linear donor fragment released
in situ from a vector improve the efficiency of zinc-finger nuclease-mediated genome
editing. *Gene Ther* 21 (3):282–8. doi: 10.1038/gt.2013.83.

Zheng, C., B. J. Baum, M. J. Iadarola, and B. C. O'Connell. 2000. Genomic integration and
gene expression by a modified adenoviral vector. *Nat Biotechnol* 18 (2):176–80. doi:
10.1038/72628.

6 Adenovirus Vectors for Genome Editing Involving Engineered Endonucleases

Kamola Saydaminova, Maximilian Richter,
Philip Ng, Anja Ehrhardt, and André Lieber

CONTENTS

ABSTRACT

We discuss advantages and disadvantages of using adenovirus vectors (Ads) expressing engineered endonucleases (ENs) for targeted genome engineering *in vitro* and *in vivo*, in mice. For *in vitro* application of Ad-ENs we focus on hematopoietic stem cells, mesenchymal stem cells, and induced pluripotent stem cells. Among the advantages of using Ad vectors for EN delivery are the ability to transduce nondividing cells, the episomal nature of Ad genomes, the easiness to modify vector tropism, and the relatively low cost of vector manufacturing, which is particularly relevant for *in vivo* genome editing. Disadvantages include toxicity to stem cells and immunogenicity of first-generation vectors; problems that can largely be circumvented by helper-dependent Ad vectors. Problems specific to EN-expressing Ad vectors are the risk of vector genome rearrangements and loss of EN expression during vector production, cyto-/genotoxicity associated with extended EN expression from episomal Ad vector genomes in slow or nonproliferating cells, and immune responses against ENs *in vivo*. We review examples for *in vitro* and *in vivo* Ad-EN-mediated genome editing and conclude that more studies on the efficacy and the safety (specifically after *in vivo* Ad-EN application) are required to assess the utility of this vector system for genome editing in humans.

6.1 GENOME ENGINEERING IN STEM CELLS: OVERVIEW

6.1.1 TARGET CELLS

Our review of *in vitro* applications of Ad-EN vectors focuses on genome engineering in hematopoietic stem cells (HSCs), mesenchymal stem cells (MSCs), and induced pluripotent stem (iPS) cells for the purpose of gene therapy and generation of disease models. Two central features characterize HSCs, the ability to self-renew through asymmetric cell division and the ability to replenish all blood cell types [1,2]. HSCs are therefore an important gene therapy target and recent clinical trials of HSC gene therapy have shown clear therapeutic benefits for blood diseases that are otherwise incurable (for a review: [3]). MSCs are multipotent stromal cells that can differentiate into a variety of cell types, including osteoblasts, chondrocytes, myocytes, and adipocytes. In the context of gene therapy, MSCs have been used as a source for producing secreted proteins. iPS cells can propagate indefinitely and give rise to every other cell type in the body (such as neurons, heart, pancreatic, and liver cells). iPS cells therefore represent an important source for regenerative medicine and disease modeling.

6.1.2 SITE-SPECIFIC DNA BREAKS MEDIATED BY ENDONUCLEASES

A major tool for genome engineering is engineered endonucleases (ENs) that target a DNA break to preselected genomic sites (for a review of ENs: [4]). ENs are employed to knock-out genes, correct frame shift mutations, delete or rearrange chromosomal regions, or knock-in wild type or mutated cDNAs into the endogenous site or a heterologous sites. There are now a number of different EN platforms to generate site-specific DNA breaks in the genome [5]. One group of ENs contains DNA-binding protein domains. This group includes meganucleases [6] and megaTALs [4] with DNA binding and nuclease properties as well as zinc-finger nucleases (ZFNs) and transcription activator-like effector nucleases (TALENs) in which the DNA binding domain is fused with the bacterial EN *Fok*I. Because DNA cleavage by *Fok*I requires two *Fok*I molecules bound to each of the DNA strands, two subunits of the *Fok*I containing EN have to be expressed [7]. A second group of ENs is based on RNA-guided DNA recognition and utilizes the CRISPR/Cas9 bacterial system [8]. Because they are easy to design, CRISPR/Cas9 nucleases have found widespread application during the last 2 years. Side-by-side comparisons of different EN platforms are still rare and the verdict as to which EN platform is the best for genome engineering, is still out. Most likely this has to be evaluated for each target cell type and genomic target site as it was recently exemplary done for T-cell receptor gene targeting [9].

Several approaches have been used to deliver EN expression cassettes to HSCs, MSCs, and iPS cells. Because it is thought that the ENs need to be expressed only for a short time to achieve permanent modification of the target genomic sequence, most of the EN delivery approaches focused on transient expression of ENs without integration of the EN gene into the host genome. Among these delivery systems are (i) electroporation of plasmid, minicircle, or mRNA-encoding ENs. While avoiding the problems associated with viral gene delivery vectors, electroporation of plasmid DNA can be associated with cytotoxicity in primary cells especially in stem cells [10,11]. (ii) Infection with integrase-defective, non-integrating lentivirus vectors [10]. Limitations of this approach can include relatively low EN expression levels and epigenetic silencing of the EN cassette [12,13], limited insert capacity (<8 kb) of lentivirus vectors, and the potential risk of recombination between identical sequences when two EN-*Fok*I units are expressed from the same vector. (iii) Transduction with Ad vectors which will be discussed in detail below. For a review on EN delivery technologies see Reference [14].

6.1.3 GENE TARGETING BY HOMOLOGY DIRECTED REPAIR INVOLVING DONOR VECTORS

It has been shown that site-specific DNA breaks stimulate homology-directed repair (HDR) pathways and gene addition using homologous donor templates [15–19]. For most gene targeting purposes, these donors contain DNA sequences that are homologous to genomic sequences flanking the EN cleavage site. Platforms to deliver donor DNA include (i) synthetic single-stranded oligonucleotides with a length of up to 300 nucleotides [20,21], linearized circular plasmids [22] that are electroporated together with the nuclease-encoding mRNA; (ii) non-integrating lentivirus vectors [23]; (iii) rAAV vectors [24]; and (iv) adenovirus vectors [25]. Potential advantages of adenovirus vectors as for delivery of homology donor templates are discussed below.

6.2 ADENOVIRUS VECTORS FOR GENOME EDITING—ADVANTAGES

6.2.1 GENE DELIVERY TO NONDIVIDING CELLS

HSC, MSC, and iPS cells are, *per se*, nondividing cells. Proliferation of these cells is generally associated with differentiation and loss of pluripotency [26]. Ad serotype

5-derived vectors (Ad5) can efficiently transduce nondividing cells *in vitro* and *in vivo* (for a review, see Reference [27]). The efficiency of Ad5 infection relies to a large degree on efficient targeting of the Ad genome to the host cell nucleus. Ad vector DNA, including helper-dependent adenoviral (HDAd) vector DNA is packaged together with viral core proteins and pTP/TP, the terminal protein, into virions [28]. After entry into the host cell, the virion is uncoated and the Ad DNA is transported into the nucleus. It is generally thought that the nuclear localization domain in the pTP/TP and the core protein V play a crucial role in directing this complex to the nucleus. Efficient transduction of nondividing HSCs with capsid-chimeric Ad5/35 vectors was demonstrated for example by Nilsson et al. [29]. Viral replication studies with other Ad serotypes, including serotypes 11 and 35, also indicate that Ad genomes are efficiently delivered to the nucleus of nondividing CD34+ cells [30].

6.2.2 Non-Integrating Nature of Most Ad Serotypes

As outlined above, EN expression is only needed for a short time period. Therefore, viral vectors that do not integrate into the host genome are the vectors of choice for EN gene delivery. Stable integration of Ad DNA into the host genome has been reported only for wild type forms of specific Ad serotypes, for example, Ad12, and appears not to occur in a detectable manner with the Ad5 vectors widely used for gene transfer *in vitro* and *in vivo* [31]. When sequences with homology to chromosomal DNA are inserted into Ad5 vector genomes, chromosomal Ad vector integration frequencies of only 10^{-5} to 10^{-7} were reported, with a large fraction of integration involving homologous recombination [32,33]. To minimize the integration of HDAd vectors, stuffer DNA from scrambled human DNA fragments was used to avoid large stretches of homology with the human genome [34].

6.2.3 Insert Capacity

The packaging capacity of rAAV vectors or lentivirus vectors limits the ability of incorporating ZFNs and TALENs which have a length of 5 to 6 kb [35]. Notably, two ZFN and TALEN subunits are required for DNA cleavage. Furthermore, for a number of applications, for example inactivation of human immunodeficieny virus (HIV) genomes, for rearrangement/deletion of chromosomal regions, or for correction of multigenic diseases, the simultaneous expression of multiple ENs is desirable. The size limitation of rAAV or lentivirus vectors is not as acute for the CRSIPR/Cas9 system. With about 4.2 kb, the cDNA of the most commonly used Cas9 is small enough to fit into these vectors as long as a short Pol II promoter is used to drive Cas9 expression (for review: [36]).

HDAds can accommodate transgene cassettes of up to 36 kb. For example, we generated an HDAd vector that contained about 27 kb of the human globin locus control region [37]. This large packaging capacity is also relevant for using HDAd vectors as donor vectors. It is thought that the efficiency of homologous recombination directly correlates with the length of the homology region. This has also been demonstrated in gene correction studies with HDAd vectors [38,32,39]. A drawback of using large homology regions for gene targeting in patient cells is that this requires the sequencing of the target loci in patient cells to exclude small-nucleotide polymorphisms that can decrease the efficacy of HDR. Cathomen's group was the first to use second-generation Ad vectors as donors in combination with CRISPR/Cas9-induced HDR [25]. They compared the specificity and accuracy with non-integrating lentivirus vector donor templates. They found that lentiviral vectors showed substantial random integration outside the targeted region. In contrast, HDAd delivery of the repair template yielded a more precise insertion profile. One possible explanation for this reduced promiscuity is that the ends of adenoviral genomes are capped with a terminal protein, TP [40], which may favor HDR over random integration of linear DNAs with free ends. On the other hand, other studies have shown that HDR is more efficient when the donor template contains free DNA ends [17]. Specifically, it has been shown that *in vivo* cleavage of donor DNA plasmids promotes EN-mediated targeted integration [16].

6.2.4 CAPSID MODIFICATION TO IMPROVE TROPISM TO TARGET CELLS

The most commonly used Ad vectors belong to species C serotype 5 Ads (Ad5). Vectors of this serotype utilize the coxsackie–adenovirus receptor (CAR) as the receptor for primary attachment to target cells. On human HSCs, CAR is only marginally expressed [41,42], making these cells virtually refractory to Ad5 transduction. Similarly, low expression levels were observed on MSCs [43,44]. In order to allow transduction of these target populations, the tropism of the Ad vector has to be modified. For this, several ways can be envisioned: switching the serotype of the whole virus or of just the fiber shaft and knob, the insertion of a binding motif into the fiber knob, or the coupling of the vector to a targeting molecule through a bridging molecule (for review: [45]).

While Ad5 vectors use CAR for attachment to target cells, species B Ads utilize either desmoglein 2 (DSG2) [46] or CD46 [47] as primary receptors. While CAR expression on HSCs and MSCs is only marginal, expression levels of DSG2 and CD46 are more promising. Both HSCs and MSCs have been shown to express CD46 [48,49,29], while HSCs also have been shown to express DSG2 [50]. In accordance with this, HSCs can be readily transduced by members of the group B adenoviruses targeting either DSG2 [51] or CD46 [30], namely Ad35, Ad11, and Ad3. In order to be able to harness both, the stem cell transduction potential of serotypes 3 and 35 and the well understood vector system represented by Ad5, fiber-chimeric adenovirus vectors can be employed. To generate these vectors, Ad5 is equipped with either just the fiber knob or the fiber knob and shaft of other serotypes. This has been accomplished for several serotypes resulting in Ad5/3 [52], Ad5/35 [30], Ad5/11 [53], and Ad5/50 [54] fiber-chimeric vectors. Consequently, work done by us and others demonstrated that fiber-chimeric Ad5/3 and Ad5/35 vectors could be used to efficiently transduce HSCs of both humans and nonhuman primates as well as human MSCs [51,55,56]. Specifically, Ad5/35 vectors have been widely used for gene transfer into human HSCs [57,58,29,30,42] and human lymphocytes [59–65]. Ad5/35 vectors have also been employed in clinical trials to treat viral infections in immunocompromised individuals [66,64].

With regard to iPS cell transduction, Mitani's group was the first to demonstrate the suitability of HDAd5 vectors for gene targeting purposes in iPS cells [67,32].

More recently, fiber-chimeric (Ad5/35) HDAd vectors found application for gene targeting in iPS cells [68–70]. Notably, iPS cells in compact colonies are difficult to transduce. In most of the studies listed above, iPS cell colonies were dissociated with trypsin and single cells were transduced in the presence of ROCK inhibitor Y-27632 to prevent apoptosis.

6.2.5 *In Vivo* Application

The ability of cost-efficient production of high yields of Ad vectors was one of the reasons why this vector system was widely used for systemic application in humans, specifically for virotherapy of cancer. *In vivo* Ad vector application is also relevant for targeting HSCs and MSCs. A major disadvantage of Ad5-based gene therapy vectors *in vivo* stems from its interaction with the liver upon systemic application as well as innate immune responses to the vector. Upon systemic injection of Ad5 vectors, up to 98% of injected vector will be sequestered by the liver within 30 minutes of virus injection [71]. The majority of virus is taken up by liver macrophages, that is, Kupffer cells. This uptake of viral particles not only leads to a loss of massive amounts of vector in this "sink" but also causes destruction of Kupffer cells [72] and a massive inflammatory response [73]. Another part of virus lost in the liver is taken up by hepatocytes. Hepatocyte transduction after intravenous Ad5 vector injection involved the interaction of the Ad capsid protein hexon with vitamin K-dependent coagulation factors, for example factor X (FX), and the uptake of this complex by cellular heparan sulfate proteoglycans (HSPGs) [74,75]. Recent research however suggests the existence of alternative entry pathways for Ad5–FX complexes [76]. In addition to the liver acting as a significant virus sink, it has also been shown that erythrocytes express CAR leading to further loss of Ad5-based vectors after intravascular injection [77].

In order to ameliorate both mentioned problems, that is, the loss of virus through sinks throughout the body as well as innate immune responses, a fiber-chimerism strategy can be pursued. On one hand, replacement of the CAR-interacting Ad5 fiber knob with species B knobs will abrogate the interaction of the vector with erythrocytes in the blood stream. To reduce liver uptake of the virus, focusing on the fiber shaft rather than the knob appears to be effective. In a study comparing short- and long-shafted Ad vectors with Ad9 (CAR-tropic) and Ad35 (CD46-tropic) knobs, it was shown that the binding of FX to short-shafted vectors was compromised, most likely due to a sterical block of the FX-interacting domains within the Ad5 hexon [78]. Furthermore, cytokine responses and resulting liver damage were much milder in animals injected with the short-shafted Ads [79]. Along the same line, another study found that short-shafted Ads were not efficiently taken up by Kupffer cells, most likely contributing to the milder innate immune responses toward the vector [80].

In recent research from our group, we compared the influence of the fiber shaft length in the context of DSG2-targeting Ad5/3 fiber-chimeric vectors. In accordance with other studies, the short-shafted Ad5/3S vector was not able to interact with FX to increase HSPG-mediated transduction of cells *in vitro*. In contrast, long-shafted Ad5/3L transduction was greatly improved through the addition of FX (Figure 6.1a).

When both vectors were administrated systemically *in vivo* in a DSG2-transgenic mouse model [50], a marked decrease in liver transduction could be observed when the long fiber shaft was replaced with a shorter version (Figure 6.1b and c). In the context of using Ad5/3 and Ad5/35 vectors for *in vivo* HSC transduction it is notable that both receptors in normal epithelial tissues are mostly trapped epithelial junctions and therefore not accessible to Ad vector transduction [50].

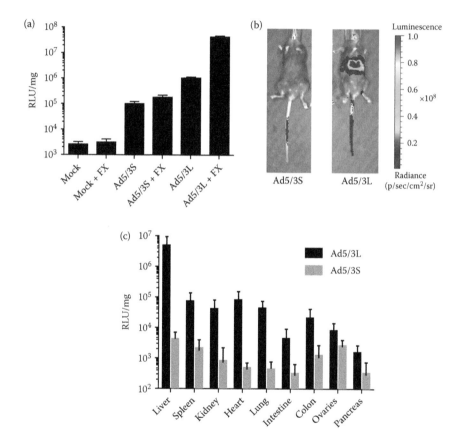

FIGURE 6.1 Influence of Ad fiber length on blood factor interaction and liver uptake. (a) HSPG-expressing CHO-K1 cells were infected with long-shafted Ad5/3L or short-shafted Ad5/3S in the presence or absence of FX *in vitro*. The long fiber shaft of Ad5/3L allows for interaction with FX and increased cell transduction through uptake of virus–FX complexes. This interaction is not possible in the context of a short virus fiber, therefore preventing increased transduction in the presence of FX for Ad5/3S. (b) *In vivo* imaging of DSG2-transgenic mice injected with luciferase-expressing Ad5/3 carrying a short or a long fiber. For the short-shafted Ad5/3S, liver uptake is clearly diminished. (c) Biodistribution of luciferase-expressing Ad5/3S and Ad5/3L after systemic administration. Tissues were collected and luciferase expression in tissue lysates was analyzed. The liver transduction of Ad5/3S was clearly lower than for Ad5/3L. However, transduction levels in other tissues were also lower compared to the long-shafted variant.

6.2.6 COST OF VECTOR MANUFACTURING

A major obstacle for a widespread clinical application of gene therapy is the cost associated with clinical grade vector production. This is particularly the case for the production of rAAV [81] and lentivirus vectors [82], which require, for each new vector preparation, large-scale plasmid transfection or more recently, transduction with baculovirus vectors that express corresponding viral proteins. On the other hand, once an Ad vector preparation is generated and characterized, it can be used as a seed stock for further amplification. The relatively low costs for producing high yields of Ad vectors are particularly relevant for *in vivo* gene therapy. Based on our experience, the costs for laboratory-grade HDAd vector is about $2000 per 1×10^{13} viral genomes. For intravenous injection this would correspond to a cost of $10–$20 per adult mouse. In comparison, for *in vivo* delivery of rAAV, $0.5 - 1 \times 10^{13}$ viral genomes per adult mouse are routinely used, which would be a cost in the range of $500–$1000 per mouse (Jeff Chamberlain, personal communication).

6.3 ADENOVIRUS VECTORS FOR GENOME EDITING—DISADVANTAGES

6.3.1 CYTOTOXICITY OF FIRST-GENERATION VECTORS IN STEM CELLS

In the past, most Ad vector transduction studies of HSCs, MSCs, and iPS cells were done with first-generation (E1/E3-deleted) Ad vectors. Despite the absence of trans-activating E1 gene products, first-generation vectors express low levels of early (E2A and E4) and late (pIX, fiber, and hexon) genes in transduced cells, which is associated with cytotoxicity, especially in primary cells [83,84]. We and others have shown that upon transduction of CD34+ cells with first-generation Ad5/35 vectors, the ability to form progenitor colonies is greatly diminished [57,85]. Toxicity related to leaky viral gene expression from first-generation Ad vectors can be circumvented by the use of helper-dependent (HD) Ad5/35 vectors that lack all viral genes [38,86,87,37]. We and others have shown that HDAd5/35 vectors efficiently transduce human CD34+ cells *in vitro* without signs of cytotoxicity [87,88,37].

6.3.2 LOSS OF EN EXPRESSION CASSETTES FROM VIRAL GENOMES DURING Ad VECTOR PRODUCTION

During Ad vector production, transgene products are produced at massive amounts in Ad producer cells. This is a problem if the transgene product is potentially cytotoxic. High levels of EN expression are poorly tolerated in Ad producer 293 cells, which prevent the rescue of vectors or selects for recombined vector genomes and deletion of EN expression cassettes. Early production of EN-expressing first-generation Ad vectors often involved either the separation of the two EN subunits into two separate vectors which are then co-infected to reconstitute EN activity [89] or the suppression of EN expression in 293 cells using inducible systems [64,55]. Among our early attempts to produce CCR5 ZFN-expressing HDAd vectors was a vector that allowed for Tet-inducible transgene expression using a fusion of the Krüppel-associated box (KRAB) domain and the tetracycline repressor. We produced GFP-expressing HDAd5/35 vectors and showed that background expression in 293 cells with Tet induction was suppressed [85]. However, when we replaced that GFP gene with the CCR5 ZFN gene, the resulting HDAd genomes isolated from purified particles demonstrated genomic rearrangements and a deletion of parts of the ZFN cassette. This supports the conclusion that Ad genomes encoding a pair of EN sequences may be unstable, a notion which is further supported by the existence of two identical copies of the *Fok*I gene as well as several DNA binding motifs in the ZFN or TALEN vector genomes, which can serve as substrates for intramolecular vector genome recombination. Notably, Ad replicates through a single-stranded DNA intermediate which enables efficient intramolecular recombination between repeated sequences in the viral genome [40,90].

The production of these HDAd vectors required that ZFN and TALEN expression in HDAd producer 293-Cre or 116 cells was suppressed. To generate HDAd5/35 vectors that would express ENs in CD34+ cells, we used a micro-RNA (miRNA)-regulated gene expression system [91]. If the mRNA of a transgene contains a target site for an miRNA that is expressed at high levels in a given cell type, the mRNA will be degraded and transgene expression avoided in this cell type. The power of this approach has been documented in a number of recent studies [92,93]. Using miR-183-5p and miR-218-5p based regulation of transgene gene expression, we successfully produced HDAd5/35++ vectors-expressing ZFNs and TALENs [91].

Maggio et al. [94] expressed the Cas9 and gRNA (against *AAVS1* or an integrated *gfp* gene) from two separate adenoviral vectors. However, more recent studies used first-generation Ad vectors that contained both a Pol-III-driven sRNA cassette and a Pol-II-driven Cas9 cassette (all-in-one vector), specifically for *in vivo* application [95–99]. We succeeded in generating a functional HDAd5/35-CRISPR/Cas9 vector without the use of the miRNA regulation system. We found however that the production yield of this vector was significantly lower compared to a corresponding vector in which the CRISPR/Cas9 was under miR183/218 control (KS unpublished). This indicates that the problem of EN transgene cassette rearrangements during Ad vector production appears to be less pronounced for the CRISPR/Cas9 system. There are also reports suggesting that TALENs are better tolerated in mammalian cells than ZFNs [100,101].

6.3.3 CYTOTOXICITY ASSOCIATED WITH EXTENDED EN EXPRESSION

Short-term EN expression causes permanent genetic modifications that are carried on to progeny cells, regardless of continued vector presence. Considering that extended EN expression may increase the risk of cytotoxicity and off-site cleavage, it may be safest for many applications, specifically therapeutic use in humans, if viral EN vectors did not integrate and the EN was expressed only for a short time period. At this point, this is best achieved by electroporated EN mRNA. However, this approach is not applicable for *in vivo* genome engineering. Electroporation is also associated with significant cytotoxicity, specifically in HSCs and iPS cells. As outline above this justifies the use of Ad-EN vectors.

While we showed efficient DNA cleavage with HDAd5/35 vectors-expressing ENs *in vitro* (Figure 6.2a), we also noted cytotoxicity upon vector transduction,

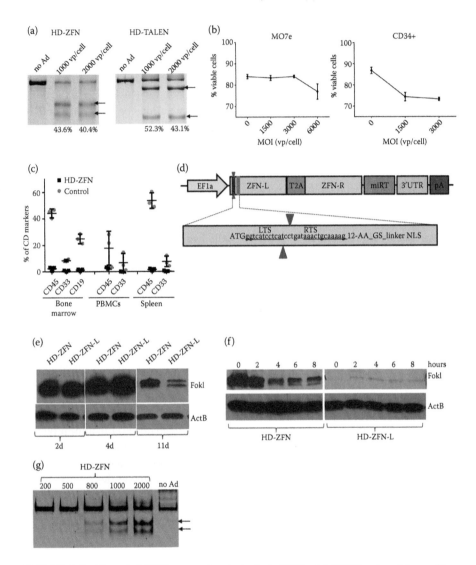

FIGURE 6.2 Extended ZFN expression is toxic to HSCs. (a) CCR5 mismatch sensitive assay in MO7e cells infected with a CCR5 ZFN-expressing HDAd5/35 vector (HD-ZFN (91)) (left panel) or a β-globin LCR-specific TALEN-expressing HDAd5/35 vector (HD-TALEN (91)). MO7e cells are a model HSC cell line (85). Genomic DNA was isolated and analyzed with T7E1 nuclease assay (91). Specific PCR products generated by ZFN cleavage are indicated by arrows. The percentage of ccr5 gene knockout is shown at the bottom. (b) Cytotoxicity in MO7e and CD34+ cells infected with HDAd5/35-ZFN. The percentage of viable cells was measured 48 hours after transduction at increasing MOIs by flow cytometry using an Annexin V/apoptosis kit. (c) Analysis of engraftment of human CD45+, CD33+, and CD19+ cells after transplantation of human CD34+ cells that were *ex vivo* transduced with an HDAd5/35-GFP vector (control) or with HD-ZFN at an MOI of 2000 vp/cell. Transduced cells were transplanted into irradiated immunodeficient NOG mice.   *(Continued)*

which was particularly pronounced in CD34+ cells (Figure 6.2b). We found that HD-ZFN transduction decreased the engraftment rate, survival, and/or expansion of CD34+ cells in NOG mice (Figure 6.2c). This was not due to HDAd5/35 transduction and vector-associated toxicity *per se*, because engraftment rates were higher with cells that were transduced with a GFP-expressing HDAd5/35 vector (Figure 6.2c). We therefore speculated that this is related to ZFN expression over an extended time period. Non-integrating HDAd vector genomes are lost after several rounds of cell division, however persist longer in nondividing cells such as hepatocytes [102]. Because HSCs are low proliferative, HDAd5/35 genomes could be maintained for longer time periods and thus express ENs.

Several techniques have been employed to limit EN expression levels and/or duration from transfected plasmids. Pruett-Miller et al. described two techniques for ZFN regulation that allowed for reduction of ZFN cytotoxicity while maintaining its DNA cleavage activity [103]. The authors first showed in 293 cells that ZFNs need to be only expressed for 32 hours after transfection to attain maximal gene targeting efficacy. To regulate ZFN half-life and the duration of expression they first destabilized ZFNs by linking an ubiquitin moiety to the N-terminus and regulated ZFN levels using a proteasome inhibitor. In a second approach, they destabilized ZFNs by linking a modified destabilizing FKBP12 domain to the N-terminus and regulated ZFN levels by using a small molecule that blocks the destabilization effect of the N-terminal domain [103].

To shorten nuclease expression, we designed and tested a novel autocatalytic loop expression system (Figure 6.2d). The ZFN DNA recognition sequence was fused in-frame to the N-terminus of the first ZFN subunit. If the ZFN is expressed in CD34+ cells, it cleaves its own gene inside the HDAd vector genome thereby inactivating it. Note that this will not occur during HDAd production in 293 cells because (EN) expression is suppressed in 293 cells through miR183/218 regulation [104]. Figure 6.2e shows that the ZFN is still detectable by Western blot 11 days after transduction of slow-proliferating MO7e cells. The half-life of *de novo* produced ZFN was determined to be 4 hours in MO7e cells (Figure 6.2f). These data underscore the problem

---

FIGURE 6.2 (Continued) Bone marrow, PBMCs, and spleen cells were harvested 6 weeks after transplantation and analyzed for lineage markers by flow cytometry. N = 3. (d) Schematic of the autocatalytic loop containing HD-ZFN-L vector. Left and right ZFN DNA target sequences (LTS, RTS) are inserted after the start codon at the N-terminus of the first ZFN subunit. The LTS and RTS sequence is followed by a glycine/serine linker and a nuclear localization signal (NLS). (e) Western blot for FokI after transduction of MO7e cells with HD-ZFN or HD-ZFN-L vector at an MOI of 400 vp/cell. Cells were exposed to HDAd for 2 hours and then washed to remove virus. Cells were harvested at day 2, 4, and 11 after transduction. β-actin (ActB) serves as a loading control. (f) The ZFN half-life was measured in HDAd treated MO7 cells in the presence of 200 μM cycloheximide to block *de novo* protein expression. MO7e cells were treated with HDAds at 400 vp/cell. Cells were infected with HDAd vectors and 2 days later cycloheximide was added (time point "0"). Cells were collected 2, 4, 6, and 8 hours later. (g) T7E1 nuclease mismatch sensitive assay in MO7e cells infected with HD-ZFN-L at different MOIs. Genomic DNA was analyzed 48 hours after infection.

to shorten ZFN expression. ZFN expression from the autocatalytic loop containing HD-ZFN-L vector was greatly attenuated while the ZFN cleavage activity was not compromised in MO7e cells (Figure 6.2g). However, a study with *ex vivo* transduced CD34+ cells that were transplanted into irradiated mice did not show an advantage of HD-ZFN-L over HD-ZFN with regard to improved engraftment rates or CCR5 knockout in engrafted cells, indicating that further control of ZFN expression is required. Zhang et al. [105] used a similar ZFN suicide strategy in the context of plasmids and showed lower ZFN-related toxicity in 293 cells.

Inducible expression systems could potentially be used to regulate EN expression. We and others however found that in the context Ad5/35 vectors doxycycline inducible Tet-on or autoregulated rtTA systems either displayed high levels of background expression without doxycycline (most likely due to activity of viral promoters present in the vector genome) or that the induction of the transgene was restricted to only a small subset of CD34+ cells [57,85].

6.4 ADENOVIRUS VECTOR FOR GENOME EDITING—EXAMPLES

Table 6.1 gives an overview on studies that used EN-expressing Ad vectors for genome editing.

6.4.1 EN-BASED ANTIVIRAL THERAPIES

The first published study on an Ad5/35 vector-expressing a ZFN for genome editing in primary human T-cells is from Perez et al. [64]. They demonstrated inhibition of HIV-1 infection in T-cells transduced *ex vivo* with a first-generation Ad5/35

TABLE 6.1

Overview of Studies That Used EN-Expressing Ad Vectors for Genome Editing

Ad Capsid	Ad Vector Type	EN Type	Targeted Gene	Cell Type	Reference
Ad5/F50	Second-generation	TALEN	AAVS1	AAVS1	54
Ad5/F35	Second-generation	CRISPR/Cas9, TALEN	AAVS1	HeLa	25
Ad5/F35	First-generation	ZFN	CCR5,CXCR4	HSCs, CD4+ T-cells	57,106,107
Ad5/F50	Second-generation	CRISPR/Cas9, TALEN	AAVS1	HeLa, U2OS, myoblasts, hMSCs	94
Ad5	First-generation	TALEN	D1Pas1, Ddx3x, Ddx3y	ratC6, mouse neura-2a	108
Ad5/F35	First-generation	CRISPR/Cas9	CCR5	CD4	109
Ad5	First-generation	CRISPR/Cas9	dystrophin	*In vivo* mouse skeletal muscle	99
Ad5	First-generation	CRISPR/Cas9	Pcsk9, Cebpα	*In vivo* mouse liver	95,110
Ad5/35	HDAd	ZFN, TALEN	CCR5, β-globin	HSCs, Mo7e	91
Ad5, Ad5/35	HDAd	ZFN, TALEN	β-globin	hiPSCs	111

vector expressing a *ccr5*-specific ZFN. The approach is currently being tested in clinical trials at U Penn and Sangamo Biosciences demonstrating good safety and the first promising efficacy data [106]. This therapy requires harvesting of peripheral blood mononuclear cells (PBMCs), *in vitro* culture, Ad5/35.ZFN transduction, and T-cell retransplantation. Efficient *ccr5* disruption in primary CD4+ T-cells was also achieved by an Ad5/35 vector expressing a *ccr5*-targeted CRISPR/Cas9 [109]. In this study, both the gRNA and Cas9 expression cassettes were incorporated into one Ad5/35 vector.

More recent attempts have focused on *ccr5*-knockout in HSCs. Li et al. [57] used a first-generation Ad5/35 vector that expressed a ccr5 ZFN. While efficient in T-cells, the Ad5/35 vector achieved less than 5% *ccr5* gene disruption in HSCs. Reasons for this poor efficacy could include suboptimal expression of ZFN from the promoter used (a minimal CMV promoter linked to tet-operator sequences) and cytotoxicity associated with the use of first-generation Ad5/35 vectors. A limitation of *ccr5*-targeted EN-based antiviral gene therapy is the risk of emergence of HIV strains that use the other HIV co-receptor CXCR4 [112]. A double-knockout of both co-receptors, CCR5 and CXCR4 using ZFNs, while being performed in T-cells [113,114], would most likely not be possible in HSCs because of the critical role of CXCR4 in HSC biology [115,116].

Other EN-based antiviral therapies involved ZFNs that directly target the viral genome, as shown by the Jerome's group for human hepatitis virus (HBV), Herpes simplex virus (HSV), and HIV with rAAV vectors [117,119]. The authors also found in their HIV studies that using a ZFN against only one target site can trigger the development of HIV escape mutants, implying that the use of multiple ENs against different sites may be needed to prevent the emergence of resistance [118]. As noted above, the large insert capacity of HDAd vectors allows accommodating several EN-expression cassettes. In the case of ZFNs and TALENs this would require breaking up large stretches of DNA homology through the introduction of silent mutations. In this context, CRISPR studies using several different gRNAs (under different Pol III promoters) and one Cas9 unit in an all-in-one vector are also noteworthy [120].

6.4.2 Gene Targeting in Cell Lines and Immortalized Primary Cells

Major contributions to the use of Ad vectors for genome editing in cell lines stem from Goncalves' and Cathomen's groups. Their early attempts focused on the production of first- and second-generation Ad5/50 TALEN vectors, whereby the two TALEN subunits were split up into two different Ad vectors (see Section 3.2) [54,25,89]. The TALEN used in their studies was directed against the AAVS1 site in the human genome, a site that is thought to be a "safe harbor", although this, so far, has not been proven in clinical trials [121]. They showed that Ad-TALEN vectors mediated efficient ds-DNA cleavage in HeLa cells and in immortalized human myoblasts. These studies also indicated that Ad vectors are superior to non-integrating lentivirus vectors in delivery of functional TALENs. More recently, they focused on Ad vectors expressing CRISPR/Cas9 specific to AAVS1 [25,122,94]. Again, in these studies, the gRNA and catalytic Cas9 units were split into two vectors. CRSIPR/Cas9 delivery with second-generation Ad5/50 vectors resulted in gene disruption frequencies in various cell types (HeLa, human osteosarcoma U2OS, human MSCs, and immortalized human myoblasts) ranging from 18% to 65% [94]. In these studies, no significant difference in the efficacy of HDR-directed AAVS1 site targeting between TALEN and CRISPR/Cas9 platforms was seen [25,94]. Notably the necessity for co-infection with two Ad vectors could be a limitation in situations when the multiplicity of infection (MOI) is low, for example after intravenous injection.

Also of note is the relative ease of producing Ad vector libraries [123,124]. This could theoretically be beneficial for high-throughput genome-wide screening with CRSIPR/Cas9 libraries to investigate the function of large gene numbers. This has, so far, only been done with lentivirus vector-based CRISPR/Cas9 libraries [125–128].

6.4.3 Genome Editing in iPS Cells to Introduce or Correct Disease Mutations

Given their role in regenerative research, iPS cells have become a crucial platform for disease modeling and potential applications in personalized medicine. Pivotal work on Ad-mediated genome editing in iPS cells was done by the Izpisua Belmonte group [68,69,70,111]. Their earlier work demonstrated the utility of HDAd vectors in introducing mutations to specific genes (FANCA, HBB, and LMNA) in iPS cells without ENs. To increase the efficacy of gene correction, this group also employed TALEN-expressing HDAd5 and HDAd5/35 vectors in iPS cells in three different disease models [111]. They constructed two vectors each containing one of the TALEN subunits. Each vector also contained a donor homology sequence with silent mutations to avoid TALEN cleavage. Gene targeting rates were superior to those achieved by HDAd-donor vectors alone or those achieved with bacterial artificial chromosome (BAC) transgenesis. Genome analysis in modified iPS cells revealed a very low rate of off-target effects, which is important for therapeutic applications. Notably, Mitani et al. [129] also showed that gene targeting in iPS cells can be achieved with HDAd vectors without ENs. However, the frequency of this approach was low and required drug selection of gene corrected clones.

6.4.4 Genome Editing in HSCs

Gene targeting in HSCs remains challenging, given the rare nature of the cells, the low frequency of dsDNA break repair, and low transduction efficiency. As outlined above, efficient transduction of HSCs without toxicity can be achieved with HDAd vectors that target CD46, for example chimeric vectors containing Ad35, Ad11, or Ad50 fibers. We have shown that transduction of CD34+ cells with HDAd5/35 vectors-expressing GFP or β-globin did not decrease their ability to form multi-lineage progenitor colonies or engraftment in irradiated recipients [91,85,37]. HDAd5/35 vectors or Ad5/35 vectors expressing a *ccr5* ZFN mediated efficient target site cleavage in CD34+ cells [57,91].

The relatively low activity of the cellular DNA repair and recombination machinery in HSCs (which are, *per se,* quiescent) is a major limitation of gene targeting strategies involving nonhomologous-end-joining or homologous recombination [130]. For example, Naldini's group found that HDR-mediated gene targeting using lentivirus vectors was 20-fold lower in primitive HSCs than in more differentiated cells [10]. They were able to increase gene targeting in HSCs by stimulating cell cycle progression through prolonged incubation of cells with tailored cytokine cocktails under conditions that inhibited differentiation.

While HDAd5/35 vectors were used for gene addition in HSCs without the use of ENs [37,131], we are not aware of studies involving HDR stimulated by ENs expressed from Ad vectors.

6.4.5 Genome Editing in MSCs

MSCs can be found in adipose tissue and bone marrow. Their ability to differentiate into various tissue types as well as to migrate to tumors or other inflammatory sites makes them a relevant target for *ex vivo* or *in vivo* genome editing. The vast majority of published studies involve genetically modified MSCs to express proteins capable of attenuating tumor growth or stimulating antitumor immune responses (for example, interferon β, TRAIL, or HSV thymidine kinase) [132,134]. An advantage of using MSCs is that, due to low expression of HLA MHC-I molecules, MSCs can, to a certain degree, escape immune surveillance and destruction by T-cells specific to the transgene product. We found only one example for targeted gene addition to HSCs using Ad-ENs [135]. Benabdallah et al. [136] used a *ccr5*-specific ZFN to target the integration of an erythropoietin expression cassette to ccr5 in MSCs. The authors used the same approach for targeted dystrophin gene addition in human myoblasts.

6.4.6 In Vivo Genome Editing

To achieve efficient *in vivo* transduction after systemic application, high vector doses are required (based on our experience, $2-5 \times 10^{10}$ viral genomes per adult mouse for Ad vectors and $0.5-1 \times 10^{13}$ viral genomes per mouse for rAAV vectors). As outlined above, compared to other viral gene transfer vectors, Ad vectors are superior for *in vivo* application due to the ease and relatively low costs of vector production.

In vivo genome engineering using EN-expressing Ads has been used in mice to create disease models or correct mutations.

Liver: Three studies capitalized on liver transduction of intravenously injected Ad5 vectors. Wang et al. [98] used an all-in-one, first-generation Ad5 vector to deliver a CRISPR/Cas9 specific to *Pten* to the liver after intravenous injection. Pten is a negative regulator of the PI3K-AKT pathway and, in the liver, its mutation is involved in the development of nonalcoholic steatohepatitis (NASH). The authors showed site-specific gene knockout frequencies in the liver in the range of 20%. Four months after vector infusion, mice that received the *Pten* gene-editing Ad vector showed massive hepatomegaly and features of NASH. This study also noted humoral and cellular immune responses against Cas9 as well as transaminitis. Cheng et al. [95] used an all-in-one, first-generation Ad5 to express a CRISPR/Cas9 against the transcription factor *Cebpα*, a gene that the authors previously identified in a proteome screen to be critical for hepatocyte function. They found a gene knockout frequency of 90% in the liver after intravenous Ad-EN injection. Efficient

Cebpα knockout was validated by detecting dysregulation of selected *Cebpα* targets. Further analyses revealed no off-target activity and confirmed stable editing even under conditions of forced liver regeneration. As expected from a first-generation Ad vector, Cas9 expression dropped within a few days after injection due to an immune-mediated clearance of transduced cells. Notably, T-cell responses against Ad transduced hepatocytes can vary between different mouse strains [137]. Ding et al. [110] used a similar approach to knockout PCSK9, an enzyme produced in and secreted from the liver which antagonizes the low-density lipoprotein (LDL) receptor. Remarkably, within 1 week after vector administration, mice displayed up to 50% on-target gene editing, an increase in LDL-receptor, as well as reductions in plasma PCSK9 and lipoprotein.

Muscle: Xu et al. [99] generated all-in-one, first-generation Ad5 vectors to deliver two different CRISPR/Cas9 nucleases specific to different sites in the mouse dystrophin gene to muscle in *mdx* mice after intramuscular injection. They showed that CRISPR/Cas9-mediated genome editing efficiently excised a 23-kb genomic region on the X-chromosome covering the mutant exon 23 in a mouse model of Duchenne muscular dystrophy, and restored dystrophin expression and the dystrophin–glycoprotein complex in the sarcolemma of skeletal muscles in live *mdx* mice. This study is an impressive example for the *in vivo* use of Ad-Ens to remove deleterious mutations.

Lung: Maddalo et al. [96] demonstrated *in vivo* engineering of oncogenic chromosomal rearrangements with CRISPR/Cas9-expressing Ad vectors. Their goal was to generate a lung cancer model through creating *Eml4-Alk* fusions, an intrachromosomal inversion recurrently found in non-small-cell lung cancers (NSCLC). They used two Pol-III promoters to drive the expression of the two anti-*Eml4* or anti-*Alk* gRNAs on an all-in-one Ad5 vector. Starting at 1 month after intratracheal vector application, lungs of treated mice invariably displayed signs of hyperplasia, up to bilateral tumors showing histopathological and molecular features typical of human NSCLCs.

Problem—EN Immunogenicity: The studies by Maddalo et al. [96] and Xu et al. [99] demonstrate the power of *in vivo* application of Ad-EN vectors to create chromosomal deletions or rearrangements. Notably, these approaches did not require a donor vector and an HDR. All of the above listed studies used first-generation vectors, which bears the risk of losing gene corrected cells due to viral gene expression from Ad-transduced cells and subsequent anti-Ad T-cell responses. This problem can be circumvented by the use of HDAd vectors. A more critical obstacle to *in vivo* genome engineering with Ad-EN vectors is the immunogenicity of the bacterial EN subunits, that is, *Fok*I and Cas9, specifically in low proliferative cells (hepatocytes, myocytes, etc.) in which the episomal Ad vector genome is not rapidly lost due to cell division. As outlined above, approaches to restrict the duration of EN expression might lessen the problem of *Fok*1 or Cas9 immunogenicity. Theoretically, recently developed plant-based oral tolerance approaches could be applied to induce tolerance to bacterial ENs [138]. Furthermore immunosuppressive approaches, including approaches that have been tested clinically, for example, pentostatin, cyclophosphamide, prednisolone [139,140], or are currently under development [141,142], can be considered.

Notably, EN immunogenicity is not an acute problem in approaches that involve EN genome edited HSCs that were transplanted after *ex vivo* transduction. The genetic modification of only a fraction of HSCs can result in a microchimerism with regard to transgene product production in the hematopoietic system and can establish peripheral tolerance, most likely through generation of tolerogenic T-cells in the thymus [143,144].

Problem—Efficacy: The above discussed studies also indicate that the efficacy of current Ad-EN based approaches for therapeutic *in vivo* genome editing may not be high enough to achieve a phenotypic cure of a genetic disease. In our opinion, attempts to incorporate both homology donor sequences and EN expression cassettes into one HDAd vector are required to increase the efficacy of *in vivo* genome engineering using this vector system.

Problem—EN safety: The safety of systemically injected Ad-ENs has, to our knowledge, not been addressed so far in published studies. Even with tropism-optimized Ad vectors, it is not likely that only the target cell and tissues types are transduced and genetically modified. Therefore, future studies should focus on detailed analysis of Ad-EN induced genomic rearrangements in off-target tissues, specifically germ line tissues.

ACKNOWLEDGMENTS

We thank Jeff Chamberlain (University of Washington) for providing valuable information on rAAV production. We also thank Roma Yumul for editing the manuscript.

REFERENCES

1. Krause, D. S., N. D. Theise, M. I. Collector, O. Henegariu, S. Hwang, R. Gardner, S. Neutzel, and S. J. Sharkis. 2001. Multi-organ, multi-lineage engraftment by a single bone marrow-derived stem cell. *Cell*, **105**:369–377.
2. Notta, F., S. Doulatov, E. Laurenti, A. Poeppl, I. Jurisica, and J. E. Dick. 2011. Isolation of single human hematopoietic stem cells capable of long-term multilineage engraftment. *Science*, **333**:218–221.
3. Naldini, L. 2015. Gene therapy returns to centre stage. *Nature*, **526**:351–360.
4. Boissel, S., J. Jarjour, A. Astrakhan, A. Adey, A. Gouble, P. Duchateau, J. Shendure et al. 2014. megaTALs: A rare-cleaving nuclease architecture for therapeutic genome engineering. *Nucl Acids Res*, **42**:2591–2601.
5. Kim, H. and J. S. Kim. 2014. A guide to genome engineering with programmable nucleases. *Nat Rev Genet*, **15**:321–334.
6. Silva, G., L. Poirot, R. Galetto, J. Smith, G. Montoya, P. Duchateau, and F. Paques. 2011. Meganucleases and other tools for targeted genome engineering: Perspectives and challenges for gene therapy. *Curr Gene Ther*, **11**:11–27.
7. Gaj, T., C. A. Gersbach, and C. F. Barbas, 3rd. 2013. ZFN, TALEN, and CRISPR/Cas-based methods for genome engineering. *Trends Biotechnol*, **31**:397–405.
8. Wright, A. V., J. K. Nunez, and J. A. Doudna. 2016. Biology and applications of CRISPR systems: Harnessing nature's toolbox for genome engineering. *Cell*, **164**:29–44.
9. Osborn, M. J., B. R. Webber, F. Knipping, C. L. Lonetree, N. Tennis, A. P. DeFeo, A. N. McElroy et al. 2015. Evaluation of TCR gene editing achieved by TALENs, CRISPR/Cas9, and megaTAL nucleases. *Mol Ther*, **24**:570–581.

10. Genovese, P., G. Schiroli, G. Escobar, T. Di Tomaso, C. Firrito, A. Calabria, D. Moi et al. 2014. Targeted genome editing in human repopulating haematopoietic stem cells. *Nature*, **510**:235–240.

11. Holt, N., J. Wang, K. Kim, G. Friedman, X. Wang, V. Taupin, G. M. Crooks et al. 2010. Human hematopoietic stem/progenitor cells modified by zinc-finger nucleases targeted to CCR5 control HIV-1 *in vivo*. *Nat Biotechnol*, **28**:839–847.

12. Pelascini, L. P. and M. A. Goncalves. 2014. Lentiviral vectors encoding zinc-finger nucleases specific for the model target locus HPRT1. *Methods Mol Biol*, **1114**:181–199.

13. Pelascini, L. P., I. Maggio, J. Liu, M. Holkers, T. Cathomen, and M. A. Goncalves. 2013. Histone deacetylase inhibition rescues gene knockout levels achieved with integrase-defective lentiviral vectors encoding zinc-finger nucleases. *Hum Gene Ther Methods*, **24**:399–411.

14. Ul Ain, Q., J. Y. Chung, and Y. H. Kim. 2015. Current and future delivery systems for engineered nucleases: ZFN, TALEN and RGEN. *J Control Release*, **205**:120–127.

15. Carbery, I. D., D. Ji, A. Harrington, V. Brown, E. J. Weinstein, L. Liaw, and X. Cui. 2010. Targeted genome modification in mice using zinc-finger nucleases. *Genetics*, **186**:451–459.

16. Cristea, S., Y. Freyvert, Y. Santiago, M. C. Holmes, F. D. Urnov, P. D. Gregory, and G. J. Cost. 2013. *In vivo* cleavage of transgene donors promotes nuclease-mediated targeted integration. *Biotechnology and Bioengineering*, **110**:871–880.

17. Wang, J., G. Friedman, Y. Doyon, N. S. Wang, C. J. Li, J. C. Miller, K. L. Hua et al. 2012. Targeted gene addition to a predetermined site in the human genome using a ZFN-based nicking enzyme. *Genome Res*, **22**:1316–1326.

18. Wefers, B., M. Meyer, O. Ortiz, M. Hrabe de Angelis, J. Hansen, W. Wurst, and R. Kuhn. 2013. Direct production of mouse disease models by embryo microinjection of TALENs and oligodeoxynucleotides. *Proc Natl Acad Sci U S A*, **110**:3782–3787.

19. Zou, J., M. L. Maeder, P. Mali, S. M. Pruett-Miller, S. Thibodeau-Beganny, B. K. Chou, G. Chen et al. 2009. Gene targeting of a disease-related gene in human induced pluripotent stem and embryonic stem cells. *Cell Stem Cell*, **5**:97–110.

20. Chen, F., S. M. Pruett-Miller, and G. D. Davis. 2015. Gene editing using ssODNs with engineered endonucleases. *Methods Mol Biol*, **1239**:251–265.

21. Schumann, K., S. Lin, E. Boyer, D. R. Simeonov, M. Subramaniam, R. E. Gate, G. E. Haliburton et al. 2015. Generation of knock-in primary human T cells using Cas9 ribonucleoproteins. *Proc Natl Acad Sci U S A*, **112**:10437–10442.

22. Urnov, F. D., J. C. Miller, Y. L. Lee, C. M. Beausejour, J. M. Rock, S. Augustus, A. C. Jamieson et al. 2005. Highly efficient endogenous human gene correction using designed zinc-finger nucleases. *Nature*, **435**:646–651.

23. Lombardo, A., D. Cesana, P. Genovese, B. Di Stefano, E. Provasi, D. F. Colombo, M. Neri et al. 2011. Site-specific integration and tailoring of cassette design for sustainable gene transfer. *Nat Methods*, **8**:861–869.

24. Wang, J., C. M. Exline, J. J. DeClercq, G. N. Llewellyn, S. B. Hayward, P. W. Li, D. A. Shivak et al. 2015. Homology-driven genome editing in hematopoietic stem and progenitor cells using ZFN mRNA and AAV6 donors. *Nat Biotechnol*, **33**:1256–1263.

25. Holkers, M., I. Maggio, S. F. Henriques, J. M. Janssen, T. Cathomen, and M. A. Goncalves. 2014. Adenoviral vector DNA for accurate genome editing with engineered nucleases. *Nat Methods*, **11**:1051–1057.

26. Glimm, H., I. H. Oh, and C. J. Eaves. 2000. Human hematopoietic stem cells stimulated to proliferate *in vitro* lose engraftment potential during their S/G(2)/M transit and do not reenter G(0). *Blood*, **96**:4185–4193.

27. Hitt, M. M., C. L. Addison, and F. L. Graham. 1997. Human adenovirus vectors for gene transfer into mammalian cells. *Adv Pharmacol*, **40**:137–206.

28. Lieber, A., M. A. Kay, and Z. Y. Li. 2000. Nuclear import of moloney murine leukemia virus DNA mediated by adenovirus preterminal protein is not sufficient for efficient retroviral transduction in nondividing cells. *J Virol*, **74**:721–734.

29. Nilsson, M., S. Karlsson, and X. Fan. 2004. Functionally distinct subpopulations of cord blood CD34+ cells are transduced by adenoviral vectors with serotype 5 or 35 tropism. *Mol Ther*, **9**:377–388.

30. Shayakhmetov, D. M., T. Papayannopoulou, G. Stamatoyannopoulos, and A. Lieber. 2000. Efficient gene transfer into human CD34(+) cells by a retargeted adenovirus vector. *J Virol*, **74**:2567–2583.

31. Doerfler, W. 1993. Adenoviral DNA integration and changes in DNA methylation patterns: A different view of insertional mutagenesis. *Prog Nucl Acid Res Mol Biol*, **46**:1–36.

32. Ohbayashi, F., M. A. Balamotis, A. Kishimoto, E. Aizawa, A. Diaz, P. Hasty, F. L. Graham, C. T. Caskey, and K. Mitani. 2005. Correction of chromosomal mutation and random integration in embryonic stem cells with helper-dependent adenoviral vectors. *Proc Natl Acad Sci U S A*, **102**:13628–13633.

33. Wang, Q. and M. W. Taylor. 1993. Correction of a deletion mutant by gene targeting with an adenovirus vector. *Mol Cell Biol*, **13**:918–927.

34. Sandig, V., R. Youil, A. J. Bett, L. L. Franlin, M. Oshima, D. Maione, F. Wang, M. L. Metzker, R. Savino, and C. T. Caskey. 2000. Optimization of the helper-dependent adenovirus system for production and potency *in vivo*. *Proc Natl Acad Sci U S A*, **97**:1002–1007.

35. Grieger, J. C. and R. J. Samulski. 2005. Packaging capacity of adeno-associated virus serotypes: Impact of larger genomes on infectivity and postentry steps. *J Virol*, **79**:9933–9944.

36. Schmidt, F. and D. Grimm. 2015. CRISPR genome engineering and viral gene delivery: A case of mutual attraction. *Biotechnol J*, **10**:258–272.

37. Wang, H., D. M. Shayakhmetov, T. Leege, M. Harkey, Q. Li, T. Papayannopoulou, G. Stamatoyannopolous, and A. Lieber. 2005. A capsid-modified helper-dependent adenovirus vector containing the beta-globin locus control region displays a nonrandom integration pattern and allows stable, erythroid-specific gene expression. *J Virol*, **79**:10999–11013.

38. Balamotis, M. A., K. Huang, and K. Mitani. 2004. Efficient delivery and stable gene expression in a hematopoietic cell line using a chimeric serotype 35 fiber pseudotyped helper-dependent adenoviral vector. *Virology*, **324**:229–237.

39. Suzuki, K., K. Mitsui, E. Aizawa, K. Hasegawa, E. Kawase, T. Yamagishi, Y. Shimizu, H. Suemori, N. Nakatsuji, and K. Mitani. 2008. Highly efficient transient gene expression and gene targeting in primate embryonic stem cells with helper-dependent adenoviral vectors. *Proc Natl Acad Sci U S A*, **105**:13781–13786.

40. van der Vliet. 1995. Adenovirus DNA replication. In: a. P. B. W. Doerfler (ed.) *The Molecular Repertoire of Adenoviruses*, Springer-Verlag, Berlin, vol. 2., p. 1–31.

41. Rebel, V. I., S. Hartnett, J. Denham, M. Chan, R. Finberg, and C. A. Sieff. 2000. Maturation and lineage-specific expression of the coxsackie and adenovirus receptor in hematopoietic cells. *Stem Cells*, **18**:176–182.

42. Yotnda, P., H. Onishi, H. E. Heslop, D. Shayakhmetov, A. Lieber, M. Brenner, and A. Davis. 2001. Efficient infection of primitive hematopoietic stem cells by modified adenovirus. *Gene Ther*, **8**:930–937.

43. Olmsted-Davis, E. A., Z. Gugala, F. H. Gannon, P. Yotnda, R. E. McAlhany, R. W. Lindsey, and A. R. Davis. 2002. Use of a chimeric adenovirus vector enhances BMP2 production and bone formation. *Hum Gene Ther*, **13**:1337–1347.

44. Suzuki, T., K. Kawamura, Q. Li, S. Okamoto, Y. Tada, K. Tatsumi, H. Shimada, K. Hiroshima, N. Yamaguchi, and M. Tagawa. 2014. Mesenchymal stem cells are efficiently transduced with adenoviruses bearing type 35-derived fibers and the transduced cells with the IL-28A gene produces cytotoxicity to lung carcinoma cells co-cultured. *BMC Cancer*, **14**:713.

45. Yamamoto, M. and D. T. Curiel. 2010. Current issues and future directions of oncolytic adenoviruses. *Mol Ther*, **18**:243–250.

46. Wang, H., Z. Y. Li, Y. Liu, J. Persson, I. Beyer, T. Moller, D. Koyuncu et al. 2011. Desmoglein 2 is a receptor for adenovirus serotypes 3, 7, 11 and 14. *Nat Med*, **17**:96–104.

47. Gaggar, A., D. Shayakhmetov, and A. Lieber. 2003. CD46 is a cellular receptor for group B adenoviruses. *Nature Medicine*, **9**:1408–1412.

48. Manchester, M., K. A. Smith, D. S. Eto, H. B. Perkin, and B. E. Torbett. 2002. Targeting and hematopoietic suppression of human CD34+ cells by measles virus. *J Virol*, **76**:6636–6642.

49. Mizuguchi, H., T. Sasaki, K. Kawabata, F. Sakurai, and T. Hayakawa. 2005. Fiber-modified adenovirus vectors mediate efficient gene transfer into undifferentiated and adipogenic-differentiated human mesenchymal stem cells. *Biochem Biophys Res Commun*, **332**:1101–1106.

50. Wang, H., I. Beyer, J. Persson, H. Song, Z. Li, M. Richter, H. Cao et al. 2012. A new human DSG2-transgenic mouse model for studying the tropism and pathology of human adenoviruses. *J Virol*, **86**:6286–6302.

51. Tuve, S., H. Wang, C. Ware, Y. Liu, A. Gaggar, K. Bernt, D. Shayakhmetov et al. 2006. A new group B adenovirus receptor is expressed at high levels on human stem and tumor cells. *J Virol*, **80**:12109–12120.

52. Krasnykh, V. N., G. V. Mikheeva, J. T. Douglas, and D. T. Curiel. 1996. Generation of recombinant adenovirus vectors with modified fibers for altering viral tropism. *J Virol*, **70**:6839–6846.

53. Stecher, H., D. M. Shayakhmetov, G. Stamatoyannopoulos, and A. Lieber. 2001. A capsid-modified adenovirus vector devoid of all viral genes: Assessment of transduction and toxicity in human hematopoietic cells. *Mol Ther*, **4**:36–44.

54. Holkers, M., T. Cathomen, and M. A. Goncalves. 2014. Construction and characterization of adenoviral vectors for the delivery of TALENs into human cells. *Methods*, **69**:179–187.

55. van Rensburg, R., I. Beyer, X. Y. Yao, H. Wang, O. Denisenko, Z. Y. Li, D. W. Russell et al. 2013. Chromatin structure of two genomic sites for targeted transgene integration in induced pluripotent stem cells and hematopoietic stem cells. *Gene Ther*, **20**:201–214.

56. Yu, D., C. Jin, M. Ramachandran, J. Xu, B. Nilsson, O. Korsgren, K. Le Blanc et al. 2013. Adenovirus serotype 5 vectors with Tat-PTD modified hexon and serotype 35 fiber show greatly enhanced transduction capacity of primary cell cultures. *PLoS One*, **8**:e54952.

57. Li, L., L. Krymskaya, J. Wang, J. Henley, A. Rao, L. F. Cao, C. A. Tran et al. 2013. Genomic editing of the HIV-1 coreceptor CCR5 in adult hematopoietic stem and progenitor cells using zinc finger nucleases. *Mol Ther*, **21**:1259–1269.

58. Lu, Z. Z., F. Ni, Z. B. Hu, L. Wang, H. Wang, Q. W. Zhang, W. R. Huang, C. T. Wu, and L. S. Wang. 2006. Efficient gene transfer into hematopoietic cells by a retargeting adenoviral vector system with a chimeric fiber of adenovirus serotype 5 and 11p. *Exp Hematol*, **34**:1171–1182.

59. Chen, Z., M. Ahonen, H. Hamalainen, J. M. Bergelson, V. M. Kahari, and R. Lahesmaa. 2002. High-efficiency gene transfer to primary T lymphocytes by recombinant adenovirus vectors. *J Immunol Methods*, **260**:79–89.

60. Cho, H. I., H. J. Kim, S. T. Oh, and T. G. Kim. 2003. *In vitro* induction of carcinoembryonic antigen (CEA)-specific cytotoxic T lymphocytes by dendritic cells transduced with recombinant adenoviruses. *Vaccine*, **22**:224–236.

61. Hurez, V., R. D. Hautton, J. Oliver, R. J. Matthews, and C. K. Weaver. 2002. Gene delivery into primary T cells: Overview and characterization of a transgenic model for efficient adenoviral transduction. *Immunol Res*, **26**:131–141.

62. Jung, D., S. Neron, M. Drouin, and A. Jacques. 2005. Efficient gene transfer into normal human B lymphocytes with the chimeric adenoviral vector Ad5/F35. *J Immunol Methods*, **304**:78–87.
63. Leen, A. M., U. Sili, B. Savoldo, A. M. Jewell, P. A. Piedra, M. K. Brenner, and C. M. Rooney. 2004. Fiber-modified adenoviruses generate subgroup cross-reactive, adenovirus-specific cytotoxic T lymphocytes for therapeutic applications. *Blood*, **103**:1011–1019.
64. Perez, E. E., J. Wang, J. C. Miller, Y. Jouvenot, K. A. Kim, O. Liu, N. Wang et al. 2008. Establishment of HIV-1 resistance in CD4+ T cells by genome editing using zinc-finger nucleases. *Nat Biotechnol*, **26**:808–816.
65. Schroers, R., Y. Hildebrandt, J. Hasenkamp, B. Glass, A. Lieber, G. Wulf, and M. Piesche. 2004. Gene transfer into human T lymphocytes and natural killer cells by Ad5/F35 chimeric adenoviral vectors. *Exp Hematol*, **32**:536–546.
66. Leen, A. M., G. D. Myers, U. Sili, M. H. Huls, H. Weiss, K. S. Leung, G. Carrum et al. 2006. Monoculture-derived T lymphocytes specific for multiple viruses expand and produce clinically relevant effects in immunocompromised individuals. *Nat Med*, **12**:1160–1166.
67. Aizawa, E., Y. Hirabayashi, Y. Iwanaga, K. Suzuki, K. Sakurai, M. Shimoji, K. Aiba et al. 2012. Efficient and accurate homologous recombination in hESCs and hiPSCs using helper-dependent adenoviral vectors. *Mol Ther*, **20**:424–431.
68. Li, M., K. Suzuki, J. Qu, P. Saini, I. Dubova, F. Yi, J. Lee, I. Sancho-Martinez, G. H. Liu, and J. C. Izpisua Belmonte. 2011. Efficient correction of hemoglobinopathy-causing mutations by homologous recombination in integration-free patient iPSCs. *Cell Res*, **21**:1740–1744.
69. Liu, G. H., K. Suzuki, M. Li, J. Qu, N. Montserrat, C. Tarantino, Y. Gu et al. 2014. Modelling Fanconi anemia pathogenesis and therapeutics using integration-free patient-derived iPSCs. *Nat Commun*, **5**:4330.
70. Liu, G. H., K. Suzuki, J. Qu, I. Sancho-Martinez, F. Yi, M. Li, S. Kumar et al. 2011. Targeted gene correction of laminopathy-associated LMNA mutations in patient-specific iPSCs. *Cell Stem Cell*, **8**:688–694.
71. Alemany, R., K. Suzuki, and D. T. Curiel. 2000. Blood clearance rates of adenovirus type 5 in mice. *J Gen Virol*, **81**:2605–2609.
72. Manickan, E., J. S. Smith, J. Tian, T. L. Eggerman, J. N. Lozier, J. Muller, and A. P. Byrnes. 2006. Rapid Kupffer cell death after intravenous injection of adenovirus vectors. *Mol Ther*, **13**:108–117.
73. Lieber, A., C. Y. He, L. Meuse, D. Schowalter, I. Kirillova, B. Winther, and M. A. Kay. 1997. The role of Kupffer cell activation and viral gene expression in early liver toxicity after infusion of recombinant adenovirus vectors. *J Virol*, **71**:8798–8807.
74. Parker, A. L., S. N. Waddington, C. G. Nicol, D. M. Shayakhmetov, S. M. Buckley, L. Denby, G. Kemball-Cook et al. 2006. Multiple vitamin K-dependent coagulation zymogens promote adenovirus-mediated gene delivery to hepatocytes. *Blood*, **108**:2554–2561.
75. Shayakhmetov, D. M., A. Gaggar, S. Ni, Z. Y. Li, and A. Lieber. 2005. Adenovirus binding to blood factors results in liver cell infection and hepatotoxicity. *J Virol*, **79**:7478–7491.
76. Zaiss, A. K., E. M. Foley, R. Lawrence, L. S. Schneider, H. Hoveida, P. Secrest, A. B. Catapang et al. 2015. Hepatocyte heparan sulfate is required for adeno-associated virus 2 but dispensable for adenovirus 5 liver transduction *in vivo*. *J Virol*, **90**:412–420.
77. Carlisle, R. C., Y. Di, A. M. Cerny, A. F. Sonnen, R. B. Sim, N. K. Green, V. Subr et al. 2009. Human erythrocytes bind and inactivate type 5 adenovirus by presenting coxsackievirus-adenovirus receptor and complement receptor 1. *Blood*, **13**:1909–1918.

78. Liu, Y., H. Wang, R. Yumul, W. Gao, A. Gambotto, T. Morita, A. Baker, D. Shayakhmetov, and A. Lieber. 2009. Transduction of liver metastases after intravenous injection of Ad5/35 or Ad35 vectors with and without factor X-binding protein pretreatment. *Hum Gene Ther*, **20**:621–629.

79. Shayakhmetov, D. M., Z. Y. Li, S. Ni, and A. Lieber. 2004. Analysis of adenovirus sequestration in the liver, transduction of hepatic cells, and innate toxicity after injection of fiber-modified vectors. *J Virol*, **78**:5368–5381.

80. Bernt, K., S. Ni, Z.-Y. Li, D. M. Shayakhmetov, and A. Lieber. 2003. The effect of sequestration by nontarget tissues on anti-tumor efficacy of systemically applied, conditionally replicating adenovirus vectors. *Mol Therapy*, **8**:746–755.

81. Song, L., X. Li, G. R. Jayandharan, Y. Wang, G. V. Aslanidi, C. Ling, L. Zhong et al. 2013. High-efficiency transduction of primary human hematopoietic stem cells and erythroid lineage-restricted expression by optimized AAV6 serotype vectors *in vitro* and in a murine xenograft model *in vivo*. *PLoS One*, **8**:e58757.

82. van Til, N. P. and G. Wagemaker. 2014. Lentiviral gene transduction of mouse and human hematopoietic stem cells. *Methods Mol Biol*, **1185**:311–319.

83. Lieber, A., C. Y. He, I. Kirillova, and M. A. Kay. 1996. Recombinant adenoviruses with large deletions generated by Cre-mediated excision exhibit different biological properties compared with first-generation vectors *in vitro* and *in vivo*. *J Virol*, **70**:8944–8960.

84. Shimizu, K., F. Sakurai, M. Machitani, K. Katayama, and H. Mizuguchi. 2011. Quantitative analysis of the leaky expression of adenovirus genes in cells transduced with a replication-incompetent adenovirus vector. *Mol Pharm*, **8**:1430–1435.

85. Wang, H., H. Cao, M. Wohlfahrt, H. P. Kiem, and A. Lieber. 2008. Tightly regulated gene expression in human hematopoietic stem cells after transduction with helper-dependent Ad5/35 vectors. *Exp Hematol*, **36**:823–831.

86. Morral, N., R. J. Parks, H. Zhou, C. Langston, G. Schiedner, J. Quinones, F. L. Graham, S. Kochanek, and A. L. Beaudet. 1998. High doses of a helper-dependent adenoviral vector yield supraphysiological levels of alpha1-antitrypsin with negligible toxicity. *Hum Gene Ther*, **9**:2709–2716.

87. Wang, H., H. Cao, M. Wohlfahrt, H. P. Kiem, and A. Lieber. 2008. Tightly regulated gene expression in human hematopoietic stem cells after transduction with helper-dependent Ad5/35 vectors. *Exp Hematol* **36**:823–831.

88. Wang, H. and A. Lieber. 2006. A helper-dependent capsid-modified adenovirus vector expressing adeno-associated virus rep78 mediates site-specific integration of a 27-kilobase transgene cassette. *J Virol*, **80**:11699–11709.

89. Holkers, M., I. Maggio, J. Liu, J. M. Janssen, F. Miselli, C. Mussolino, A. Recchia, T. Cathomen, and M. A. Goncalves. 2013. Differential integrity of TALE nuclease genes following adenoviral and lentiviral vector gene transfer into human cells. *Nucl Acids Res*, **41**:e63.

90. Steinwaerder, D. S., C. A. Carlson, and A. Lieber. 1999. Generation of adenovirus vectors devoid of all viral genes by recombination between inverted repeats. *J Virol*, **73**:9303–9313.

91. Saydaminova, K., X. Ye, H. Wang, M. Richter, M. Ho, H. Chen, N. Xu et al. 2015. Efficient genome editing in hematopoietic stem cells with helper-dependent Ad5/35 vectors expressing site-specific endonucleases under microRNA regulation. *Mol Ther Methods Clin Dev*, **1**:14057.

92. Brown, B. D., M. A. Venneri, A. Zingale, L. Sergi Sergi, and L. Naldini. 2006. Endogenous microRNA regulation suppresses transgene expression in hematopoietic lineages and enables stable gene transfer. *Nat Med*, **12**:585–591.

93. Papapetrou, E. P., D. Kovalovsky, L. Beloeil, D. Sant'angelo, and M. Sadelain. 2009. Harnessing endogenous miR-181a to segregate transgenic antigen receptor expression in developing versus post-thymic T cells in murine hematopoietic chimeras. *J Clin Invest*, **119**:157–168.

94. Maggio, I., M. Holkers, J. Liu, J. M. Janssen, X. Chen, and M. A. Goncalves. 2014. Adenoviral vector delivery of RNA-guided CRISPR/Cas9 nuclease complexes induces targeted mutagenesis in a diverse array of human cells. *Sci Rep*, **4**:5105.
95. Cheng, R., J. Peng, Y. Yan, P. Cao, J. Wang, C. Qiu, L. Tang et al. 2014. Efficient gene editing in adult mouse livers via adenoviral delivery of CRISPR/Cas9. *FEBS Lett*, **588**:3954–3958.
96. Maddalo, D., E. Manchado, C. P. Concepcion, C. Bonetti, J. A. Vidigal, Y. C. Han, P. Ogrodowski et al. 2014. *In vivo* engineering of oncogenic chromosomal rearrangements with the CRISPR/Cas9 system. *Nature*, **516**:423–427.
97. Nissim, L., S. D. Perli, A. Fridkin, P. Perez-Pinera, and T. K. Lu. 2014. Multiplexed and programmable regulation of gene networks with an integrated RNA and CRISPR/Cas toolkit in human cells. *Mol Cell*, **54**:698–710.
98. Wang, D., H. Mou, S. Li, Y. Li, S. Hough, K. Tran, J. Li et al. 2015. Adenovirus-mediated somatic genome editing of pten by CRISPR/Cas9 in mouse liver in spite of Cas9-specific immune responses. *Hum Gene Ther*, **26**:432–442.
99. Xu, L., K. H. Park, L. Zhao, J. Xu, M. El Refaey, Y. Gao, H. Zhu, J. Ma, and R. Han. 2015. CRISPR-mediated genome editing restores dystrophin expression and function in mdx mice. *Mol Ther*, **24**:564–569.
100. Mussolino, C., J. Alzubi, E. J. Fine, R. Morbitzer, T. J. Cradick, T. Lahaye, G. Bao, and T. Cathomen. 2014. TALENs facilitate targeted genome editing in human cells with high specificity and low cytotoxicity. *Nucl Acids Res*, **42**:6762–6773.
101. Tesson, L., C. Usal, S. Menoret, E. Leung, B. J. Niles, S. Remy, Y. Santiago et al. 2011. Knockout rats generated by embryo microinjection of TALENs. *Nat Biotechnol*, **29**:695–696.
102. Ehrhardt, A., H. Xu, and M. A. Kay. 2003. Episomal persistence of recombinant adenoviral vector genomes during the cell cycle *in vivo*. *J Virol*, **77**:7689–7695.
103. Pruett-Miller, S. M., D. W. Reading, S. N. Porter, and M. H. Porteus. 2009. Attenuation of zinc finger nuclease toxicity by small-molecule regulation of protein levels. *PLoS Genet*, **5**:e1000376.
104. Saydaminova, K., X. Ye, H. Wang, M. Richter, M. Ho, C. HongZhuan, N. Xu et al. 2015. Efficient genome editing in hematopoietic stem cells with helper-dependent Ad5/35 vectors expressing site-specific endonucleases under microRNA regulation. *Mol Ther Methods Clin Dev*, **1**:14057.
105. Zhang, C., K. Xu, L. Hu, L. Wang, T. Zhang, C. Ren, and Z. Zhang. 2015. A suicidal zinc finger nuclease expression coupled with a surrogate reporter for efficient genome engineering. *Biotechnol Lett*, **37**:299–305.
106. Maier, D. A., A. L. Brennan, S. Jiang, G. K. Binder-Scholl, G. Lee, G. Plesa, Z. Zheng et al. 2013. Efficient clinical scale gene modification via zinc finger nuclease-targeted disruption of the HIV co-receptor CCR5. *Hum Gene Ther*, **24**:245–258.
107. Yuan, J., J. Wang, K. Crain, K. Fearns, K. A. Kim, K. L. Hua, P. D. Gregory, M. C. Holmes, and B. E. Torbett. 2012. Zinc-finger nuclease editing of human cxcr4 promotes HIV-1 CD4(+) T cell resistance and enrichment. *Mol Ther*, **20**:849–859.
108. Zhang, Z., S. Zhang, X. Huang, K. E. Orwig, and Y. Sheng. 2013. Rapid assembly of customized TALENs into multiple delivery systems. *PLoS One*, **8**:e80281.
109. Li, C., X. Guan, T. Du, W. Jin, B. Wu, Y. Liu, P. Wang et al. 2015. Inhibition of HIV-1 infection of primary CD4+ T-cells by gene editing of CCR5 using adenovirus-delivered CRISPR/Cas9. *J Gen Virol*, **96**:2381–2393.
110. Ding, Q., A. Strong, K. M. Patel, S. L. Ng, B. S. Gosis, S. N. Regan, C. A. Cowan, D. J. Rader, and K. Musunuru. 2014. Permanent alteration of PCSK9 with *in vivo* CRISPR-Cas9 genome editing. *Circ Res*, **115**:488–492.
111. Suzuki, K., C. Yu, J. Qu, M. Li, X. Yao, T. Yuan, A. Goebl et al. 2014. Targeted gene correction minimally impacts whole-genome mutational load in human-disease-specific induced pluripotent stem cell clones. *Cell Stem Cell*, **15**:31–36.

112. Hutter, G., J. Bodor, S. Ledger, M. Boyd, M. Millington, M. Tsie, and G. Symonds. 2015. CCR5 targeted cell therapy for HIV and prevention of viral escape. *Viruses*, **7**:4186–4203.

113. Didigu, C. A., C. B. Wilen, J. Wang, J. Duong, A. J. Secreto, G. A. Danet-Desnoyers, J. L. Riley et al. 2014. Simultaneous zinc-finger nuclease editing of the HIV coreceptors ccr5 and cxcr4 protects CD4+ T cells from HIV-1 infection. *Blood*, **123**:61–69.

114. Wilen, C. B., J. Wang, J. C. Tilton, J. C. Miller, K. A. Kim, E. J. Rebar, S. A. Sherrill-Mix et al. 2011. Engineering HIV-resistant human CD4+ T cells with CXCR4-specific zinc-finger nucleases. *PLoS Pathog* **7**:e1002020.

115. Moll, N. M. and R. M. Ransohoff. 2010. CXCL12 and CXCR4 in bone marrow physiology. *Expert Rev Hematol*, **3**:315–322.

116. Tachibana, K., S. Hirota, H. Iizasa, H. Yoshida, K. Kawabata, Y. Kataoka, Y. Kitamura et al. 1998. The chemokine receptor CXCR4 is essential for vascularization of the gastrointestinal tract. *Nature*, **393**:591–594.

117. Aubert, M., N. M. Boyle, D. Stone, L. Stensland, M. L. Huang, A. S. Magaret, R. Galetto, D. J. Rawlings, A. M. Scharenberg, and K. R. Jerome. 2014. *In vitro* inactivation of latent HSV by targeted mutagenesis using an HSV-specific homing endonuclease. *Mol Ther Nucl Acids*, **3**:e146.

118. De Silva Feelixge, H. S., D. Stone, H. L. Pietz, P. Roychoudhury, A. L. Greninger, J. T. Schiffer, M. Aubert, and K. R. Jerome. 2016. Detection of treatment-resistant infectious HIV after genome-directed antiviral endonuclease therapy. *Antiviral Res*, **126**:90–98.

119. Weber, N. D., D. Stone, R. H. Sedlak, H. S. De Silva Feelixge, P. Roychoudhury, J. T. Schiffer, M. Aubert, and K. R. Jerome. 2014. AAV-mediated delivery of zinc finger nucleases targeting hepatitis B virus inhibits active replication. *PLoS One*, **9**:e97579.

120. Maeder, M. L. and C. A. Gersbach. 2016. Genome editing technologies for gene and cell therapy. *Mol Ther* **24**:430–446.

121. Sadelain, M., E. P. Papapetrou, and F. D. Bushman. 2012. Safe harbours for the integration of new DNA in the human genome. *Nat Rev Cancer*, **12**:51–58.

122. Maggio, I. and M. A. Goncalves. 2015. Genome editing at the crossroads of delivery, specificity, and fidelity. *Trends Biotechnol*, **33**:280–291.

123. Choi, E. W., D. S. Seen, Y. B. Song, H. S. Son, N. C. Jung, W. K. Huh, J. S. Hahn, K. Kim, J. Y. Jeong, and T. G. Lee. 2012. AdHTS: A high-throughput system for generating recombinant adenoviruses. *J Biotechnol*, **162**:246–252.

124. Yamamoto, Y., N. Goto, K. Miura, K. Narumi, S. Ohnami, H. Uchida, Y. Miura, M. Yamamoto, and K. Aoki. 2014. Development of a novel efficient method to construct an adenovirus library displaying random peptides on the fiber knob. *Mol Pharm*, **11**:1069–1074.

125. Koike-Yusa, H., Y. Li, E. P. Tan, C. Velasco-Herrera Mdel, and K. Yusa. 2014. Genome-wide recessive genetic screening in mammalian cells with a lentiviral CRISPR-guide RNA library. *Nat Biotechnol*, **32**:267–273.

126. Peng, J., Y. Zhou, S. Zhu, and W. Wei. 2015. High-throughput screens in mammalian cells using the CRISPR-Cas9 system. *FEBS J*, **282**:2089–2096.

127. Wang, T., J. J. Wei, D. M. Sabatini, and E. S. Lander. 2014. Genetic screens in human cells using the CRISPR-Cas9 system. *Science*, **343**:80–84.

128. Zhou, Y., S. Zhu, C. Cai, P. Yuan, C. Li, Y. Huang, and W. Wei. 2014. High-throughput screening of a CRISPR/Cas9 library for functional genomics in human cells. *Nature*, **509**:487–491.

129. Mitani, K. 2014. Gene targeting in human-induced pluripotent stem cells with adenoviral vectors. *Methods Mol Biol*, **1114**:163–167.

130. Beerman, I., J. Seita, M. A. Inlay, I. L. Weissman, and D. J. Rossi. 2014. Quiescent hematopoietic stem cells accumulate DNA damage during aging that is repaired upon entry into cell cycle. *Cell Stem Cell*, **15**:37–50.

131. Yamamoto, H., M. Ishimura, M. Ochiai, H. Takada, K. Kusuhara, Y. Nakatsu, T. Tsuzuki, K. Mitani, and T. Hara. 2016. BTK gene targeting by homologous recombination using a helper-dependent adenovirus/adeno-associated virus hybrid vector. *Gene Ther,* **23**:205–213.

132. Grisendi, G., R. Bussolari, L. Cafarelli, I. Petak, V. Rasini, E. Veronesi, G. De Santis et al. 2010. Adipose-derived mesenchymal stem cells as stable source of tumor necrosis factor-related apoptosis-inducing ligand delivery for cancer therapy. *Cancer Res,* **70**:3718–3729.

133. Nakamura, K., Y. Ito, Y. Kawano, K. Kurozumi, M. Kobune, H. Tsuda, A. Bizen, O. Honmou, Y. Niitsu, and H. Hamada. 2004. Antitumor effect of genetically engineered mesenchymal stem cells in a rat glioma model. *Gene Ther,* **11**:1155–1164.

134. Studeny, M., F. C. Marini, R. E. Champlin, C. Zompetta, I. J. Fidler, and M. Andreeff. 2002. Bone marrow-derived mesenchymal stem cells as vehicles for interferon-beta delivery into tumors. *Cancer Res,* **62**:3603–3608.

135. Benabdallah, B. F., E. Allard, S. Yao, G. Friedman, P. D. Gregory, N. Eliopoulos, J. Fradette et al. 2010. Targeted gene addition to human mesenchymal stromal cells as a cell-based plasma-soluble protein delivery platform. *Cytotherapy,* **12**:394–399.

136. Benabdallah, B. F., A. Duval, J. Rousseau, P. Chapdelaine, M. C. Holmes, E. Haddad, J. P. Tremblay, and C. M. Beausejour. 2013. Targeted gene addition of microdystrophin in mice skeletal muscle via human myoblast transplantation. *Mol Ther Nucl Acids,* **2**:e68.

137. Barr, D., J. Tubb, D. Ferguson, A. Scaria, A. Lieber, C. Wilson, J. Perkins, and M. A. Kay. 1995. Strain related variations in adenovirally mediated transgene expression from mouse hepatocytes *in vivo*: Comparisons between immunocompetent and immunodeficient inbred strains. *Gene Ther,* **2**:151–155.

138. Wang, X., J. Su, A. Sherman, G. L. Rogers, G. Liao, B. E. Hoffman, K. W. Leong, C. Terhorst, H. Daniell, and R. W. Herzog. 2015. Plant-based oral tolerance to hemophilia therapy employs a complex immune regulatory response including LAP + CD4+ T cells. *Blood,* **125**:2418–2427.

139. Hassan, R., A. C. Miller, E. Sharon, A. Thomas, J. C. Reynolds, A. Ling, R. J. Kreitman et al. 2013. Major cancer regressions in mesothelioma after treatment with an anti-mesothelin immunotoxin and immune suppression. *Sci Transl Med,* **5**:208ra147.

140. Nathwani, A. C., E. G. Tuddenham, S. Rangarajan, C. Rosales, J. McIntosh, D. C. Linch, P. Chowdary et al. 2011. Adenovirus-associated virus vector-mediated gene transfer in hemophilia B. *N Engl J Med,* **365**:2357–2365.

141. Hareendran, S., B. Balakrishnan, D. Sen, S. Kumar, A. Srivastava, and G. R. Jayandharan. 2013. Adeno-associated virus (AAV) vectors in gene therapy: Immune challenges and strategies to circumvent them. *Rev Med Virol,* **23**:399–413.

142. Pastan, I. and R. Hassan. 2014. Discovery of mesothelin and exploiting it as a target for immunotherapy. *Cancer Res,* **74**:2907–2912.

143. Bagley, J., J. L. Bracy, C. Tian, E. S. Kang, and J. Iacomini. 2002. Establishing immunological tolerance through the induction of molecular chimerism. *Front Biosci,* **7**:d1331–d1337.

144. Zakas, P. M., H. T. Spencer, and C. B. Doering. 2011. Engineered hematopoietic stem cells as therapeutics for hemophilia A. *J Genet Syndr Gene Ther,* **1**:pii, 2410.

7 Adenoviruses for Vaccination

Michael A. Barry

CONTENTS

ABSTRACT

Adenoviruses (Ads) are arguably one of the most potent vectors for in vivo gene
delivery. They are also thought to be one of the most immunogenic vectors driving
potent cellular and humoral immune responses. While this immunogenicity can be
a handicap in using Ads for gene therapy, this feature can be a potent strength when
Ads are instead used as gene-based vaccine. This chapter discussed this application
area for Ad vectors.

7.1   INTRODUCTION: BACKGROUND AND DRIVING FORCES

Vaccines are said to be the most economical medical intervention to control infectious
agents and perhaps cancer (http://www.who.int/topics/immunization/en/). Yet, traditional vaccine strategies are slow to respond to these threats and to provide effective
vaccines. While we have some potent vaccines, it has been said that all of the easy vaccines have been made and we are now faced with engineering to protect against some
of the most intractable infectious diseases and to explore our ability to combat cancer.

7.1.1   INACTIVATED/ATTENUATED PATHOGEN VACCINES

Vaccines have historically been made by inactivating or attenuating real pathogens
(i.e., polio, measles, and influenza vaccines) (1). While many attenuated pathogen
vaccines can be quite potent by virtue of being able to drive both antibody and
cellular immune responses, they also run a real risk of causing the disease that
they aim to prevent. For example, there is currently a large polio outbreak raging in
Nigeria that was actually caused by the oral polio vaccine (2).

7.1.2 Protein Vaccines

It is feasible to create vaccines from single recombinant proteins from pathogens rather than using the pathogen itself. Using one protein from a pathogen removes all risk of infection because the whole pathogen's genome is never delivered to the vaccinee. The two best protein vaccine examples are those against hepatitis B virus (HBV) and human papillomavirus (HPV). Single pathogen genes can be expressed from bacteria, yeast, mammalian cells, or even plants to produce recombinant protein vaccines. Protein vaccines are generally good at generating antibodies to neutralize pathogens, but they do not generally produce good T-cell responses to kill intracellular pathogens.

7.1.3 Genetic or DNA Vaccines or Gene-Based Vaccines

In the late 1980s and 1990s, it was shown that one could deliver simple naked plasmid DNA by injection into muscles (3) or into the skin with a gene gun to express foreign transgenes *in vivo* (4). Based on these results, it was speculated that this could be used for "...*genetic vaccination of animals against pathogenic infection by producing foreign antigens in restricted subsets of self-cells that mimics natural infections...*" (5). This so-called "genetic immunization" (5) or "genetic vaccine" (5) was followed by "DNA vaccines" (6), "DNA-based vaccines" (7), "polynucleotide vaccines" (8), or "gene-based vaccines" (9). In this case, a rose by any other name is still a rose regardless of what you "first" name it.

Importantly, unlike protein vaccines that drive primarily humoral responses, intracellular production of antigens from a plasmid not only produces antibodies, but also displays antigen peptides on MHC I and II to drive CD4 and CD8 T cell to kill intracellular pathogens. Because genetic or DNA vaccines use only one or a few pathogen genes there is no risk of infection by the original pathogen.

While naked DNA vaccines worked well in mice and other small animals, they had markedly less efficacy in primates and humans. This weakness can be mitigated to some degree by combining naked DNA injection with *in vivo* electroporation to increase transfection into the muscle or skin. An alternate approach is to bioengineer plasmids into polymers that increase DNA uptake into cells for vaccine effects.

7.1.4 Prime–Boost Combinations of Gene-Based and Protein-Based Vaccines

It should be noted that in many animal and human studies, a gene-based vaccine is used to "prime" the immune system and a protein vaccine is used to "boost" this prime. This takes advantage of the different strengths of both vaccine types and mitigates some of their weaknesses. Protein vaccines (10) are generally effective at driving antibody responses, but less potent at driving supportive T-cell responses (reviewed in References 11 and 12). Conversely, gene-based vaccines that express protein antigens *in situ* can be quite potent at priming T-cell responses, but are generally weaker at driving high-titer antibody responses. Protein vaccines are weak for T-cell response, but their mass delivery of antigen is good at boosting antibody responses. Given their differing potencies, gene-based and protein vaccines are frequently used in combination as prime–boost vaccines. In most cases, the gene-based vaccine is used first for priming, and then boosted with a protein vaccine.

7.1.5   VIRAL VECTORS AS GENE-BASED VACCINES

When plasmid-based genetic vaccines waned in efficacy in larger hosts, many investigators resurrected the use of viral vectors to deliver vaccine genes *in vivo*. In essence, this approach takes advantage of the evolved DNA or RNA delivery efficiency that is optimized by viruses.

7.1.6   ADENOVIRUSES AS VACCINES

Adenoviruses (Ads) were first defined as adenoidal-pharyngeal-conjunctival (APC) viruses in 1953 (13,14). By 1956, Ads were being used in humans to vaccinate military recruits against respiratory infections. As such, the very first work with Ads as vaccines used fully replication-competent (RC) "wild" Ads as vaccines against themselves in humans (15). Since then, thousands of military recruits have received fully wild, replication-competent Ads (RC-Ads) as vaccines.

One might ask how do you use a live virus to vaccinate without causing the disease you aim to avoid (like with the polio vaccine)? Interestingly, when live Ads were used to vaccinate military recruits, they were delivered in oral capsules (15). This was not because oral delivery works better (it does not), but was used primarily to protect health care personnel and vaccinees from respiratory infection by the RC-Ad vaccine.

7.1.7   ACCIDENTAL PROOF OF PRINCIPLE FOR THE USE
OF Ads AS GENE-BASED VACCINES

When some of the wild Ads were produced as military vaccines, they were grown in monkey cell lines. Later work testing Ads as potential cancer causing agents accidentally discovered that human Ad serotype 7 (hAd7) produced a protein that cross-reacted strongly with simian virus 40 (SV40) tumor (T) antigen (16). This cross-reactive protein was found to actually be SV40 T antigen. Surprisingly, this expression from Ad7 actually occurred because the virus had captured the SV40 T antigen gene by recombination with SV40 viral DNA present in the monkey cell line (17). These Ad7 vaccines were presumably used in humans and therefore not only vaccinated these humans, but also inoculated them with a known oncogene. Fortunately, this was not associated with any known increase in cancer in vaccinees.

This is therefore the first demonstration of gene-based vaccination provided by nature and man. As many in the research community enjoy using 293 T cells for the ease with which they can be transfected, they should note that the "T" stands for tumor antigen, so you too could be producing your own home-grown Ad-SV40 virus by accident.

7.1.8 Ads as Gene-Based Vaccines

A spectrum of viruses is being developed as genetic vaccines. Within this spectrum, Ad vectors are arguably one of the most potent (18–21). For example in our review of the 36 gene-based HIV vaccines occurring in 2011 (12), seven used protein vaccines: four used proteins with no gene-based vaccine and three used proteins as a boost for gene-based vaccines. 32 of the 36 trials used some form of gene-based vaccine. Three used plasmid DNA with no other gene-based vaccine. 13 of the 36 trials used plasmid combined with one or more other viral gene-based vaccines. 13 of the 36 used Ad vaccines, 10 used poxvirus vaccines, one used both Ad and pox, and the remainder used measles, vesicular stomatitis virus (VSV), or simian immunodeficiency virus (SIV) vectors. Therefore, DNA and Ad vaccines are at the top of human trials, followed by poxviruses, protein, measles, VSV, and SIV vectors. A further indication of Ad potency as a gene-based vaccine comes from the response to the 2014 Ebola outbreak. In this case, two gene-based vaccines were tested in humans: a RC VSV vectors and a replication-defective chimpanzee Ad vector cAd3-EBO-Z (22–24).

7.2 ADPPLES AND ORANGES

Ads are one of the most potent gene-based vaccines available today. In head to head comparisons in nonhuman primates, replication-defective Ad (RD-Ad) vaccines were more robust than DNA or vaccinia vaccines (25). These vaccine comparisons are generally not well controlled. In most cases, two vectors are not compared based on the number of transgenes they deliver, but instead are compared at some favored dose for each different vector. For example, you might vaccinate a macaque with 1 mg of a plasmid vaccine and 10^{10} viral particles of Ad. While these might be good doses for each vaccine, you are delivering wildly different numbers of the test transgene. 1 mg of a 6000 base pair plasmid would deliver about 5×10^{13} vaccine transgenes as compared to 1×10^{10} transgenes in the Ad vaccine. Likewise, if you compare to the popular adeno-associated virus (AAV) vector platform, one typically delivers 10–100-times as many viral genomes in a typical application other than Ad. If you are comparing Ad to an enveloped virus vector, these are not expressed in genomes, but usually plaque-forming units (PFU), infectious units (IFU), or tissue culture infectious dose 50% (TCID50) units based on wildly different biological tests for different viruses performed in different labs. For some enveloped viruses, particle, or genome to IFU ratios can be as low as 10 or as high as 10^6 (Apples and oranges).

While Ads appear potent even in these apples and oranges comparison, it should be noted that we are probably not actually comparing our most robust gene-based Ad vaccines to other vector systems. The vast majority of Ad vaccines are actually drastically weakened RD-Ads and they are sometimes being compared to RC viral vaccines. Work by our lab and others are exploring re-harnessing Ads with different DNA replication and viral replication capacities to amplify vaccine efficacy.

7.3   Ad VECTORS WITH DIFFERENT DNA AND VIRUS REPLICATION CAPACITIES

7.3.1   RC-Ad VACCINES AMPLIFY ANTIGEN TRANSGENES

Fully wild RC-Ads have been used in thousands of humans as anti-Ad vaccines. Some of the earliest Ad gene-based vaccines used RC-Ads based on human serotypes 4, 5, and 7 (hAd4, 5, and 7) in nonhuman primates against HBV and HIV (26–33). These RC-Ad vaccines can infect a cell, replicate their genome *and also replicate the transgene* up to 100,000-fold in each cell (Figures 7.1 and 7.2). This produces massively more transgene protein to stimulate immune responses. For example, if you compare luciferase expression by RD- and RC-Ad, 100–1000 times as much transgene activity is generated by the replicating virus than the standard ΔE1 vector (34). If the host is permissive for the full Ad life cycle, an RC-Ad can also generate infectious progeny virions that can infect even more cells to expand antigen expression. This will of course be countered by intracellular and immune system responses that will amplify in parallel to suppress this adenovirus infection.

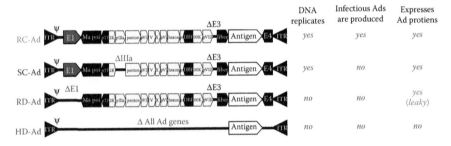

FIGURE 7.1   Cartoon of current and new Ad vectors being tested as gene-based vaccines.

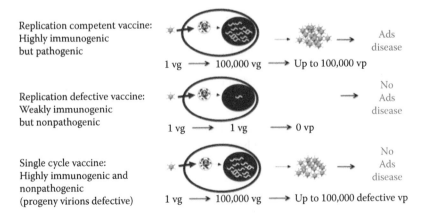

FIGURE 7.2   Ads with different replication and progeny virion production. Vg = viral genomes. *Note*: Replicating the viral genome also replicates the vaccine antigen gene.

7.3.2 RC-Ad VACCINES AMPLIFY INFECTIOUS PROGENY VIRUSES AND CAN RISK ADENOVIRUS INFECTIONS

Wild RC-Ads naturally cause a number of respiratory, ocular, digestive, and in some case internal infections. An RC-Ad vaccine can cause the same infections. When live RC-Ad4 and Ad7 were used to vaccinate military recruits, they were used in oral capsules specifically to protect health care personnel and vaccinees from respiratory infection by Ads (Robert Couch, Baylor College of Medicine, personal communication and (15)). Oral RC-Ad delivery avoided the side effects of these wild viruses, but still allowed mucosal vaccination to occur to protect from natural RC-Ad respiratory infection.

Therefore, while RC-Ad gene-based vaccines can be potent, they also have a finite risk of causing adenovirus infections. This could be an infection in the person who actually gets the vaccine, the health care worker, or to family members later if the vaccine is shed. One solution is to lyophilize the virus in capsules and swallow them. Of course, this will work only if the virus causes diseases outside of the gastrointestinal tract. Considering that some Ads cause digestive infections (Ad40 and 41), that many Ads are shed in the feces after initial ocular or respiratory infections, and others are persistent in the gut, oral delivery may or may not be a robust way to avoid all negative effects of the use of a fully RC-Ad vaccine.

7.3.3 RD-Ads: FIRST GENERATION ΔE1, ΔE3 VECTORS

The vast majority of Ad vaccines in use today are actually engineered as RD-Ads by deleting the pivotal viral E1 region (Figure 7.1). This renders them RD for both viral DNA replication and infectious progeny virus production. This prevents them from causing Ad infections. Most of these are based on human Ad serotype 5 (Ad5) background owing in part to the commercial availability of kits to make these E1, E3-deleted viruses.

Thus RD-Ads are largely safe, at least from an Ad infection risk perspective. However, they are not particularly potent because key transgene amplification features have been inactivated to make them safe. An E1-deleted RD-Ad vaccine will infect a cell, deliver its one copy of a vaccine transgene and express "1X" of the resulting antigen protein without the virus generating progeny virions (Figure 7.2). In contrast, the RC-Ad amplifies the same transgene up to 100,000-fold. Therefore, while RD-Ads are potent relative to many gene-based vaccines, but they are engineered to favor safety over potency. This may be an Achille's heel in human Ad-based vaccines.

7.3.4 RD HELPER-DEPENDENT Ad VECTORS TO EVADE VECTOR-SPECIFIC T-CELL RESPONSES

Most RD-Ad vectors are deleted for E1 and E3, but still carry all other Ad protein open reading frames (ORFs) that can be expressed in vaccinated cells and targeted by anti-Ad T-cell responses ((35,36) and Figure 7.1). In contrast, all viral ORFs are deleted in helper-dependent adenoviral (HD-Ad) vectors ((37–39) and Figure 7.1). No Ad ORFs mean no Ad proteins can be produced in the HD-Ad-transduced cell. This reduces T-cell responses that can kill transduced cells (37–39). In gene therapy applications, this reduced immunogenicity and reduced liver damage allows for transgene expression in rodents and in nonhuman primates over several years ((40–42) and see Chapter 3).

Based on this unique feature, we tested HD-Ads expressing GFP-luciferase (GL) or HIV antigens in mice and in rhesus macaques (43,44). In head to head comparisons with ΔE1ΔE3 RD-Ad, HD-Ads do indeed generate better immune responses than RD-Ad. However, unlike in gene therapy applications, HD-Ads expressing antigens do not give persistent gene expression over months or years. Rather, this expression is quenched nearly as quickly as occurs with RD-Ad (43,44). This difference in HD-Ad expression duration makes perfect sense, since we are making this stealthy vector express highly immunogenic vaccine antigens. Therefore, while Ad ORFs are not being targeted by immune cells, the vaccine antigen still is. Despite this relatively rapid loss of expression, HD-Ads still mediate better antibody responses against antigens (43,44) perhaps by slowing the loss of transduced cells or by reducing antiviral inflammatory responses that may attenuate antigen expression.

7.3.5   HEAD TO HEAD COMPARISONS OF RC AND RD-Ad VACCINES

Dr. Marjorie Guroff's group has compared Ad vaccines with differing replication capacities (11,45–56). In nonhuman primates, they showed that RC-Ads are superior to RD-Ads as gene-based vaccines against HIV, particularly by the oral route. For example, when chimpanzees were vaccinated with vectors expressing the HIV envelope, they showed that priming with RD-Ad5 followed by a boost with RD-Ad7 generated no detectable antibodies against the protein (48). In contrast, a prime–boost with RC-Ad5 and RC-Ad7 generated significant anti-HIV titers. This demonstrates the efficacy benefit of allowing an Ad to amplify antigen transgenes and/or progeny virions. Considering that chimps can be as large as humans, this emphasizes that we may need Ads to replicate their antigen transgenes to generate relevant vaccine effects in humans.

7.3.6   HEAD TO HEAD COMPARISONS OF RC, RD, AND HD-Ad VACCINES

We later compared RD-Ad5 versus RC-Ad5 in mice, but also included HD-Ad5 in the mix (57). Mice cannot support the full life cycle of Ads, but they do support some level of Ad DNA replication, so the pros and cons of Ad vaccine DNA replication could still be assessed. RC, RD, and HD-Ad5 vectors expressing the GL transgene were injected intramuscular (i.m.) and intravenous routes in BALB/c mice (57). Notably, RC-Ad and HD-Ad both induced significantly higher anti-luciferase antibodies than conventional RD-Ad vectors. From this, immunologically "stealthy" HD-Ads seem favorable over conventional ΔE1 Ad vectors. Likewise, RC-Ads that can amplify transgenes are also better than current RD-Ad vectors. Thus, the vast majority of Ad vaccines in use today are not as good as they could be. If they fail in humans, this may well be the reason. We are therefore likely to be underestimating the true potential of Ad vaccines in preclinical and clinical tests.

Case in point: The "best" Ad vaccine that went into humans for Ebola virus vaccination was a chimpanzee RD-Ad called cAd3-EBO-Z (22–24). Consider how much more potent these could have been if they were able to replicate and amplify their antigen genes. Consider also how much better the responses might have been if a stealthy HD-Ad vector was used to avoid strong pre-existing and vector-induced anti-Ad T-cell responses.

Two caveats need to be considered for those who may consider exploring replicating Ads. First, testing RD versus RC-Ads in mouse models works, but not well. Mice can amplify Ad genomes *up to* 10-fold in certain tissue, but not nearly 10,000-fold as can be observed in human cells. Mouse cells also do not produce functional virions for second waves of vaccine infection. Syrian hamsters support more of species C Ad DNA and virus replication and are considered a better model for replicating Ads (58,59). Three caveats occur with hamsters. First, it may be that only species C Ads will replicate in their cells (60), so if you want to test other serotypes, they may not work. Second, we have never observed a second "round" of infection in terms of reporter gene expression in hamsters even with species C Ads (data not shown). Third, immunological reagents are almost impossible to find for hamsters. If permissive animals or humans are tested, one must realize that a full RC-Ad may well induce significant side effects including overt Ad infections that will be unobserved and unappreciated in mice. The primary caveat with HD-Ads is that there are more difficult to produce than run of the mill RD-Ad vectors. You may need to recruit HD-Ad "artists" like the editors of this book to make these vectors for you.

7.3.7 SINGLE-CYCLE Ad VACCINES: TRANSGENE REPLICATION WITHOUT VIRUS PRODUCTION

We recently engineered so-called "single-cycle" Ad (SC-Ad) vectors to try to take advantage of the amplification potential of a DNA RC virus, but without the risk of causing frank Ad infections (Figures 7.1 and 7.2 and (34,59)). To do this we retained the pivotal E1 gene, but instead deleted key late genes that are required *after* viral DNA replication occurs (34,59). This allows the virus to still replicate its genome (and transgene), but blocks the production of functional progeny (Figure 7.3). We generated SC-Ads by deleting several Ad genes including those for protein V, fiber, and pIIIa. Our best SC-Ad has the gene for the pIIIa capsid cement protein deleted ((34,59) and Figure 7.1). We demonstrated that delta IIIa SC-Ads replicate their genomes as well as RC-Ad markedly outproducing transgene proteins over RD-Ad (34,59). Without IIIa, adenoviral genomes are not packaged into virions (34). This unpackaged DNA appears to be expressed somewhat longer and may better stimulate Toll-like receptor 9 (TLR9) resulting in SC-Ads generating more robust and more persistent immune responses than RD-Ads in permissive Syrian hamsters and in rhesus macaques. Surprisingly, SC-Ad was more potent than RC-Ad at generating immune responses in systemic and mucosal immune compartments ((34,59) and data not shown). This effect may actually be related to antiviral responses against RC-Ad, since it drives more strong activation of interferon-stimulated genes than SC-Ad (data not shown).

7.3.8 AMPLIFIED PRODUCTION OF FUNCTIONAL INFLUENZA ANTIBODIES BY SC-Ad

To test a legitimate pathogen vaccine, we inserted the gene for influenza A/PR/8/34 hemagglutinin (HA) into RD- and SC-Ad6 in permissive Syrian hamsters. Consistent

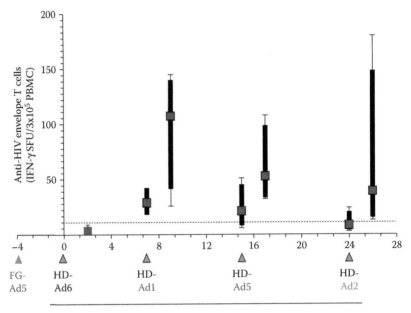

FIGURE 7.3 Repeated prime–boost with HD-Ad vaccines in macaques in the face of anti-Ad immune responses. Macaques were pre-immunized with first-generation Ad5 and then immunized with the indicated HD-Ad serotypes expressing the HIV envelope. Peripheral mononuclear cells (PBMCs) were collected before and after each immunization and anti-HIV T-cell responses were measured by ELISPOT (from reference in ELISPOT plates were stimulated with 3 pools of 50–70 peptides spanning the gp140 region of the SF162 envelope for 36 h and IFN-γ secreting spots were detected and counted. Responses in terms of IFN-γ spot forming cells (SFC) for 10^5 total input $CD8^+$ T cells were determined for individual monkeys after subtracting background values of cells cultured in the medium.

with the anti-GFP data, SC-Ad6 generated markedly higher anti-HA binding and influenza hemagglutination inhibition (HAI) antibodies than RD-Ad6 (data not shown). Notably, after only a single immunization, a 100-fold lower dose of SC-Ad could be used to generate HAI titers that were equal to a large dose of RD-Ad. This is consistent with SC-Ad transducing primary human lung cells with GFP with 33-fold less vector than RD-Ad (34). This suggests that one viral production of an SC-Ad vaccine could immunize 33–100 times as many people as the same production size of current RD-Ad vectors. We believe that SC-Ads are a Goldilocks vector for gene-based vaccines having the ability to amplify antigen genes without the risk of Ad infections.

7.4 PRE-EXISTING AND VECTOR-INDUCED ANTI-Ad IMMUNE RESPONSES

The immune system has evolved to repel viruses, so it is not surprising that these responses impact Ad-based vaccines.

7.4.1 Innate Immune Responses

Innate immune responses occur in a dose-dependent fashion in mice, nonhuman primates, and humans after intravenous injection of Ad vectors. Uptake of Ad virions by immune and non-immune cells precipitates the release of massive amounts of inflammatory cytokines including IL-6 and TNF-α within 3–24 hours of intravenous injection (reviewed in Reference 61).

These events produced by intravenous injection of large Ad doses (e.g., 10^{13} virus particles) can lead to lethal events in nonhuman primates (62). Innate responses to Ad also likely played a significant role in the unfortunate death of Jessie Gelsinger in the ornithine transcarbamylase deficiency gene therapy trial (63). While these events are a problem after intravenous administration, oral administration is in contrast remarkably safe, as evidenced by the safe administration of live RC-Ad to thousands of military recruits (15). These innate responses tend to be markedly weaker when Ads are applied by intranasal routes (data not shown). They can be quite strong after i.m. injection of Ads since a substantial portion of the vector will actually leak into the blood and infect the liver.

7.4.2 T-Cell Responses

Once innate responses resolve within a few days of injection, the host immune system mounts adaptive CD4 and CD8 T cells against Ad vector proteins and transgenes. RD-Ads are ΔE1 and this reduces, but does not eliminate expression from Ad ORFs and these antigens can be presented by MHC I and MHC II molecules to drive T-cell responses. The resulting production of cytotoxic T lymphocytes (CTLs) against RD-Ad modified cells results in the destruction of these cells within 2–3 weeks after vector administration (35,36). These CTL responses can target leaky Ad antigens (36) or can be directed against the transgene products (35).

7.4.3 Neutralizing Antibodies

Neutralizing antibodies are a significant problem for Ad vectors. These neutralizing antibodies are generally serotype-specific. These include pre-existing neutralizing antibodies produced by previous infection with the same Ad serotype. For example, as much as 27%–100% of humans have pre-existing neutralizing antibodies against Ad5 (64,65). These pre-existing antibodies in humans can blunt (but not ablate) even the first inoculation of vector into patients. Neutralizing antibodies are also produced after one inoculation of the vector into a naive host. These vector-induced antibodies can attenuate the level of gene delivery after subsequent administration of the vector if the same serotype of Ad is used (66). Because of this, each Ad serotype vaccine can be thought of as a "one off." You can use it once, but probably not again.

7.4.4 Comparison to Other Vaccine Platforms

This "one off" problem is also true for other viral vectors. Any viral vector can be neutralized by pre-existing or vaccine-induced antibodies. This is not a problem for plasmid vaccines, since they have no capsid proteins, so they have nothing that can be neutralized. A protein vaccine generates immune responses against itself, so repeated immunization can amplify responses when they are weak, but have little effect when they are strong, because neutralizing antibodies will "quench" further immune stimulation.

With Ads, we have the luxury of human and nonhuman serotypes to use as a "palette" of immunologically distinct vaccines. For monotypic viruses like VSV or measles, they are also "one offs," but with no palette of serotypes as a backup. For measles, the problem is compounded by the fact that most humans have been actively immunized against the virus.

These monotypic viruses can be made more stealthy by pseudotyping with the glycoproteins of other viruses. For example, measles can be pseudotyped with canine distemper virus F and H proteins. This can be used to avoid their own neutralization as well as to drive immune responses against a different pathogen. For example, the VSV Ebola virus vaccine, rVSV-ZEBOV-GP (67–72), is pseudotyped with the Ebola Zaire glycoprotein, so it can escape anti-VSV antibodies and also immunize against the Zaire Ebola virus. However, vaccine-induced antibodies will largely render second immunization with VSV-ZEBOV-GP markedly less effective. One can pseudotype with other Ebola virus glycoproteins (i.e., Marburg, Sudan, etc.), but this is a limited subset of proteins.

7.5   EVADING IMMUNE RESPONSES AGAINST Ad VECTORS

7.5.1   HD-Ad VECTORS TO EVADE VECTOR-SPECIFIC T-CELL RESPONSES

As discussed above, one approach to reduce the immunogenicity of expressed Ad proteins is to delete every viral gene from the adenoviral genome to produce HD-Ad vectors ((37,38,39) and see Chapter 3). We have shown that HD-Ad vaccines are more potent than RD-Ad vaccines (57). Another benefit of HD-Ad vectors is that removal of all viral genes now allows sequences as large as 35 Kbp to be packaged allowing very large genes or multiple antigen genes to be packaged in one vaccine (e.g., *three HIV genomes could theoretically be packaged in one HD-Ad*).

7.5.2   USE LOWER SEROPREVALENCE ADS TO AVOID PRE-EXISTING Ad ANTIBODIES IN HUMANS

Most humans have previously been exposed to viruses like Ad5 and are already immune. Given this, there have been recent efforts to convert from using Ad5 to using lower seroprevalence human or nonhuman Ad serotypes to avoid pre-existing immunity. This is feasible and necessary if you want to use Ad in humans. There are approximately 60 serotypes of human Ads and many other nonhuman Ads with low seroprevalence (reviewed in Reference 73). In particular, human Ad26 and Ad35 (65,74) and Ad serotypes from nonhuman primates are currently being tested in animals and humans to avoid Ad5 immunity (75–77).

7.5.3   Serotype Switching to Evade Vector-Induced Neutralizing Antibodies

One approach to evade neutralizing antibodies is to "serotype switch" the vector. For example, administration of Ad2 in mice generated potent neutralizing antibodies against Ad2 that reduced second transduction (78). However, if an Ad5 vector is used for the second round of gene delivery, this vector is not neutralized by Ad2-specific antibodies and it is effective (42,78). This serotype-switching approach has subsequently been repeatedly demonstrated as a robust approach for RD-Ad HIV vaccines by Drs. Barouch and Ertl's groups and others (76–80).

The HD-Ad system is even better suited to serotype switching, since Ads in the same class can generally cross-package each other's genomes. For example, an HD-Ad vector bearing a packaging signal and inverted terminal repeats (ITRs) from Ad5 can be cross-packaged by other type C Ads including Ad1, Ad2, and Ad6 ((42,78) and see below).

7.5.4   PEGylation to Evade Pre-Existing Neutralizing Antibodies and Reduce Adaptive Antibody and Cellular Responses against Ad Vectors

Polyethylene glycol (PEG) is a clinically approved conjugation agent used to improve the pharmacokinetics of a variety of protein therapeutics. In these applications, the hydrophilic PEG molecule is cross-linked to the therapeutic agent to "shield" or reduce interactions of it with proteins and cells that would normally decrease the therapeutic interactions with its target. PEG has also been applied to improve the pharmacology of Ad vector. 3–5 kDa PEG molecules bearing reactive groups are randomly chemically conjugated to free amine groups on the virion surface. By this approach, as many as 15,000 PEG molecules can be added to the virion surface to shield the virus. Previous work shows that PEG can protect Ad vectors from neutralizing antibodies to allow multiple administration into immune recipients (66,81,82). PEGylation also reduces the production of new antibody and cellular immune responses against Ad proteins (66).

In practice, most random PEGylation strategies have relative weak abilities to shield at least Ad5 from neutralizing antibodies (83). In our hands, random PEGylation does rescue Ad activity, but you may still lose 95% of activity relative to the same vector acting without antibodies. Also, random PEGylation reduces CAR binding, so PEGylated vectors are weaker by the intramuscular and intranasal route (83). PEG affects Ad5 vaccines less by the intranasal route (83), perhaps because the vectors may be relying on penton base-integrin interactions more than by fiber–CAR interactions. Using PEGs with glucose or galactose on their ends can rescue some of this lost activity by the intranasal route (83), perhaps by retargeting to receptors that bind these sugars.

7.5.5   Examples: Serotype-Switching HD-Ad HIV Vaccines

Barouch and Ertl's groups have performed a number of studies serotype-switching ΔE1 RD-Ad vectors for HIV vaccines (76,77,79,80). We used HD-Ads for this same purpose, but using their lower immunogenicity and the unique ability of viruses in one Ad species to cross-package one vaccine vector.

Ad5 is a species C Ad. Other members of species C include Ad1, 2, and 6. We generated an HD-Ad5 vector expressing the HIV envelope (44,57). We then packaged this one vector with different species C helper viruses: Ad1, Ad2, Ad5, and Ad6. We next pre-immunized macaques with Ad5 and then vaccinated them with either only HD-Ad5 three times or by serotype switching with HD-Ad6, 1, and 2 each expressing the HIV envelope (44,57). Three rounds of HD-Ad5 barely generated anti-envelope antibodies in the animals. In contrast, serotype switching with HD-Ad6, 1, and 2 generated stronger T cells and antibodies at every immunization than by HD-Ad5 in Ad5-immune animals (57).

When these animals were challenged rectally with SHIV-SF$_{162P3}$, both vaccine groups reduced SHIV viremia versus controls, but the serotype-switched group mediated better protection than the HD-Ad5 vaccine group. These results are significant as these species C HD-Ad vectors were applied in macaques that were previously immunized against species C Ad5. These four serotypes of HD-Ads were administered for four rounds of immunization without overt reduction in boosting at any immunization due to anti-vector immune responses. Even HD-Ad5 worked in the face of prior immunization with Ad5.

In a second macaque study, three groups of four macaques were immunized intranasally with Ad5 to mimic natural infection in humans (Figure 7.3 and (43)). Four weeks later, eight of the animals were then immunized with HD-Ads expressing only the HIV envelope. Four animals were immunized i.m. with HD-Ad6 to repeat the previous study. Another set of four were immunized with HD-Ad6 by the intravaginal route to compare mucosal immunization with systemic (i.m.) immunization. A third set of animals were unvaccinated controls. The HD-Ad groups were then boosted by the same routes with other Ad serotypes over 24 weeks. The animals were then mucosally challenged by the rectal route with SHIV-SF162P3.

At week 15, we diverged from the previous study to explore the question of exactly how problematic anti-Ad5 immune responses are to these HD-Ad vaccines. Given the animals had already received Ad5 and were boosted by two rounds of vaccination with Ad6 and Ad1 (which should boost immune responses to conserved species C Ad epitopes), these animals should be profoundly resistant to boosting with Ad5. To test this, the two vaccine groups were boosted with the HD-Ad5-env. Finally, at week 24 they received their third Ad boost in the form of HD-Ad1. Therefore, these Ad5-immune animals received prime–boost–boost–boost with HD-Ads (Figure 7.3 and (43)). Relevant to serotype switching and HD-Ads, it was notable that every Ad boost, including that by HD-Ad generated detectable T-cell responses. Even HD-Ad5 boosted T-cell responses in the animals after they had already been exposed to RD-Ad5, HD-Ad6, and HD-Ad1, all species C Ads.

7.5.6 PEG versus Serotype Switching

When we have compared PEG versus serotype switching, the latter is generally markedly more robust for prime–boost immunizations. However, from a clinical vaccine product standpoint, serotype switching requires making and testing more than one GMP-grade viral vector for FDA approval. For the PEG approach, GMP-grade PEG can be purchased, but then you are using this as a chemical reactant with Ads and will have to prove that this is reproducible and will likely have to perform additional safety studies for PEG, for the Ad, and for the combination.

7.6 OPPORTUNITIES AND CHALLENGES

7.6.1 CHALLENGES: BAD PUBLIC RELATIONS: THE HIV STEP VACCINE TRIAL

It has long been recognized that antibody and T-cell responses against Ad can attenuate gene therapy or vaccination. However, data from the STEP HIV vaccine trial in humans suggested that these responses may have additional unexpected side effects.

First generation RD-Ad5 vaccines expressing HIV antigens were tested in the STEP and HVTN-505 trials. These trials were halted due to concerns of possible increased HIV-1 infection particularly in men with higher pre-existing anti-Ad5 antibodies before vaccination (84–87). It was hypothesized that pre-existing antibodies against Ad5 may be forming immune complexes that enhance HIV infection (88). An alternate hypothesis is that amplified numbers of CD4 T cells that are generated against HIV in the course of Ad (or any) vaccination may provide a reservoir of cells for HIV to infect thereby increasing the risk of HIV infection rather than reducing it with the vaccine. This was a big blow to the Ad vaccine field as a vaccine intended to protect against HIV seemed to actually make you more likely to get infected.

7.6.2 REALITY CHECK

While many Ad "haters" jumped on this bandwagon, it has subsequently become clear that this side effect is not unique to Ad vaccines. It is instead unique to HIV vaccines and may apply to vaccines against other pathogens that infect immune cells. Within the STEP trial itself, people were not only at increased risk if they had high anti-Ad5 antibodies, but they were also at increased risk if they had a concomitant herpes simplex infection. This suggested an unspecific immune effect rather than one housed solely in anti-Ad immunity.

A subsequent review stated "...*Ad5-specific immunity did not alter HIV susceptibility in vaccinees, and strongly suggest that HIV-specific T-cell responses elicited by vaccination enhanced HIV acquisition in the STEP and HVTN505 trials.*" (89). Subsequent work reinforced this concept by demonstrating that many HIV or SIV vaccines can induce increased viral acquisition or the so-called "STEP Effect." For example, live and attenuated SIV vaccine increases SIV acquisition (90). Similarly, an attenuated varicella zoster virus (VZV) gene-based vaccine also increased SIV acquisition in nonhuman primates (91). Finally, even simple naked plasmid DNA vaccines combined with simple protein vaccines also increase SIV acquisition relative to vaccine controls (92). Therefore, it is not Ad that has a problem; it is vaccines in general that have a problem with HIV and its analogs.

These data indicate that any HIV vaccine can increase HIV, SIV, or SHIV acquisition. This side effect is likely due to the fact that a potent vaccine will generate

CD4 T cells in the natural course of educating the immune system. Thus a vaccine will amplify CD4 cells that happen to be the favored cell that HIV and SIV infect. Therefore, any potent vaccine may increase the number of at risk CD4 cells to increase the odds that SIV or HIV can start an infection (89). Any increase in the odds of an initial infection in CD4 T cells will increase the odds that an animal or person will become systemically infected with SIV or HIV.

This increased infection risk is a transient phenomenon. Risk of viral acquisition is increased soon after immunization (e.g., 2 weeks), but then declines within months (90). After this decline, protective immune responses prevail and the vaccines appear effective. These kinetics of increased acquisition may be due to a peak of at risk CD4 T cells in mucosal tissues soon after immunization, but then decline in these numbers over time after immunization. Considering that it is likely impossible to avoid driving CD4 T cells during any HIV vaccination, the path forward will likely involve avoiding HIV exposure at times soon after vaccination.

One can extrapolate this immunization side effect to any pathogen that infects immune cells if the vaccine amplifies the number of at risk cells. One can also mostly ignore it for the vast majority of pathogens that do not infect immune cells and that will not benefit by amplification of CD4 and other immune cells during active vaccination.

7.6.3   OPPORTUNITY: AN IMPROVING RD-Ad VACCINE RESULT LANDSCAPE

In head to head comparisons in nonhuman primates, Ad vaccines are more robust than DNA or vaccinia vaccines (25). Recently, RD-Ad26 and RD-Ad35 or RD-Ad25 and MVA vaccines have been shown to mediate protection against SIVmac251 (93). Vaccines that carried only SIV gag–pol reduced the infection hazard ratio to 0.71 whereas vaccines with gag–pol and env reduced risk to 0.2. More recent work by Barouch et al. builds on these results (94). Rhesus macaques were primed with RD-Ad26 vectors expressing SIV gag, pol, and env twice and boosted three times with AS01B-adjuvanted SIV env gp140 protein. This prime-boost regimen mediated complete protection in 50% of the animals against repeated intrarectal challenges with heterologous SIVmac251. These data suggest that two RD-Ad primes and 3 protein boost can protect against low dose stringent challenge.

7.6.4   OPPORTUNITY: RD-Ad MOVED FORWARD FOR
EBOLA VIRUS VACCINATION IN HUMANS

The fact that a chimpanzee RD-Ad (cAd3-EBO-Z) moved forward as one of the two Ebola vaccines for accelerated testing in humans supports that Ads are robust vaccine platforms (22–24). In the initial report of cAd2-EBO-Z testing in 20 humans, an i.m. dose of 2×10^{10} or 2×10^{11} virus particles (vp) generated anti-Ebola glycoprotein antibodies in all recipients (23). Antibody and T-cell responses were higher in the 2×10^{11} group than in the 2×10^{10} group. Extrapolating the 2×10^{11} vp dose to an average 10^{13} vp from a Cell Factory, this would equate to getting 50 human doses from an average RD-Ad preparation.

Subsequent follow up to these studies that included 1000 participants noted no serious adverse events with cAd3-EBO-Z (24). The authors report that the immune responses observed in the phase 1 trial would be protective in nonhuman primates. Chimpanzee Ads have been championed as more likely to avoid pre-existing immunity in humans (77,95,96). However, high levels of neutralizing immune responses against chimpanzee Ads have been observed in humans (up to 85%) and this was noted as a concern regarding the cAd3-EBO-Z vaccine (24). In response to this criticism, the study's authors noted that there was no correlation between pre-existing anti-cAd3 antibody levels of the production of antibodies or CD4 T cells against Ebola glycoprotein or CD4+ T-cell responses (24). However, anti-cAd3 antibodies did correlate with reduced CD8+ T-cell responses. The study's authors also emphasized that anti-cAd3 antibodies are still markedly lower in humans than anti-Ad5 responses (24).

7.6.5 ADPPLES AND ORANGES CHALLENGE: VSV EBOLA VACCINE VERSUS RD-Ad EBOLA VACCINE?

RD-cAd3-EBO-Z is a replication-defective chimpanzee Ad 3 vector with the E1 and E4 genes deleted. The other tested Ebola vaccine was rVSV-ZEBOV-GP, an attenuated VSV where its glycoprotein gene was deleted and replaced with the Ebola glycoprotein gene rVSV-ZEBOV-GP ((67–72) and http://www.who.int/vaccine_safety/committee/topics/ebola/Jun_2015/en/)). Whereas RD-cAd3-EBO-Z is replication-defective, rVSV-ZEBOV-GP is attenuated but remains RC. A*d*pples and Oranges.

The World Health Organization (WHO) reported in July 2015 that 271 adults were vaccinated with ChAd3-EBO-Z in Phase 1 studies in the United States, the United Kingdom, Mali, and Switzerland with more in Uganda. No serious adverse events ascribed to the Ad vaccine (http://www.who.int/vaccine_safety/committee/topics/ebola/Jun_2015/en/).

Phase 1 and 1b studies of the RC rVSV-ZEBOV-GP vaccine involved approximately 500 volunteers. Administration of rVSV vaccine resulted in vaccine viremia for up to 2 weeks after vaccination (http://www.who.int/vaccine_safety/committee/topics/ebola/Jun_2015/en/). The vaccine was detected in the saliva and urine in less than 10% of vaccinees. No vaccine-related serious adverse reactions, but vaccine virus replication was observed in the joints and the skin and this was associated with arthralgia, arthritis, dermatitis, rash, and cutaneous vasculitis. For those that suffered vaccine-induced arthritis, this pain lasted 2–3 weeks, but occasionally more than 3 months, and was most common in older subjects. Skin vesicles and mouth ulcers were also reported. Despite wildly different levels of vaccine side effects, the WHO concluded that "Safety data from Phase 1 studies of both ChAd3 and rVSV vaccines indicate an acceptable safety profile in healthy adults."

7.6.6 OPPORTUNITY AND CHALLENGE: RC-Ad VACCINES

As we can see from the Ebola vaccine trial, we are comparing an RD-Ad vaccine to an RC VSV vaccine. This apples to oranges comparison is unlikely to come out in favor of Ad vaccines unless the VSV side effects become quite dangerous (they are already alarming…). We believe that RC-Ads or single-cycle Ads are a more equal competitor to VSV and other RC vaccines. Whether the risk of an Ad infection by RC-Ad vaccine (like the risk of VSV infection) is too big a risk for general vaccination remains to be tested in non-military applications. As discussed, live RC-Ad4 and Ad7 have been used in thousands of military personnel and their risk of infection was mitigated elegantly by containing them in oral capsules (15).

Our single-cycle Ads may have as good of potency as RC-Ads, but without risk of infection by genetic control. In animals, SC-Ad seems better than RC-Ad, but it is uncertain if this will hold up in humans. Also, SC-Ads do not benefit from a second or third wave of vaccine infection like fully wild RC-Ads. To what degree these waves favor RC-Ad or SC-Ad versus the stronger antiviral response against RC-Ad versus SC-Ad occurring in humans needs to be tested. Regardless, one path forward to more robust Ad vaccines is to give them the ability to amplify transgenes and themselves to equal this capacity already present in their competitors like VSV.

REFERENCES

1. Lycke N. Recent progress in mucosal vaccine development: Potential and limitations. *Nature Reviews Immunology.* 2012;12(8):592–605. Epub 2012/07/26. doi: 10.1038/nri3251. PubMed PMID: 22828912.
2. Burns CC, Shaw J, Jorba J, Bukbuk D, Adu F, Gumede N, Pate MA et al. Multiple independent emergences of type 2 vaccine-derived polioviruses during a large outbreak in northern Nigeria. *Journal of Virology.* 2013;87(9):4907–22. Epub 2013/02/15. doi: 10.1128/JVI.02954-12. PubMed PMID: 23408630; PMCID: 3624331.
3. Wolff JA, Malone RW, Williams P, Chong W, Acsadi G, Jani A, Felgner PL. Direct gene transfer into mouse muscle *in vivo. Science.* 1990;247:1465–8.
4. Williams RS, Johnston SA, Riedy M, DeVit MJ, McElligott SG, Sanford JC. Introduction of foreign genes into tissues of living mice by DNA-coated microprojectiles. *Proceedings of the National Academy of Sciences of the United States of America.* 1991;88:2726–30.
5. Tang D, DeVit M, Johnston SA. Genetic immunization is a simple method for eliciting an immune response. *Nature.* 1992;356:152–4.
6. Fynan EF, Webster RG, Fuller DH, Haynes JR, Santoro JC, Robinson HL. DNA vaccines: Protective immunizations by parenteral, mucosal, and gene-gun inoculations. *Proceedings of the National Academy of Sciences of the United States of America.* 1993;90:11478–82.
7. Davis HL, Michel M-L, Whalen RG. DNA-based immunization induces continuous secretion of hepatitis B surface antigen and high levels of circulating antibody. *Human Molecular Genetics.* 1993;2:1847–51.
8. Conry RM, LoBuglio AF, Kantor J, Schlom J, Loechel F, Moore SE, Sumerel LA, Barlow DL, Abrams S, Curiel DT. Immune response to a carcinoembryonic antigen polynucleotide vaccine. *Cancer Research.* 1994;54:1164–8.
9. Han R, Cladel NM, Reed CA, Peng X, Christensen ND. Protection of rabbits from viral challenge by gene gun-based intracutaneous vaccination with a combination of cottontail rabbit papillomavirus E1, E2, E6, and E7 genes. *Journal of Virology.* 1999;73(8):7039–43. PubMed PMID: 10400806.
10. Berns KI, Byrne BJ, Flotte TR, Gao G, Hauswirth WW, Herzog RW, Muzyczka N et al. Adeno-associated virus Type 2 and hepatocellular carcinoma? *Human Gene Therapy.* 2015;26(12):779–81. doi: 10.1089/hum.2015.29014.kib. PubMed PMID: 26690810.

11. Demberg T, Robert-Guroff M. Mucosal immunity and protection against HIV/SIV infection: Strategies and challenges for vaccine design. *International Reviews of Immunology*. 2009;28(1):20–48. PubMed PMID: 19241252.

12. Barry MA. Recombinant vector vaccines for the prevention and treatment of HIV infection. *Drugs of the Future*. 2011;36(11):833.

13. Rowe WP, Huebner RJ, Gilmore LK, Parrott RH, Ward TG. Isolation of a cytopathogenic agent from human adenoids undergoing spontaneous degeneration in tissue culture. *Proceedings of the Society for Experimental Biology and Medicine Society for Experimental Biology and Medicine (New York, NY)*. 1953;84(3):570–3. PubMed PMID: 13134217.

14. Bell JA, Huebner RJ, Paffenbarger RS, Jr., Rowe WP, Suskind RG, Ward TG. Studies of adenoviruses (APC) in volunteers. *American Journal of Public Health and the Nation's Health*. 1956;46(9):1130–46. Epub 1956/09/01. PubMed PMID: 13354831; PMCID: 1624049.

15. Couch RB, Chanock RM, Cate TR, Lang DJ, Knight V, Huebner RJ. Immunization with types 4 and 7 adenovirus by selective infection of the intestinal tract. *The American Review of Respiratory Disease*. 1963;88(Suppl):394–403.

16. Huebner RJ, Chanock RM, Rubin BA, Casey MJ. Induction by adenovirus Type 7 of tumors in hamsters having the antigenic characteristics of Sv40 virus. *Proceedings of the National Academy of Sciences of the United States of America*. 1964;52:1333–40. PubMed PMID: 14243505; PMCID: PMC300449.

17. Rapp F, Melnick JL, Butel JS, Kitahara T. The incorporation of Sv40 material into adenovirus 7 as measured by intranuclear synthesis of Sv40 tumor antigen. *Proceedings of the National Academy of Sciences of the United States of America*. 1964;52:1348–52. PubMed PMID: 14243507; PMCID: PMC300451.

18. Ulmer JB, Wahren B, Liu MA. Gene-based vaccines: Recent technical and clinical advances. *Trends of Molecular Medicine*. 2006;12(5):216–22. PubMed PMID: 16621717.

19. Lasaro MO, Ertl HC. New insights on adenovirus as vaccine vectors. *Molecular Therapy*. 2009;17(8):1333–9. PubMed PMID: 19513019.

20. Shiver JW, Emini EA. Recent advances in the development of HIV-1 vaccines using replication-incompetent adenovirus vectors. *Annual Review of Medicine*. 2004;55:355–72. PubMed PMID: 14746526.

21. Barouch DH. Novel adenovirus vector-based vaccines for HIV-1. *Current Opinion in HIV and AIDS*. 2010;5(5):386–90. doi: 10.1097/COH.0b013e32833cfe4c. PubMed PMID: 20978378; PMCID: PMC2967414.

22. Cohen J. Infectious disease. Ebola vaccines racing forward at record pace. *Science*. 2014;345(6202):1228–9. Epub 2014/09/13. doi: 10.1126/science.345.6202.1228. PubMed PMID: 25214582.

23. Ledgerwood JE, Sullivan NJ, Graham BS. Chimpanzee adenovirus vector ebola vaccine—Preliminary report. *The New England Journal of Medicine*. 2015;373(8):776. doi: 10.1056/NEJMc1505499. PubMed PMID: 26287857.

24. Zhang Q, Seto D. Chimpanzee adenovirus vector ebola vaccine—Preliminary report. *The New England Journal of Medicine*. 2015;373(8):775–6. doi: 10.1056/NEJMc1505499#SA1. PubMed PMID: 26287858.

25. Shiver JW, Fu TM, Chen L, Casimiro DR, Davies ME, Evans RK, Zhang ZQ et al. Replication-incompetent adenoviral vaccine vector elicits effective anti-immunodeficiency-virus immunity. *Nature*. 2002;415(6869):331–5. PubMed PMID: 11797011.

26. Gurwith MJ, Horwith GS, Impellizzeri CA, Davis AR, Lubeck MD, Hung PP. Current use and future directions of adenovirus vaccine. *Seminars in Respiratory Infections*. 1989;4(4):299–303.

27. Lubeck MD, Davis AR, Chengalvala M, Natuk RJ, Morin JE, Molnar-Kimber K, Mason BB et al. Immunogenicity and efficacy testing in chimpanzees of an oral hepatitis B vaccine based on live recombinant adenovirus. *Proceedings of the National Academy of Sciences of the United States of America.* 1989;86(17):6763–7. Epub 1989/09/01. PubMed PMID: 2570422; PMCID: 297926.

28. Natuk RJ, Chanda PK, Lubeck MD, Davis AR, Wilhelm J, Hjorth R, Wade MS et al. Adenovirus-human immunodeficiency virus (HIV) envelope recombinant vaccines elicit high-titered HIV-neutralizing antibodies in the dog model. *Proceedings of the National Academy of Sciences of the United States of America.* 1992;89(16):7777–81. PubMed PMID: 1502197.

29 Natuk RJ, Lubeck MD, Chanda PK, Chengalvala M, Wade MS, Murthy SC, Wilhelm J et al. Immunogenicity of recombinant human adenovirus-human immunodeficiency virus vaccines in chimpanzees. *AIDS Research and Human Retroviruses.* 1993;9(5):395–404. PubMed PMID: 8318268.

30. Lubeck MD, Natuk RJ, Chengalvala M, Chanda PK, Murthy KK, Murthy S, Mizutani S et al. Immunogenicity of recombinant adenovirus-human immunodeficiency virus vaccines in chimpanzees following intranasal administration. *AIDS Research and Human Retroviruses.* 1994;10(11):1443–9. PubMed PMID: 7888199.

31. Natuk RJ, Davis AR, Chanda PK, Lubeck MD, Chengalvala M, Murthy SC, Wade MS et al. Adenovirus vectored vaccines. *Developments in Biological Standardization.* 1994;82:71–7. PubMed PMID: 7958485.

32. Buge SL, Richardson E, Alipanah S, Markham P, Cheng S, Kalyan N, Miller CJ et al. An adenovirus-simian immunodeficiency virus env vaccine elicits humoral, cellular, and mucosal immune responses in rhesus macaques and decreases viral burden following vaginal challenge. *Journal of Virology.* 1997;71(11):8531–41. PubMed PMID: 9343211.

33. Lubeck MD, Natuk R, Myagkikh M, Kalyan N, Aldrich K, Sinangil F, Alipanah S et al. Long-term protection of chimpanzees against high-dose HIV-1 challenge induced by immunization. *Nature Medicine.* 1997;3(6):651–8.

34. Crosby CM, Barry MA. IIIa deleted adenovirus as a single-cycle genome replicating vector. *Virology.* 2014;462–463(462–463):158–65. doi: 10.1016/j.virol.2014.05.030. PubMed PMID: 24996029; PMCID: 4125442.

35. Tripathy SK, Black HB, Goldwasser E, Leiden JM. Immune responses to transgene-encoded proteins limit the stability of gene expression after injection of replication-defective adenovirus vectors. *Nature Medicine.* 1996;2(5):545–50.

36. Yang Y, Su Q, Wilson JM. Role of viral antigens in destructive cellular immune responses to adenovirus vector-transduced cells in mouse lungs. *Journal of Virology.* 1996;70(10):7209–12.

37. Mitani K, Graham FL, Caskey CT, Kochanek S. Rescue, propagation, and partial purification of a helper virus-dependent adenovirus vector. *Proceedings of the National Academy of Sciences of the United States of America.* 1995;92(9):3854–8.

38. Clemens PR, Kochanek S, Sunada Y, Chan S, Chen HH, Campbell KP, Caskey CT. *In vivo* muscle gene transfer of full-length dystrophin with an adenoviral vector that lacks all viral genes. *Gene Therapy.* 1996;3(11):965–72.

39. Fisher KJ CH, Burda J, Chen SJ, Wilson JM. Recombinant adenovirus deleted of all viral genes for gene therapy of cystic fibrosis. *Virology.* 1996;217(1):11–22.

40. Chen HH, Mack LM, Kelly R, Ontell M, Kochanek S, Clemens PR. Persistence in muscle of an adenoviral vector that lacks all viral genes. *Proceedings of the National Academy of Sciences of the United States of America.* 1997;94(5):1645–50.

41. Morral N, Parks RJ, Zhou H, Langston C, Schiedner G, Quinones J, Graham FL, Kochanek S, Beaudet AL. High doses of a helper-dependent adenoviral vector yield supraphysiological levels of a1-antitrypsin with negligible toxicity. *Human Gene Therapy.* 1998;9:2709–16.

42. Morral N, O'Neal NW, Rice K, Leland M, Kaplan J, Piedra PA, Zhou H et al. Administration of helper-dependent adenoviral vectors and sequential delivery of different vector serotype for long-term liver-directed gene transfer in baboons. *Proceedings of the National Academy of Sciences of the United States of America.* 1999;96(22):12816–21.

43. Weaver EA, Nehete PN, Nehete BP, Yang G, Buchl SJ, Hanley PW, Palmer D et al. Comparison of systemic and mucosal immunization with helper-dependent adenoviruses for vaccination against mucosal challenge with SHIV. *PLoS One.* 2013;8(7):e67574. Epub 2013/07/12. doi: 10.1371/journal.pone.0067574. PubMed PMID: 23844034; PMCID: 3701068.

44. Weaver EA, Nehete PN, Nehete BP, Buchl SJ, Palmer D, Montefiori DC, Ng P, Sastry KJ, Barry MA. Protection against mucosal SHIV challenge by peptide and helper-dependent adenovirus vaccines. *Viruses.* 2009;1(3):920. Epub 2010/01/29. doi: 10.3390/v1030920. PubMed PMID: 20107521; PMCID: 2811377.

45. Malkevitch N, Patterson LJ, Aldrich K, Richardson E, Alvord WG, Robert-Guroff M. A replication competent adenovirus 5 host range mutant-simian immunodeficiency virus (SIV) recombinant priming/subunit protein boosting vaccine regimen induces broad, persistent SIV-specific cellular immunity to dominant and subdominant epitopes in Mamu-A*01 rhesus macaques. *Journal of Immunology.* 2003;170(8):4281–9. PubMed PMID: 12682263.

46. Zhao J, Lou Y, Pinczewski J, Malkevitch N, Aldrich K, Kalyanaraman VS, Venzon D et al. Boosting of SIV-specific immune responses in rhesus macaques by repeated administration of Ad5hr-SIVenv/rev and Ad5hr-SIVgag recombinants. *Vaccine.* 2003;21(25–26):4022–35. PubMed PMID: 12922139.

47. Patterson LJ, Malkevitch N, Venzon D, Pinczewski J, Gomez-Roman VR, Wang L, Kalyanaraman VS, Markham PD, Robey FA, Robert-Guroff M. Protection against mucosal simian immunodeficiency virus SIV(mac251) challenge by using replicating adenovirus-SIV multigene vaccine priming and subunit boosting. *Journal of Virology.* 2004;78(5):2212–21. PubMed PMID: 14963117.

48. Peng B, Wang LR, Gomez-Roman VR, Davis-Warren A, Montefiori DC, Kalyanaraman VS, Venzon D et al. Replicating rather than nonreplicating adenovirus-human immunodeficiency virus recombinant vaccines are better at eliciting potent cellular immunity and priming high-titer antibodies. *Journal of Virology.* 2005;79(16):10200–9. PubMed PMID: 16051813.

49. Zhao J, Voltan R, Peng B, Davis-Warren A, Kalyanaraman VS, Alvord WG, Aldrich K et al. Enhanced cellular immunity to SIV Gag following co-administration of adenoviruses encoding wild-type or mutant HIV Tat and SIV Gag. *Virology.* 2005;342(1):1–12. PubMed PMID: 16109434.

50. Gomez-Roman VR, Florese RH, Patterson LJ, Peng B, Venzon D, Aldrich K, Robert-Guroff M. A simplified method for the rapid fluorometric assessment of antibody-dependent cell-mediated cytotoxicity. *Journal of Immunological Methods.* 2006;308(1–2):53–67. PubMed PMID: 16343526.

51. Gomez-Roman VR, Florese RH, Peng B, Montefiori DC, Kalyanaraman VS, Venzon D, Srivastava I, Barnett SW, Robert-Guroff M. An adenovirus-based HIV subtype B prime/boost vaccine regimen elicits antibodies mediating broad antibody-dependent cellular cytotoxicity against non-subtype B HIV strains. *Journal of Acquired Immune Deficiency Syndromes.* 2006;43(3):270–7. PubMed PMID: 16940858.

52. Gomez-Roman VR, Grimes GJ, Jr., Potti GK, Peng B, Demberg T, Gravlin L, Treece J et al. Oral delivery of replication-competent adenovirus vectors is well tolerated by SIV- and SHIV-infected rhesus macaques. *Vaccine.* 2006;24(23):5064–72. PubMed PMID: 16621178.

53. Peng B, Voltan R, Cristillo AD, Alvord WG, Davis-Warren A, Zhou Q, Murthy KK, Robert-Guroff M. Replicating Ad-recombinants encoding non-myristoylated rather than wild-type HIV Nef elicit enhanced cellular immunity. *AIDS*. 2006;20(17):2149–57. PubMed PMID: 17086054.

54 Demberg T, Florese RH, Heath MJ, Larsen K, Kalisz I, Kalyanaraman VS, Lee EM et al. A replication-competent adenovirus-human immunodeficiency virus (Ad-HIV) tat and Ad-HIV env priming/Tat and envelope protein boosting regimen elicits enhanced protective efficacy against simian/human immunodeficiency virus SHIV89.6P challenge in rhesus macaques. *Journal of Virology*. 2007;81(7):3414–27. PubMed PMID: 17229693.

55. Hidajat R, Xiao P, Zhou Q, Venzon D, Summers LE, Kalyanaraman VS, Montefiori DC, Robert-Guroff M. Correlation of vaccine-elicited systemic and mucosal non-neutralizing antibody activities with reduced acute viremia following intrarectal SIVmac251 challenge of rhesus macaques. *Journal of Virology*. 2009;83(2):791–801. PubMed PMC2612365.

56. Morgan C, Marthas M, Miller C, Duerr A, Cheng-Mayer C, Desrosiers R, Flores J et al. The use of nonhuman primate models in HIV vaccine development. *PLoS Medicine*. 2008;5(8):e173. PubMed PMID: 18700814.

57. Weaver EA, Nehete PN, Buchl SS, Senac JS, Palmer D, Ng P, Sastry KJ, Barry MA. Comparison of replication-competent, first generation, and helper-dependent adenoviral vaccines. *PLoS One*. 2009;4(3):e5059. doi: 10.1371/journal.pone.0005059. PubMed PMID: 19333387; PMCID: 2659436.

58. Thomas MA, Spencer JF, Wold WS. Use of the Syrian hamster as an animal model for oncolytic adenovirus vectors. *Methods in Molecular Medicine*. 2007;130:169–83. PubMed PMID: 17401172.

59. Crosby CM, Nehete P, Sastry KJ, Barry MA. Amplified and persistent immune responses generated by single-cycle replicating adenovirus vaccines. *Journal of Virology*. 2015;89(1):669–75. Epub 2014/10/31. doi: 10.1128/JVI.02184-14. PubMed PMID: 25355873; PMCID: 4301142.

60. Chen CY, Weaver EA, Khare R, May SM, Barry MA. Mining the adenovirus virome for oncolytics against multiple solid tumor types. *Cancer Gene Therapy*. 2011;18(10):744–50. PubMed PMID: 21886190.

61. Liu Q, Muruve DA. Molecular basis of the inflammatory response to adenovirus vectors. *Gene Therapy*. 2003;10(11):935–40. PubMed PMID: 12756413.

62. Brunetti-Pierri N, Palmer DJ, Beaudet AL, Carey KD, Finegold M, Ng P. Acute toxicity after high-dose systemic injection of helper-dependent adenoviral vectors into nonhuman primates. *Human Gene Therapy*. 2004;15(1):35–46. PubMed PMID: 14965376.

63. Marshall E. Gene therapy death prompts review of adenovirus vector. *Science*. 1999;286(5448):2244–5.

64. Piedra PA, Poveda GA, Ramsey B, McCoy K, Hiatt PW. Incidence and prevalence of neutralizing antibodies to the common adenoviruses in children with cystic fibrosis: Implication for gene therapy with adenovirus vectors. *Pediatrics*. 1998;101(6):1013–9. PubMed PMID: 9606228.

65. Abbink P, Lemckert AA, Ewald BA, Lynch DM, Denholtz M, Smits S, Holterman L et al. Comparative seroprevalence and immunogenicity of six rare serotype recombinant adenovirus vaccine vectors from subgroups B and D. *Journal of Virology*. 2007;81(9):4654–63. PubMed PMID: 17329340.

66. Croyle MA, Chirmule N, Zhang Y, Wilson JM. "Stealth" adenoviruses blunt cell-mediated and humoral immune responses against the virus and allow for significant gene expression upon readministration in the lung. *Journal of Virology*. 2001;75(10):4792–801. PubMed PMID: 11312351.

67. Takada A, Feldmann H, Stroeher U, Bray M, Watanabe S, Ito H, McGregor M, Kawaoka Y. Identification of protective epitopes on ebola virus glycoprotein at the single amino acid level by using recombinant vesicular stomatitis viruses. *Journal of Virology.* 2003;77(2):1069–74. PubMed PMID: 12502822; PMCID: PMC140786.

68. Garbutt M, Liebscher R, Wahl-Jensen V, Jones S, Moller P, Wagner R, Volchkov V, Klenk HD, Feldmann H, Stroher U. Properties of replication-competent vesicular stomatitis virus vectors expressing glycoproteins of filoviruses and arenaviruses. *Journal of Virology.* 2004;78(10):5458–65. PubMed PMID: 15113924; PMCID: PMC400370.

69. Geisbert TW, Daddario-DiCaprio KM, Williams KJ, Geisbert JB, Leung A, Feldmann F, Hensley LE, Feldmann H, Jones SM. Recombinant vesicular stomatitis virus vector mediates postexposure protection against Sudan Ebola hemorrhagic fever in nonhuman primates. *Journal of Virology.* 2008;82(11):5664–8. doi: 10.1128/JVI.00456-08. PubMed PMID: 18385248; PMCID: PMC2395203.

70. Geisbert TW, Daddario-Dicaprio KM, Geisbert JB, Reed DS, Feldmann F, Grolla A, Stroher U et al. Vesicular stomatitis virus-based vaccines protect nonhuman primates against aerosol challenge with Ebola and Marburg viruses. *Vaccine.* 2008;26(52):6894–900. doi: 10.1016/j.vaccine.2008.09.082. PubMed PMID: 18930776; PMCID: PMC3398796.

71. Geisbert TW, Daddario-Dicaprio KM, Lewis MG, Geisbert JB, Grolla A, Leung A, Paragas J et al. Vesicular stomatitis virus-based ebola vaccine is well-tolerated and protects immunocompromised nonhuman primates. *PLoS Pathogens.* 2008;4(11):e1000225. doi: 10.1371/journal.ppat.1000225. PubMed PMID: 19043556; PMCID: PMC2582959.

72. Geisbert TW, Geisbert JB, Leung A, Daddario-DiCaprio KM, Hensley LE, Grolla A, Feldmann H. Single-injection vaccine protects nonhuman primates against infection with marburg virus and three species of ebola virus. *Journal of Virology.* 2009;83(14):7296–304. doi: 10.1128/JVI.00561-09. PubMed PMID: 19386702; PMCID: PMC2704787.

73. Khare R, Chen CY, Weaver EA, Barry MA. Advances and future challenges in adenoviral vector pharmacology and targeting. *Current Gene Therapy.* 11(4):241–58. PubMed PMID: 21453281.

74. Thorner AR, Vogels R, Kaspers J, Weverling GJ, Holterman L, Lemckert AA, Dilraj A et al. Age dependence of adenovirus-specific neutralizing antibody titers in individuals from sub-Saharan Africa. *Journal of Clinical Microbiology.* 2006;44(10):3781–3. PubMed PMID: 17021110.

75. Vogels R, Zuijdgeest D, van Rijnsoever R, Hartkoorn E, Damen I, de Bethune MP, Kostense S et al. Replication-deficient human adenovirus type 35 vectors for gene transfer and vaccination: Efficient human cell infection and bypass of preexisting adenovirus immunity. *Journal of Virology.* 2003;77(15):8263–71.

76. Pinto AR, Fitzgerald JC, Giles-Davis W, Gao GP, Wilson JM, Ertl HC. Induction of CD8(+) T cells to an HIV-1 antigen through a prime boost regimen with heterologous E1-deleted adenoviral vaccine carriers. *Journal of Immunology.* 2003;171(12):6774–9.

77. McCoy K, Tatsis N, Korioth-Schmitz B, Lasaro MO, Hensley SE, Lin SW, Li Y et al. Effect of preexisting immunity to adenovirus human serotype 5 antigens on the immune responses of nonhuman primates to vaccine regimens based on human- or chimpanzee-derived adenovirus vectors. *Journal of Virology.* 2007;81(12):6594–604. PubMed PMID: 17428852.

78. Parks R, Evelegh C, Graham F. Use of helper-dependent adenoviral vectors of alternative serotypes permits repeat vector administration. *Gene Therapy.* 1999;6(9):1565–73. PubMed PMID: 10490766.

79. Barouch DH, Pau MG, Custers JH, Koudstaal W, Kostense S, Havenga MJ, Truitt DM et al. Immunogenicity of recombinant adenovirus serotype 35 vaccine in the presence of pre-existing anti-Ad5 immunity. *Journal of Immunology.* 2004;172(10):6290–7. PubMed PMID: 15128818.

80. Lemckert AA, Sumida SM, Holterman L, Vogels R, Truitt DM, Lynch DM, Nanda A et al. Immunogenicity of heterologous prime-boost regimens involving recombinant adenovirus serotype 11 (Ad11) and Ad35 vaccine vectors in the presence of anti-ad5 immunity. *Journal of Virology.* 2005;79(15):9694–701. PubMed PMID: 16014931.

81. O'Riordan CR, Lachapelle A, Delgado C, Parkes V, Wadsworth SC, Smith AE, Francis GE. PEGylation of adenovirus with retention of infectivity and protection from neutralizing antibody *in vitro* and *in vivo*. *Human Gene Therapy.* 1999;10(8):1349–58.

82. Croyle MA, Chirmule N, Zhang Y, Wilson JM. PEGylation of E1-deleted adenovirus vectors allows significant gene expression on readministration to liver. *Human Gene Therapy.* 2002;13(15):1887–900. PubMed PMID: 12396620.

83. Weaver EA, Barry MA. Effects of shielding adenoviral vectors with polyethylene glycol on vector-specific and vaccine-mediated immune responses. *Human Gene Therapy.* 2008;19(12):1369–82. Epub 2008/09/10. doi: 10.1089/hgt.2008.091. PubMed PMID: 18778197; PMCID: 2922072.

84. STEP study: Disappointing, but not a failure. *Lancet.* 2007;370(9600):1665. PubMed PMID: 18022020.

85. Sekaly RP. The failed HIV Merck vaccine study: A step back or a launching point for future vaccine development? *The Journal of Experimental Medicine.* 2008;205(1):7–12. PubMed PMID: 18195078.

86. Steinbrook R. One step forward, two steps back—will there ever be an AIDS vaccine? *The New England Journal of Medicine.* 2007;357(26):2653–5. PubMed PMID: 18160684.

87. Gray G, Buchbinder S, Duerr A. Overview of STEP and Phambili trial results: Two phase IIb test-of-concept studies investigating the efficacy of MRK adenovirus type 5 gag/pol/nef subtype B HIV vaccine. *Current Opinion in HIV and AIDS.* 5(5):357–61. PubMed PMID: 20978374.

88. Liu J, O'Brien KL, Lynch DM, Simmons NL, La Porte A, Riggs AM, Abbink P et al. Immune control of an SIV challenge by a T-cell-based vaccine in rhesus monkeys. *Nature.* 2008. PubMed PMID: 18997770.

89. McChesney MB, Miller CJ. New directions for HIV vaccine development from animal models. *Current Opinion in HIV and AIDS.* 2013;8(5):376–81. doi: 10.1097/COH.0b013e328363d3a2. PubMed PMID: 23836045.

90. Byrareddy SN, Ayash-Rashkovsky M, Kramer VG, Lee SJ, Correll M, Novembre FJ, Villinger F et al. Live attenuated rev-independent Nef‾SIV enhances acquisition of heterologous SIVsmE660 in acutely vaccinated rhesus macaques. *PLoS One.* 2013;*in press.*

91. Staprans SI, Feinberg MB. The roles of nonhuman primates in the preclinical evaluation of candidate AIDS vaccines. *Expert Review of Vaccines.* 2004;3(4 Suppl):S5–32. Epub 2004/08/03. PubMed PMID: 15285703.

92. Tenbusch M, Ignatius R, Nchinda G, Trumpfheller C, Salazar AM, Topfer K, Sauermann U et al. Immunogenicity of DNA vaccines encoding simian immunodeficiency virus antigen targeted to dendritic cells in rhesus macaques. *PLoS One.* 2012;7(6):e39038. Epub 2012/06/22. doi: 10.1371/journal.pone.0039038. PubMed PMID: 22720025; PMCID: 3373620.

93. Barouch DH, Liu J, Li H, Maxfield LF, Abbink P, Lynch DM, Iampietro MJ et al. Vaccine protection against acquisition of neutralization-resistant SIV challenges in rhesus monkeys. *Nature.* 2012;482(7383):89–93. doi: 10.1038/nature10766. PubMed PMID: 22217938; PMCID: 3271177.

94. Barouch DH, Alter G, Broge T, Linde C, Ackerman ME, Brown EP, Borducchi EN et al. Protective efficacy of adenovirus-protein vaccines against SIV challenges in rhesus monkeys. *Science.* 2015. doi: 10.1126/science.aab3886. PubMed PMID: 26138104.

95. Santra S, Sun Y, Korioth-Schmitz B, Fitzgerald J, Charbonneau C, Santos G, Seaman MS et al. Heterologous prime/boost immunizations of rhesus monkeys using chimpanzee adenovirus vectors. *Vaccine*. 2009;27(42):5837–45. PubMed PMID: 19660588.
96. Tatsis N, Blejer A, Lasaro MO, Hensley SE, Cun A, Tesema L, Li Y et al. A CD46-binding chimpanzee adenovirus vector as a vaccine carrier. *Molecular Therapy*. 2007;15(3):608–17. PubMed PMID: 17228314.

8 Use of Oncolytic Adenoviruses for Cancer Therapy

Cristian Capasso, Manlio Fusciello, Erkko Ylösmäki, and Vincenzo Cerullo

CONTENTS

ABSTRACT

Adenoviral vectors represent a reliable and safe platform to transduce genes to target cells. The refinement of gene editing techniques allowed for a more sophisticated manipulation of the viral genome leading to new applications of adenoviruses. Among these, the deletion of genes involved in the replication cycle or the insertion of tissue-specific promoters made possible the generation of vectors that could discriminate between normal and malignant cells. Oncolytic adenoviruses can efficiently infect and lyse tumor cells in a specific manner. This results in the spreading of tumor antigens and danger signals that can activate the immune system and elicit anti-tumor responses. This dual mode of action (oncolysis and immune activation) fostered innovative research and the investigation of combinational approaches.

In this chapter, we will describe the steps that led to the development of oncolytic adenoviruses and their latest applications in the translational research.

8.1   INTRODUCTION

The hypothesis that viruses might affect the growth of tumors has been at the center of first studies in the oncology field. The experiments of Peyton Rous, published at the start of the twentieth century in the *Journal of Experimental Medicine*, showed for the first time that cancer could have an infectious etiology (Rubin 2011; Weiss and Vogt 2011). The Rous sarcoma virus (RSV) could indeed infect an organism and cause the malignant transformation of healthy cells through the expression of the *src* oncogene. Rous' studies led other researchers to the discovery that different pathogens could promote tumor growth: Epstein–Barr virus (Maeda et al. 2009), Kaposi's sarcoma virus (Antman and Chang 2000), or human papilloma virus (Chai et al. 2016; Hochmann et al. 2016; Ruel et al. 2016). However, during the same period, other contrasting reports from several clinicians described tumor regression following viral infection. In fact, the earliest report of tumor cell lysis by viruses was a spontaneous regression of cervical cancer after rabies vaccination (DePace 1912). In addition, cases of remission of Burkitt's (Bluming and Ziegler 1971) and Hodgkin's (Taqi et al. 1981) lymphomas were reported after a measles infection. Following those reports, researchers started to study the ability of other viruses to infect and kill cancer cells in animal models. Oncolytic viruses gained more and more attention due to their potential, offering a selective and self-amplifying anti-cancer drug.

In this chapter, we will describe the major features of oncolytic adenoviruses (OAds) and provide an overview of the latest development of the field.

8.1.1   CANCER

Despite the growing investment of pharmaceutical industries into research and drug development, cancer remains one of the leading causes of death.

Chemotherapy and radiotherapy base their efficacy on reducing the number of tumor cells in a never-ending race against the proliferative potential of the tumor.

8.1.2   CURRENT THERAPIES

Chemotherapy is the most used therapy for cancer patients. Drugs like methotrexate or vincristine aim at inhibiting the replication of tumor cells; while molecules such as doxorubicin are able to induce apoptosis in fast-replicating cells by causing DNA damage, an aspect that is shared by radiation therapy. However, these approaches lack specificity. Kinase inhibitors represent an improvement as they show increased specificity toward specific enzymes, which drive tumor proliferation. However, in a paradoxical way, these drugs favor the appearance of resistant tumor cells due to the selective pressure applied on the tumor cell population. In addition, they also cause side effects if the target kinase shows homology with other kinases.

Some tumors express antigens or proteins that can be targeted by monoclonal antibodies (MAbs). Overexpressed receptors or embryonic antigens are often used as targets. Once the MAbs attack the tumor cells they cause antibody-dependent cellular cytotoxicity (ADCC). It is worth mentioning the case of the human epidermal growth factor receptor 2 (HER2) positive breast cancer. The presence of such specific marker drastically changed the therapy: Trastuzumab, a humanized MAb, can specifically target HER2 positive tumor cells leading to positive clinical responses in patients. However, as for kinase inhibitors, the tumor can become resistant to the therapy by downregulating the expression of the target protein.

In the last decades, several studies revealed that the tumor can evade the surveillance of immune cells. Malignant cells can establish a favorable microenvironment where antigen-presenting cells (APCs) and tumor-specific T-lymphocytes are unable to function properly. Immunotherapy aims at restoring the activity of the immune system through powerful mediators such as interferon gamma (IFN-γ), interleukin 2 (IL-2), or tumor necrosis factor alpha (TNF-α). These adjuvant therapies induce the patient's immune system to fight the cancer. However, the lack of specificity often leads to unwanted autoimmune reactions, which represent the main limitation of these approaches.

8.2 HISTORY OF OAds

Virus-based gene therapy for cancer is an attractive alternative to classic therapies; it is performed with genetically engineered viruses that are called "oncolytic viruses" that are viruses that can preferentially replicate and kill cancer cells (Figure 8.1). However, the efficiency of gene transfer represents one important limitation as, due to safety reasons, replication-deficient viral vectors have been used often. Despite the encouraging results in preclinical studies, clinical trials demonstrated that viral vectors face several additional challenges: pre-existing immunity, low tissue penetration, and rapid clearance by the immune system. For these reasons, the transfer of therapeutic genes (e.g., cytokines) remains limited for conventional non-replicating

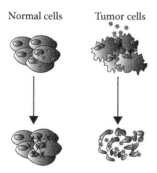

FIGURE 8.1 Oncolytic viruses. Schematic representation of the general concept of tumor-specific replication.

vectors. Hence, it appears highly desirable to have viruses that might naturally select and accurately replicate only in tumor cells.

From the beginning of the last century many observations suggested that some viruses can impact tumor growth. As early as 1922, Levatidi and Nicolau documented lysis of murine tumors by vaccinia virus (Levatidi and Nicolau 1922). Similarly, measles infection was linked to tumor regression in humans (Bluming and Ziegler 1971; Taqi et al. 1981). In the period from 1940 to 1950, several patients with different kind of cancers were treated with attenuated rabies virus and remissions were reported (Pack 1950).

Different wild type viruses show a certain degree of selectivity toward tumor cells. Vesicular stomatitis virus (VSV), for instance, can exploit defects in the IFN-γ pathway of tumor cells and replicate into the tumor tissue (Stojdl et al. 2000) and reovirus also shows to replicate with high efficiency in tumor cells (Lal et al. 2009).

8.2.1 ONCOLYTIC ADENOVIRUSES

Adenoviruses do not naturally replicate selectively in tumors cells. Therefore, genome engineering was needed to make these viruses tumor-selective. Lack of specific pathways required for the cell to eliminate virus infection (e.g., p53) or use of tumor-specific promoters to drive expression of virus replication, represent the strategies that made OAds possible (biological targeting).

8.2.1.1 Exploiting the Lack of Cell Defense Mechanisms (Biological Targeting)

As seen previously, certain wild type viruses can show some degree of selectivity toward malignant cells. This selection process is based upon the molecular differences that tumor cells have compared to healthy cells. In fact, cancer cells accumulate several mutations over time. Most of these mutations occur in genes involved in the regulation of cell proliferation or apoptosis, such as p53 or retinoblastoma protein (Rb).

Conventional adenoviruses need to inhibit the activity of such proteins in order to take control of the cell metabolism. For instance, their genome encodes for proteins that disrupt p53 (i.e., *E1B 55 KD*) or retinoblastoma function (i.e., E1A) (Choi et al. 2012; Capasso et al. 2014). Thus, mutating such genes makes the adenoviruses unable to efficiently replicate in normal cells. However, these engineered viruses would still be able to replicate into cells which show defects in p53 or Rb pathways, such as tumor cells.

ONYX-015 has been the first OAd to be tested in human patients. This virus was engineered by deleting the *E1B 55 KD* gene, thus making it able to enhance its replication in p53-defective tumors (Figure 8.2) (Heise et al. 1997; Ganly et al. 2000; Crompton and Kirn 2007). Based on similar rationale the Delta 24 (Δ24) family of oncolytic viruses was developed. These viruses feature a 24 base pair deletion into the *E1A* gene. Hence, this OAds are selective for Rb-deficient cancers (Idema et al. 2007; Kim et al. 2011, 2013). An obvious limitation of such strategy is that these OAds rely on the molecular mutations of cancer and they will not replicate efficiently in cancers with intact p53 or Rb proteins. However, these gatekeeper genes are the most commonly mutated genes among all type of cancer making this approach suitable for a wide variety of tumors.

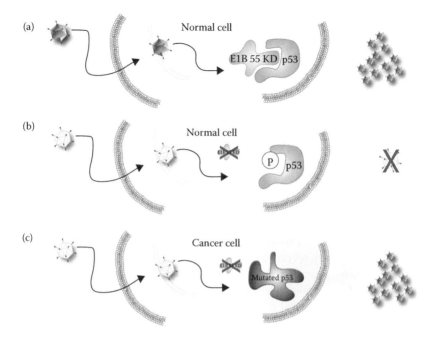

FIGURE 8.2   Schematic representation of the p53-dependent OAds. (a) A wild type adeno-virus infecting a normal cell. The El B55 KD protein of the virus abrogates the action of p53 to avoid programmed cell death allowing the virus to replicate and propagate. (b) An OAd (delta El B55 KD) infecting a normal cell. Delta El B55 KD cannot counteract the action of p53 that leads the cell to a programmed cell death and the virus cannot complete its life cycle. (c) An OAd (delta El B55 KD) infecting a cancer cell. p53 pathway is mutated and the cell does not go to a programmed cell death hence the virus is able to complete its life cycle.

8.2.1.2   Cancer Promoter-Dependent OAds (Transcriptional Targeting)

Another alternative to ensure selective replication in cancer cells rather than nor-mal tissue is using cancer-specific promoters. The initiation of the viral replication depends upon the expression of adenoviral genes, which occurs during the early phase of infection (i.e., *early phase* genes). Hence, it is possible to place such genes under the regulation of cancer-specific promoters to restrict their transcription to the tumor tissue, this is known as *transcriptional targeting* (Chu et al. 2004).

Tumor cells establish a special microenvironment and their metabolic response to several external stimuli is different compared to healthy cells. For instance, due to the limited amount of oxygen into the tumor mass, malignant cells feature a higher activity of hypoxia-related proteins and promoters, which allow tumor cells to sur-vive in low-oxygen concentrations (Kondoh et al. 2013; Ackerman and Simon 2014). Exploiting this feature was the aim of Wang et al., who generated an OAd whose *E1A* and *E1B* genes were regulated by hypoxia responsive element (*HRE*) and human telomerase reverse transcriptase (*hTERT*), respectively (Wang et al. 2008). This OAd replicated with a higher efficiency in tumor cells compared to normal cells. In addi-tion, the production of *E1A* and *E1B* proteins was weak or absent in normal cells, as assayed by Western blot, proving that the transcription of related genes was dependent on *hTERT* promoter and *HRE*. Similarly, Nokisalmi et al. (2010) built a serotype 5 adenovirus whose *expression* of the *E1A* gene was regulated by a modified *E2F* tumor-specific promoter: ICOVIR-7. This OAd, selective for Rb-p16 defective tumors, was tested in a small cohort of patients. In 5 out of 12 radiologically evaluable patients, anti-tumor activity was documented. Interestingly, one young patient experienced a partial response, with a 37% overall decrease in the sum of tumor diameters. A similar approach was used by Hemminki et al. who engineered a serotype 3 human adenovirus by placing the *E1A* gene under the control of *hTERT* promoter. The virus showed tumor-selective replication and a reduced toxicity to normal tissues, such as liver (Hemminki et al. 2012).

Expression of specific proteins is often upregulated in tumor cells. Chromogranin A (*CgA*) is a 48 kDa protein highly expressed by neuroendocrine tumors. Hence, the *CgA* human promoter can be used to drive the expression *early* genes. Leja et al. (2007) constructed an OAd based having the *E1B* gene deleted and the *E1A* gene under the control of the *CgA* promoter.

Similarly, another research group mutated OAds by making their expression dependent upon Y-box binding protein 1 (YB-1) (Rognoni et al. 2009). YB-1 is able to interact with the adenoviral *E2*-late promoter and mediates the E1-independent replication. Hence, an adenovirus lacking *E1* genes, could still be able to replicate in YB-1 positive cells, such as glioblastoma tumor cells. The dependence of such OAd (Ad-Delo3-RGD) upon YB-1 has been confirmed by observing a reduced viral replication after downregulating YB-1 by small interference RNA (siRNA) in R28 cells (Mantwill et al. 2013). In addition, this brain-tumor-selective OAds showed anti-tumor activity in preclinical studies using murine orthotopic models.

8.2.1.3 Armed OAd

The ability of OAds to replicate selectively within the tumor tissue has two main advantages (i) the lysis of cells will be tumor specific; (ii) the proteins encoded by OAds will accumulate within the tumor. The last mechanism has been exploited by several groups to foster the expression of suicide genes or immunomodulatory molecules into the tumor tissue.

8.2.1.3.1 Suicide Gene Therapy by Adenoviruses

Thymidine kinase (TK) is an enzyme of the pyrimidine salvage pathway, which phosphorylates thymidine to produce the dTMP. The TK enzyme of herpes simplex virus (HSV) is less specific, hence recognizes a very broad range of substrates including guanosine analogs such as ganciclovir or acyclovir. Once phosphorylated by the HSV-TK enzyme, these analogs can interfere with the DNA synthesis pathway leading to cell cycle arrest and cell death (Kokoris and Black 2002). The HSV-TK/ganciclovir system represents one of the most common suicide gene/prodrug combinations, hence it has been used in combination with OAds. Freytag et al. developed an OAd expressing *HSV-TK* and *cytosine deaminase* (*CD*), two suicide genes. In addition, the vector featured the expression of interleukin 12 (IL-12) to modulate the immune system. In their preclinical studies the trivalent vector (HSV-TK, CD, and IL-12) was able to increase the survival of mice, proving that the combination of the oncolytic activity, suicide gene therapy, and production of IL-2 was effective (Freytag et al. 2013). A similar approach was used by Kostova et al., who engineered a glioblastoma-selective OAd (thus dependent on YB-1 protein) to express a more active form of the HSV-TK protein: sr39TK. Treatment with the oncolytic vector in combination with ganciclovir resulted in a more efficient killing of tumor cells compared to the controls (Kostova et al. 2015).

The *CD* from *Escherichia coli* is another well-known suicide gene. Similarly to HSV-TK mechanism, CD converts prodrugs like 5-fluorocytosine into 5-fluorouracil. The latter, induces thymidylate synthase inhibition and formation of RNA–DNA complexes leading to apoptosis. This suicide gene has also been successfully cloned into oncolytic vectors for cancer treatment. Zhang et al. (2010) constructed an OAd expressing *CD* gene and observed an increased cytotoxicity *in vitro* and anti-tumor activity *in vivo*. A similar approach has been developed by Tang et al., who armed a *E1B*-55 kDa deficient adenovirus with the multisubstrate deoxyribonucleoside kinase of *Drosophila melanogaster* (*Dm*-dNK). This enzyme is able to increase to susceptibility of cells to different cytotoxic nucleoside analogs. *In vitro* and *in vivo* studies showed that the combination of this vector with nucleoside analogs achieved an improved anticancer activity compared to controls (Tang et al. 2015).

8.2.1.3.2 Expression of Immune Modulators

Recent studies have shown that the tumor microenvironment plays a key role in the regulation of local immune cells. Hence, modulating it might be an effective strategy to increase the outcome of anticancer therapies. To this end, oncolytic vectors have also been used to foster the production of immune modulating molecules within the tumor microenvironment. We have discussed earlier how the expression of IL-12 from an oncolytic vector increased the survival of mice compared to control treatment. Besides IL-12 several other potent immunological mediators have been incorporated in the virus genome to burst anticancer activity and they represent some of the most effective strategies having proved their efficacy also in clinical trials.

Granulocyte-monocyte colony-stimulating factor (GM-CSF) is a potent immune stimulating molecule capable to mature and recruit different APCs. A GM-CSF-expressing OAd proved to be effective in animal models and patients (Bramante et al. 2015). Treatment with a chimeric 5/3 GM-CSF-expressing OAd led to the rejection of tumors and animals were protected by re-challenge with tumor cells, suggesting a strong immunological effect of the cytokine produced by the vector. Interestingly, the clinical benefit rate for patients receiving the treatment was 47%, according to Response Evaluation Criteria in Solid Tumors (RECIST) criteria; in addition, patients presented immunological responses against tumor antigens after the treatment with the virus (Cerullo et al. 2010). CD40L gene therapy also proved to be effective in modulating the T-lymphocyte phenotype (Liljenfeldt et al. 2014). An OAd expressing CD40L was able to promote accumulation of macrophages and T-cells within the tumor (Diaconu et al. 2012). Tumor necrosis factor alpha (TNF-α) is another potent pro-inflammatory cytokine which mediates immune cell activation and direct cell death (Victor and Gottlieb 2002; Wohlleber et al. 2012). Different groups tried to incorporate this gene into the genome of OAds (Kurihara et al. 2000) also in combination with radiation therapy (Hirvinen et al. 2015). TNF-α expression potentiated the anti-tumor effect of the virus and modulated the immunological background positively. In a similar approach, Su et al. (2006) studied a conditionally replicative adenovirus (OAds) expressing interferon-γ (IFN-γ) and demonstrated that this vector could significantly reduce the tumor growth compared to mice treated with control and pathological analysis revealed an accumulation of CD4+ cells within the tumor. A different group studied the efficacy of an IFN-γ expressing adenovirus in clinical settings reporting local tumor responses with 3 complete and 2 partial responses (Dreno et al. 2014). The interleukin 24 (IL-24) has also been used to arm OAds since it demonstrated to have potent cell killing potential. Expression of IL-24 by an *E1B*-deleted OAd resulted in a delayed growth of tumors compared to controls (Zhao et al. 2005).

In addition to cytokines, OAds can be armed also with full MAbs. The local production of the antibody would enhance its anti-tumor activity. In fact, considering the importance of antibodies in the treatment of cancer, sustained and local expression of such compounds might be beneficial for patients, limiting side effects caused by systemic administration. Different groups constructed OAds expressing full MAbs specific for cytotoxic T-lymphocyte associated-antigen-4 (CTLA-4; CD152). These engineered vectors induced tumor cells to produce functional anti-CTLA-4 MAbs (Dias et al. 2012; Du et al. 2014). However, since the oncolytic viruses have a limited amount of deleted genes and sequences one possible limitation of this approach is the size of the expression cassettes that can be included into the viral genome.

8.2.1.4   Capsid Modification of OAds (Transductional Targeting)

The infection process of adenoviruses starts with the interaction between viral protein and cellular receptors (Wickham et al. 1993; Nemerow et al. 1994; Bergelson et al. 1997; Tomko et al. 1997). This stage (i.e., virus entry) can be affected by generating capsid-modified OAds to enhance their interaction with specific tumor antigens/receptors (Capasso et al. 2014).

Adenoviruses serotypes display different tropisms mainly due to their fiber, which is responsible for interaction with membrane receptors, such as the coxackie–adenovirus receptor (CAR) for the serotype 5. Therefore, engineering this protein allows for efficient retargeting of OAds. One of the most common modifications is replacing the fiber of a serotype 5 adenovirus with a serotype 3. This strategy exploits the specificity of the Ad3-fiber for desmoglein-2, a membrane component of tumor cells (Bramante et al. 2014; Wang et al. 2011). The chimeric OAd5/3 was evaluated in a clinical study and among the 12 patients evaluated by RECIST criteria, two experienced a minor response while six patients had stable disease (Koski et al. 2010). Further engineering of this chimeric virus was achieved by modifying the shaft region of the fiber. The KKTK domain has been described as being responsible for binding the heparin sulfate proteoglycans (HSPGs), hence favoring the liver tropism of serotype 5 Ads (Di Paolo et al. 2007). Mutation of this domain has been linked to a decreased liver tropism and an increased transduction of tumor cells. However, despite being able to transduce cancer cells *in vitro* the virus was unable to cause a sustained transgene expression *in vivo* (Koski et al. 2013) highlighting the complexity of capsid engineering strategies. An additional effect of pseudotyping viruses is the change in immunogenicity. Fibers are among the immunogenic proteins of viruses and antibodies against them naturally develop over time. Hence, swapping the fiber of Ad5 with fibers from rare serotypes might overcome the neutralization by pre-existing antibodies, which is a common limitation of the use of Ad5-based vectors in humans (Uusi-Kerttula et al. 2015).

Integrins participate in the virus entry into the cell. Hence, they can be exploited to enhance the infectivity of OAds. The Arg–Gly–Asp (RGD) peptide is known to bind integrins with high efficiency and this peptide signature has been inserted into the structure of OAds with promising results. Ovarian cancer cells usually show reduced levels of CAR, however, they could be efficiently targeted by an OAd displaying the RGD motif in both the protein IX and the fiber (Gamble et al. 2012; Avci et al. 2015) were able to increase the cytotoxicity of the virus against three-dimensional spheroid of glioblastoma cells by using an RGD-containing OAd. Similarly, Rojas et al. (2012) inserted a more complex motif (CDCRGDCFC) into the HI loop of the fiber knob achieving liver detargeting and an increased anti-tumor activity after system administration compared to the non-modified virus. The potency of the RGD motif has also been exploited to target CAR-negative cells, like bladder cancer cell lines. In this context, the RGD engineered OAd showed efficient transduction and a significantly suppressed tumor growth compared to unmodified virus when pre-infected cells were implanted in mice (Yang et al. 2015).

In a similar approach, viral vectors have been covered with proteins able to interact with specific receptors on the cell membrane. This method ensures high specificity and versatility. Both fibroblast growth factor receptor (FGFR) (Greenet al. 2008) and epidermal growth factor receptor (EFGR) (Morrison et al. 2008) have been used as targets to redirect adenoviral vectors toward tumor cells. However, these ligands are still able to activate the receptors, and they might result in proliferative stimuli for the tumor cells increasing the risk of promoting tumor growth. Therefore, an antibody against EGFR was used to avoid proliferative stimuli: Cetuximab was linked to the viral capsid through a polymer linker. This EGFR-retargeted adenovirus was able to transduce EGFR-positive cells significantly better than the non-targeted vector (Morrison et al. 2009).

Taken together, these results highlighted how the modification of the viral structure could be useful to enhance their tropism toward tumor tissue in different kind of cancers, overcoming the paucity of the primary receptor of adenoviruses CAR on tumor cells.

8.3 CURRENT TRENDS IN ADENOVIRAL VECTORS FOR CANCER THERAPY

8.3.1 Toward Oncolytic Vaccines

The lysis of tumor cells (oncolysis) was considered to be the only mode of action of OAds. However, adenoviruses are able to engage the immune system on different levels (Croyle et al. 2005; Choi et al. 2013). The presence of viral proteins activates the innate immunity resulting into the production of pro-inflammatory cytokines. Neutrophils and macrophages are activated amplifying the local inflammation while professional APCs prime virus-specific T-lymphocytes. The second phase of the antiviral response is characterized by the recruitment of CTLs that can recognize and kill infected cells. Finally, T-helper lymphocytes support the production of anti-adenovirus antibodies by B-lymphocytes that will neutralize newly produced viral particles (Wonganan and Croyle 2010). Thus, the immunogenicity of adenoviruses could play a favorable role in the treatment of cancer, which is characterized by a strong local immune suppression. Therefore, OAds can address the lack of immune activation and reshape the tumor microenvironment. Nowadays, oncolytic virus and more specifically OAds have a dual mode of action, oncolysis and vaccine effect (Figure 8.3).

The replication activity of OAds is not only important for the reduction of the number of tumor cells but also for two other reasons: (i) induction of immunogenic cell death (ICD); (ii) release of tumor antigens that APCs can process and present to T-lymphocytes.

The type of cell death is known to be an important signal to the immune system. Programmed cell death, such as apoptosis is a natural and well tolerated event, while necrosis and autophagy-induced cell death are typically associated with immunogenic signals (Michaud et al. 2011). The last two cell death modalities cause the release of danger-associated molecular patterns (DAMPs), such as heat-shock proteins, uric acid, or genomic double-stranded DNA, which alert the immune system (Kono and Rock 2008). However, the autophagy is frequently avoided by dying cancer cells, to evade immune system's surveillance. Chemotherapy and radiotherapy are able to induce the release of DAMPs, thus restoring autophagic pathways (Michaud et al. 2011). Autophagy competent tumors show an increased infiltration by dendritic cells (DCs) and a reduced growth due to recognition by immune cells (Martins et al. 2012). Interestingly, autophagy and adenoviral infection are tightly related: it is reported that adenoviruses need autophagic pathways to enhance the expression of their genes (Zeng and Carlin 2013) and they are able to induce autophagy in infected cells (Rodriguez-Rocha et al. 2011). It has been demonstrated that OAds in combination with chemotherapy can increase the three main markers of ICD: extracellular ATP, release of the nuclear protein high-mobility group box 1 (HMGB-1), and calreticulin exposure on the cell membrane (Liikanen et al. 2013). These signals are sensed by immune cells that can eventually promote local inflammation.

Induction of ICD is not the only immunogenic signal that OAds can offer. In fact, adenoviral particles are phagocytosed by resident macrophages or DCs. Once into phagosomes/endosomes the pathogen-associated molecular patterns (PAMPs) are sensed by APCs through pattern recognition receptors (PRRs), such as Toll-like receptors (Rhee et al. 2011; Cerullo et al. 2012). This process results in the production and secretion of pro-inflammatory cytokines that lead to the recruitment of T-lymphocytes and other immune cells to the infection site. Different studies have shown that macrophages produce a wide array of cytokines such as macrophage chemoattractant protein 1 (MCP-1), RANTES, IL-6, and TNF-α. Interestingly also the complement system participates in the anti-adenoviral response by attacking the viral capsid via the classic pathway (Kiang et al. 2006).

The oncolytic activity of adenoviruses promotes the release of tumor proteins and antigens that can be uptaken by surrounding APCs and used to prime tumor-specific immune responses. This dual mode of action is at the basis of a new generation of

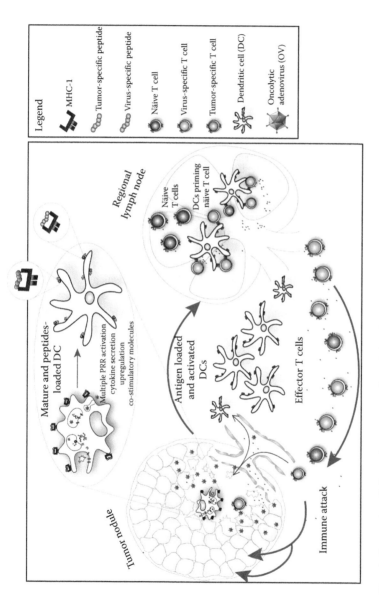

FIGURE 8.3 Oncolytic vaccine. Schematic representation of the alternative mode of action of OAds. After oncolysis tumor cells release tumor-associated antigens (TAAs) into the tumor microenvironment where they can be picked up by professional APCs such as DCs. These tumor antigen-loaded DCs are simultaneously also matured and activated by the presence of the virus so that they will migrate to the lymph nodes where they can prime naïve T cells. As a result, a tumor-specific T cell population migrates to the tumor and helps the virus in fighting against the tumor.

therapies that exploit the natural immunogenicity of OAds and their ability to drive the immune system against cancer.

We have reviewed a number of approaches where cytokine armed-OAds delivered encouraging results. Nevertheless, the immune dominance of adenoviral epitopes represents an obstacle since the activation of the immune system by encoded cytokines is broad and not specific. This problem has been addressed with the use of adenoviral vectors expressing tumor antigens or defined epitopes in addition to cytokines. Different studies suggest that the sustained expression of a defined tumor antigen together with the danger signals provided by the infection can direct the immune system against the tumor. Tumor epitopes, such as SIINFEKL or gp100$_{25}$, have been inserted into the viral gene *E3gp19 K* to enhance their transport to the endoplasmic reticulum (ER) and increase their presentation on the major histocompatibility complex (MHC) class I. This approach led to peptide-specific immune response and higher anti-tumor efficacy compared to controls (Rodriguez-Garcia et al. 2015). Similarly, Gabitzsch et al. have tested different conditionally replicating adenoviral vectors, deleted in *E1*, *E2b*, and *E3* genes, encoding three different tumor antigens: carcinoembryonic antigen (CEA), brachyury, and MUC1 (CD227). DCs, infected with the antigen-encoding vectors were able to stimulate antigen-specific T-lymphocytes and the growth of established tumors was reduced compared to mock control (Gabitzsch et al. 2015). Adenoviral-based genetic vaccines have been exploited in a prime and boost approach in combination with other viral vectors. Li et al. (2016) showed that heterologous treatment with adenovirus and modified vaccinia ankara (MVA) vectors led to sustained immunological responses against the human papilloma virus 16 antigen L1. In addition, mice that received the heterologous vaccine strategy had an increased survival compared to other controls.

One possible limitation of this approach is that the expression of the antigen is dependent upon viral replication having three consequences: (i) antigen expression depends upon the transduction efficiency; (ii) adenoviral antigens are the first to be seen, processed, and presented by APCs to T-lymphocytes; (iii) the anti-adenoviral response limits the antigen production. Taken together, these aspects represent the current challenges in the design of Ad-based genetic vaccines for cancer therapy. In our laboratory, we have tried to address some of these challenges by using the OAds as physical carriers for MHC-I tumor epitopes. The OAd would then act as an active adjuvant for the peptides, able to kill tumor cells, and mature the APCs. DCs and macrophages would sense adenoviral PAMPs and process tumor epitopes simultaneously limiting the immunodominance of OAds. In our *in vivo* studies we were able to control the growth of treated and untreated B16F10 tumors and prime human T-lymphocytes in a humanized mice model achieving the rejection of human melanomas (Capasso et al. 2015).

Although important advances have been made in this field, cancer vaccines have not been able to deliver the expected results. Among the possible limitations we would like to highlight: (i) the high mutation rate of tumor cells which might downregulate certain antigens and (ii) the immune suppressive nature of the tumor microenvironment which limits the activity of antigen-specific T-lymphocytes.

In the next section of this chapter, we will review current combinational therapies that might help overcoming these limitations.

8.3.2 OAds in Combination with Other Therapies

As described before, OAds still face several challenges to achieve full efficacy. While remarkably efficient in controlling the growth of tumors in subcutaneous and orthotopic models of human cancer, they showed only modest efficacy in clinical trials. To overcome the challenges emerged from clinical studies and to increase their efficacy, OAds have been combined with radiation therapy, chemotherapy, or other types of immunotherapy.

8.3.2.1 Combination with Chemotherapy

The cytotoxicity of OAds can be enhanced when they are combined with classic chemotherapy. Cherubini et al. developed a pro-apoptotic version of the AdΔ24 virus by deleting *E1B19 K*, a gene known to have anti-apoptotic functions. The combination of this double-deleted OAd with DNA-damaging cytotoxic drugs (e.g., gemcitabine, irinotecan or cisplatin) resulted in a significant decrease of the half-maximal effective concentration (EC50) values. In addition, *in vivo* studies showed that the combinational therapy significantly reduced the growth of tumors in two human pancreatic tumor xenografts (Cherubini et al. 2011).

In a similar approach, the ONYX-015 virus was combined with cisplatin or 5-fluorouracil showing a greater efficacy than with either agent alone (Heise et al. 1997). In a phase II trial the combination of intratumoral ONYX-015 injection with cisplatin and 5-fluorouracil was tested against cancer of the head and neck. Objective and complete responses were observed. The authors of the study report that injected lesions stabilized, while non injected lesions progressed at 6 months (Khuri et al. 2000; Lamont et al. 2000).

Many tumors respond poorly to classic chemotherapy, such as hepatocellular carcinoma (HCC). This cancer shows resistance DNA-damage inducers such as cisplatin, 5-fluorouracile (5-FU), and doxorubicin. The resistance mechanisms are based upon a low response to apoptotic stimuli. Interestingly, when HCC cells are treated with an OAds encoding apoptosis-inducing proteins such as the second mitochondria derived activator of caspases (SMAC) they become much more susceptible to chemotherapy (Pan et al. 2007).

An interesting study has been carried out with *E3*-mutants OAds in combination with paclitaxel and cisplatin. Being involved in immune regulation, the *E3* genes represent valuable candidates to enhance the efficacy of OAds when combined with drugs able to induce cell death (Wang et al. 2003). Cheong et al. investigated the role of different *E3* genes, observing that while deletion of *E3B* decreases viral efficacy, the deletion of *E3gp19 K* resulted in higher level of viral replication and enhanced anti-tumor response. Interestingly, the potency of the *E3B* could be enhanced by combining it with chemotherapy (Cheong et al. 2008).

A different study investigated whether or not the treatment schedule and sequence of the drug injection had a role into the outcome of combination therapies. Results showed that the efficacy was dependent upon the sequence of the drugs. Treatment with ONYX-015 prior to or simultaneously with chemotherapy increased the anti-tumor efficacy of the overall treatment in human tumor xenografts (Heise et al. 2000).

An interesting approach is represented by combining OAds with natural compounds. It has been reported that vitamin D3 enhances the replication of OAds at least 10-fold. Hsieh et al. constructed a prostate cancer-specific OAd having the *E1* genes under the control of the prostate-specific antigen (PSA) promoter. They observed that exposure to vitamin D3 enhances viral replication *in vitro* and *in vivo* after i.p. administration (Hsieh et al. 2002). Led by the same hypothesis, we have studied the combination of OAds with L-Carnosine, a naturally occurring dipeptide known to have anticancer properties (Iovine et al. 2014). By using the OAd as a carrier for the cytotoxic dipeptide, we were able to efficiently induce the regression of established tumors in xenograft mice models of human colorectal and lung cancer (Garofalo et al. 2016).

8.3.2.2   Combination with Radiotherapy

External beam radiotherapy (XRT) is currently used for the treatment of different types of cancer. The main mode of action is thought to be the induction of DNA damage and consequence apoptosis in tumor and tumor-stromal cells. However, accumulating evidences suggest that XRT can have immunomodulatory effects. Among these, we highlight the accumulation of CD8[+] T-lymphocytes within the tumor and stimulation of professional APCs, such as DCs (Gupta et al. 2012). Although it is believed that apoptosis following DNA damage is a form of non-ICD, certain combinations of drugs and XRT might enhance Toll-like receptor activation and induce the immune-recognition of dying cancer cells (Koji and Mimura 2013). For these reasons, the changes that XRT evokes into the tumor microenvironment might be beneficial for its combination with oncolytic virotherapy. We have recently studied the effects of combining a TNF-α-expressing OAd with radiation therapy. We found that exposure to TNF-α sensitized tumors to XRT and a higher level of necrotic (i.e., immunogenic) cells was found in irradiated mice compared to non-irradiated animals (Hirvinen et al. 2015). Contrasting reports have emerged about the combination of XRT and p53 gene therapy mediated by adenoviral vectors. In one study, combining the two treatments resulted in approximately 85% reduction in tumor size and rats receiving both radiation and Ad-p53 therapy had a significant increase in survival compared to controls (Badie et al. 1998). A similar study assessed the contribution of XRT to oncolytic virotherapy by using a conventional AdΔ24 and a p53 encoding AdΔ24. Both viruses were able to induce tumor regression *in vivo* and increase the long-term survival time when combined with XRT. However, no differences were observed between the viruses suggesting that the mechanisms might not be exclusively dependent upon p53 gene therapy (Idema et al. 2007).

A prostate-specific OAd (CG7870) has been evaluated in preclinical studies in combination with XRT. A degree of synergy was observed *in vitro* at suboptimal doses of radiation and virus. *In vivo*, while the single agents were able to stabilize or slow the growth of tumors, the combination resulted in tumor regression, with a final tumor volume of 34% compared to baseline (Dilley et al. 2005). Another variant of this prostate-specific virus was able to reduce the volume of tumors to 4% of baseline when combined with XRT (Chen et al. 2001).

8.3.2.3   Combination with Immunotherapy

As we reviewed before, the combination of OAds and MVA vectors produced a long-lasting immune response against tumor antigens, proving to be an interesting combination for immunotherapy (Li et al. 2016). Heterologous prime and boost approaches for vaccination have proved to be superior to homologous strategies. While the molecular basis of this difference remains largely unknown, several studies tried to elucidate whether or not the sequence of the vectors has a role into the efficacy of the vaccination. Interestingly, while for some viruses there seems to be no correlation between the order of the agents and the outcome, for adenoviruses the situation appears to be different. Priming with antigen-encoding adenovirus and boosting with antigen-encoding DNA resulted in a poor immune response, while priming with DNA and boosting with adenoviral vector resulted in enhanced induction of T-helper and cytotoxic T-lymphocytes (CTLs) (Lu 2009). Heterologous prime and boost strategies might also help in overcoming the decrease of efficacy of OAds due to neutralizing antibodies against the most common serotypes. Fattori et al. showed that homologous prime and boost with conventional serotypes of adenovirus lead to a poor memory response, probably due to high titers of neutralizing antibodies. On the contrary, priming with human Ad5 and boosting with a chimpanzee-derived serotype leads to high count of IFN-γ secreting T-cells at week 54, highlighting long-lasting memory response (Fattori et al. 2006).

In the context of immunotherapy, OAds have been combined also with adoptive T-cell therapy. Given the adjuvant potential of adenoviruses, they proved to increase the success of adoptive T-cell therapy, even in the absence of active replication. Mice adoptively transferred with antigen-specific T-cells experienced a reduced growth of tumors, by using B16OVA or parental B16F10 melanomas. In addition, an immune modulatory effect was observed since intratumoral adenovirus injection increased the production of pro-inflammatory cytokines, CD45[+] leukocytes, CD8[+] lymphocytes, and F4/80[+] macrophages (Tahtinen et al. 2015). OAds are not only important to provide local inflammation, but as previously described, they are used as vectors to produce cytokines and modulate the immune system. In fact, cancer gene therapy using OAds proved to enhance the outcome of CAR T-cell therapy. Nishio et al. built an OAd encoding RANTES and IL-15 to enhance the recruitment of engineered T-lymphocytes at the tumor site. Neuroblastoma-bearing mice were treated intratumorally with the oncolytic vector and then infused intravenously with CAR-T cells specific for the tumor antigen GD2. The OAd induced caspase pathways in tumor cells exposed to CAR-T cells. In addition, the local release of both RANTES and IL-15 chemokines attracted CAR-T cells and promoted their localization at the tumor site; this resulted in the increased survival of mice treated with the CAR-T/OAd therapy (Nishio et al. 2014).

8.4 CONCLUSIONS

Oncolytic and conventional adenoviruses are versatile and important tools to treat different types of cancer. Although their immunogenicity has been for a long time a limitation for their applications in humans, recent findings in the field of onco-immunology support the hypothesis that this feature might become an important advantage. The possibility to reshape the tumor microenvironment, usually characterized by local immune suppression, allows for induction of anti-tumor immune responses. The increase in ICD, the expression of pro-inflammatory molecules and the activation of innate and adaptive responses have a fundamental role in the efficacy of OAds. The combination of OAds with different therapies generated promising results, paving the way to more combinational studies that could exploit the immunological features of OAds. The recent development of immune checkpoint inhibitors gives an interesting opportunity, supported by the approval of the first oncolytic virus on the market in October 2015 by Amgen (a GM-CSF encoding HSV) (Ledford 2015). In the near future, more OAd-based therapies are expected to emerge and a greater number of combinational studies will deliver preclinical data that can be translated into new therapies for cancer patients.

REFERENCES

Ackerman, D. and M. C. Simon. 2014. Hypoxia, lipids, and cancer: Surviving the harsh tumor microenvironment. *Trends Cell Biol* 24 (8):472–8. doi: 10.1016/j.tcb.2014.06.001.

Antman, K. and Y. Chang. 2000. Kaposi's sarcoma. *N Engl J Med* 342 (14):1027–38. doi: 10.1056/NEJM200004063421407.

Avci, N. G., Y. Fan, A. Dragomir, Y. M. Akay, C. Gomez-Manzano, J. Fueyo-Margareto, and M. Akay. 2015. Delta-24-RGD induces cytotoxicity of glioblastoma spheroids in three dimensional PEG microwells. *IEEE Trans Nanobioscience* 14 (8):946–51. doi: 10.1109/TNB.2015.2499312.

Badie, B., M. H. Kramar, R. Lau, D. A. Boothman, J. S. Economou, and K. L. Black. 1998. Adenovirus-mediated p53 gene delivery potentiates the radiation-induced growth inhibition of experimental brain tumors. *J Neurooncol* 37 (3):217–22.

Bergelson, J. M., J. A. Cunningham, G. Droguett, E. A. Kurt-Jones, A. Krithivas, J. S. Hong, M. S. Horwitz, R. L. Crowell, and R. W. Finberg. 1997. Isolation of a common receptor for Coxsackie B viruses and adenoviruses 2 and 5. *Science* 275 (5304):1320–3.

Bluming, A. Z. and J. L. Ziegler. 1971. Regression of Burkitt's lymphoma in association with measles infection. *Lancet* 2 (7715):105–6.

Bramante, S., J. K. Kaufmann, V. Veckman, I. Liikanen, D. M. Nettelbeck, O. Hemminki, L. Vassilev et al. 2015. Treatment of melanoma with a serotype 5/3 chimeric oncolytic adenovirus coding for GM-CSF: Results *in vitro*, in rodents and in humans. *Int J Cancer* 137 (7):1775–83. doi: 10.1002/ijc.29536.

Bramante, S., A. Koski, A. Kipar, I. Diaconu, I. Liikanen, O. Hemminki, L. Vassilev et al. 2014. Serotype chimeric oncolytic adenovirus coding for GM-CSF for treatment of sarcoma in rodents and humans. *Int J Cancer* 135 (3):720–30. doi: 10.1002/ijc.28696.

Capasso, C., M. Garofalo, M. Hirvinen, and V. Cerullo. 2014. The evolution of adenoviral vectors through genetic and chemical surface modifications. *Viruses* 6 (2):832–55. doi: 10.3390/v6020832.

Capasso, C., M. Hirvinen, M. Garofalo, D. Romaniuk, L. Kuryk, T. Sarvela, A. Vitale et al. 2015. Oncolytic adenoviruses coated with MHC-I tumor epitopes increase the antitumor immunity and efficacy against melanoma. *OncoImmunology* 5 (4):e1105429. doi: 10.1080/2162402x.2015.1105429.

Cerullo, V., I. Diaconu, V. Romano, M. Hirvinen, M. Ugolini, S. Escutenaire, S. L. Holm, A. Kipar, A. Kanerva, and A. Hemminki. 2012. An oncolytic adenovirus enhanced for toll-like receptor 9 stimulation increases antitumor immune responses and tumor clearance. *Mol Ther* 20 (11):2076–86. doi: 10.1038/mt.2012.137.

Cerullo, V., S. Pesonen, I. Diaconu, S. Escutenaire, P. T. Arstila, M. Ugolini, P. Nokisalmi et al. 2010. Oncolytic adenovirus coding for granulocyte macrophage colony-stimulating factor induces antitumoral immunity in cancer patients. *Cancer Res* 70 (11):4297–309. doi: 10.1158/0008-5472.CAN-09-3567.

Chai, R. C., Y. Lim, I. H. Frazer, Y. Wan, C. Perry, L. Jones, D. Lambie, and C. Punyadeera. 2016. A pilot study to compare the detection of HPV-16 biomarkers in salivary oral rinses with tumour p16(INK4a) expression in head and neck squamous cell carcinoma patients. *BMC Cancer* 16 (1):178. doi: 10.1186/s12885-016-2217-1.

Chen, Y., T. DeWeese, J. Dilley, Y. Zhang, Y. Li, N. Ramesh, J. Lee et al. 2001. CV706, a prostate cancer-specific adenovirus variant, in combination with radiotherapy produces synergistic antitumor efficacy without increasing toxicity. *Cancer Res* 61 (14):5453–60.

Cheong, S. C., Y. Wang, J. H. Meng, R. Hill, K. Sweeney, D. Kirn, N. R. Lemoine, and G. Hallden. 2008. E1A-expressing adenoviral E3B mutants act synergistically with chemotherapeutics in immunocompetent tumor models. *Cancer Gene Ther* 15 (1):40–50. doi: 10.1038/sj.cgt.7701099.

Cherubini, G., C. Kallin, A. Mozetic, K. Hammaren-Busch, H. Muller, N. R. Lemoine, and G. Hallden. 2011. The oncolytic adenovirus AdDeltaDelta enhances selective cancer cell killing in combination with DNA-damaging drugs in pancreatic cancer models. *Gene Ther* 18 (12):1157–65. doi: 10.1038/gt.2011.141.

Choi, J. W., J. S. Lee, S. W. Kim, and C. O. Yun. 2012. Evolution of oncolytic adenovirus for cancer treatment. *Adv Drug Deliv Rev* 64 (8):720–9. doi: 10.1016/j.addr.2011.12.011.

Choi, J. H., S. C. Schafer, L. Zhang, T. Juelich, A. N. Freiberg, and M. A. Croyle. 2013. Modeling pre-existing immunity to adenovirus in rodents: Immunological requirements for successful development of a recombinant adenovirus serotype 5-based ebola vaccine. *Mol Pharm* 10 (9):3342–55. doi: 10.1021/mp4001316.

Chu, R. L., D. E. Post, F. R. Khuri, and E. G. Van Meir. 2004. Use of replicating oncolytic adenoviruses in combination therapy for cancer. *Clin Cancer Res* 10 (16):5299–312. doi: 10.1158/1078-0432.CCR-0349-03.

Crompton, A. M. and D. H. Kirn. 2007. From ONYX-015 to armed vaccinia viruses: The education and evolution of oncolytic virus development. *Curr Cancer Drug Targets* 7 (2):133–9.

Croyle, M. A., H. T. Le, K. D. Linse, V. Cerullo, G. Toietta, A. Beaudet, and L. Pastore. 2005. PEGylated helper-dependent adenoviral vectors: Highly efficient vectors with an enhanced safety profile. *Gene Ther* 12 (7):579–87. doi: 10.1038/sj.gt.3302441.

DePace, C. 1912. Sulla scomparsa di un enorme cancro vegetante del collo dell'utero senza cura chirurgia. *Ginecologia* 9:82–9.

Di Paolo, N. C., O. Kalyuzhniy, and D. M. Shayakhmetov. 2007. Fiber shaft-chimeric adenovirus vectors lacking the KKTK motif efficiently infect liver cells *in vivo*. *J Virol* 81 (22):12249–59. doi: 10.1128/JVI.01584-07.

Diaconu, I., V. Cerullo, M. L. Hirvinen, S. Escutenaire, M. Ugolini, S. K. Pesonen, S. Bramante et al. 2012. Immune response is an important aspect of the antitumor effect produced by a CD40L-encoding oncolytic adenovirus. *Cancer Res* 72 (9):2327–38. doi: 10.1158/0008-5472.CAN-11-2975.

Dias, J. D., O. Hemminki, I. Diaconu, M. Hirvinen, A. Bonetti, K. Guse, S. Escutenaire et al. 2012. Targeted cancer immunotherapy with oncolytic adenovirus coding for a fully human monoclonal antibody specific for CTLA-4. *Gene Ther* 19 (10):988–98. doi: 10.1038/gt.2011.176.

Dilley, J., S. Reddy, D. Ko, N. Nguyen, G. Rojas, P. Working, and D. C. Yu. 2005. Oncolytic adenovirus CG7870 in combination with radiation demonstrates synergistic enhancements of antitumor efficacy without loss of specificity. *Cancer Gene Ther* 12 (8):715–22. doi: 10.1038/sj.cgt.7700835.

Dreno, B., M. Urosevic-Maiwald, Y. Kim, J. Guitart, M. Duvic, O. Dereure, A. Khammari et al. 2014. TG1042 (adenovirus-interferon-gamma) in primary cutaneous B-cell lymphomas: A phase II clinical trial. *PLoS One* 9 (2):e83670. doi: 10.1371/journal.pone.0083670.

Du, T., G. Shi, Y. M. Li, J. F. Zhang, H. W. Tian, Y. Q. Wei, H. Deng, and D. C. Yu. 2014. Tumor-specific oncolytic adenoviruses expressing granulocyte macrophage colony-stimulating factor or anti-CTLA4 antibody for the treatment of cancers. *Cancer Gene Ther* 21 (8):340–8. doi: 10.1038/cgt.2014.34.

Fattori, E., I. Zampaglione, M. Arcuri, A. Meola, B. B. Ercole, A. Cirillo, A. Folgori et al. 2006. Efficient immunization of rhesus macaques with an HCV candidate vaccine by heterologous priming-boosting with novel adenoviral vectors based on different sero-types. *Gene Ther* 13 (14):1088–96. doi: 10.1038/sj.gt.3302754.

Freytag, S. O., K. N. Barton, and Y. Zhang. 2013. Efficacy of oncolytic adenovirus expressing suicide genes and interleukin-12 in preclinical model of prostate cancer. *Gene Ther* 20 (12):1131–9. doi: 10.1038/gt.2013.40.

Gabitzsch, E. S., K. Y. Tsang, C. Palena, J. M. David, M. Fantini, A. Kwilas, A. E. Rice et al. 2015. The generation and analyses of a novel combination of recombinant adenovirus vaccines targeting three tumor antigens as an immunotherapeutic. *Oncotarget* 6 (31):31344–59. doi: 10.18632/oncotarget.5181.

Gamble, L. J., H. Ugai, M. Wang, A. V. Borovjagin, and Q. L. Matthews. 2012. Therapeutic efficacy of an oncolytic adenovirus containing RGD ligand in minor capsid protein IX and fiber, Delta24DoubleRGD, in an ovarian cancer model. *J Mol Biochem* 1 (1):26–39.

Ganly, I., D. Kirn, G. Eckhardt, G. I. Rodriguez, D. S. Soutar, R. Otto, A. G. Robertson et al. 2000. A phase I study of Onyx-015, an E1B attenuated adenovirus, administered intratumorally to patients with recurrent head and neck cancer. *Clin Cancer Res* 6 (3):798–806.

Garofalo, M., B. Iovine, L. Kuryk, C. Capasso, M. Hirvinen, A. Vitale, M. Ylippertula, M. A. Bevilacqua, and V. Cerullo. 2016. Oncolytic adenovirus loaded with L-carnosine as novel strategy to enhance the anti-tumor activity. *Mol Cancer Ther* 15 (4):651–60. doi: 10.1158/1535-7163.MCT-15-0559.

Green, N. K., J. Morrison, S. Hale, S. S. Briggs, M. Stevenson, V. Subr, K. Ulbrich et al. 2008. Retargeting polymer-coated adenovirus to the FGF receptor allows productive infection and mediates efficacy in a peritoneal model of human ovarian cancer. *J Gene Med* 10 (3):280–9. doi: 10.1002/jgm.1121.

Gupta, A., H. C. Probst, V. Vuong, A. Landshammer, S. Muth, H. Yagita, R. Schwendener, M. Pruschy, A. Knuth, and M. van den Broek. 2012. Radiotherapy promotes tumor-specific effector CD8+ T cells via dendritic cell activation. *J Immunol* 189 (2):558–66. doi: 10.4049/jimmunol.1200563.

Heise, C., M. Lemmon, and D. Kirn. 2000. Efficacy with a replication-selective adenovirus plus cisplatin-based chemotherapy: Dependence on sequencing but not p53 functional status or route of administration. *Clin Cancer Res* 6 (12):4908–14.

Heise, C., A. Sampson-Johannes, A. Williams, F. McCormick, D. D. Von Hoff, and D. H. Kirn. 1997. ONYX-015, an E1B gene-attenuated adenovirus, causes tumor-specific cytolysis and antitumoral efficacy that can be augmented by standard chemotherapeutic agents. *Nat Med* 3 (6):639–45.

Hemminki, O., I. Diaconu, V. Cerullo, S. K. Pesonen, A. Kanerva, T. Joensuu, K. Kairemo et al. 2012. Ad3-hTERT-E1A, a fully serotype 3 oncolytic adenovirus, in patients with chemotherapy refractory cancer. *Mol Ther* 20 (9):1821–30. doi: 10.1038/mt.2012.115.

Hirvinen, M., M. Rajecki, M. Kapanen, S. Parviainen, N. Rouvinen-Lagerstrom, I. Diaconu, P. Nokisalmi, M. Tenhunen, A. Hemminki, and V. Cerullo. 2015. Immunological effects of a tumor necrosis factor alpha-armed oncolytic adenovirus. *Hum Gene Ther* 26 (3):133–44. doi: 10.1089/hum.2014.069.

Hochmann, J., J. S. Sobrinho, L. L. Villa, and L. Sichero. 2016. The Asian-American variant of human papillomavirus type 16 exhibits higher activation of MAPK and PI3 K/AKT signaling pathways, transformation, migration and invasion of primary human kerati-nocytes. *Virology* 492:145–54. doi: 10.1016/j.virol.2016.02.015.

Hsieh, C. L., L. Yang, L. Miao, F. Yeung, C. Kao, H. Yang, H. E. Zhau, and L. W. Chung. 2002. A novel targeting modality to enhance adenoviral replication by vitamin D(3) in androgen-independent human prostate cancer cells and tumors. *Cancer Res* 62 (11):3084–92.

Idema, S., M. L. Lamfers, V. W. van Beusechem, D. P. Noske, S. Heukelom, S. Moeniralm, W. R. Gerritsen, W. P. Vandertop, and C. M. Dirven. 2007. AdDelta24 and the p53-expressing variant AdDelta24-p53 achieve potent anti-tumor activity in glioma when combined with radiotherapy. *J Gene Med* 9 (12):1046–56. doi: 10.1002/jgm.1113.

Iovine, B., G. Oliviero, M. Garofalo, M. Orefice, F. Nocella, N. Borbone, V. Piccialli et al. 2014. The anti-proliferative effect of L-carnosine correlates with a decreased expression of hypoxia inducible factor 1 alpha in human colon cancer cells. *PLoS One* 9 (5):e96755. doi: 10.1371/journal.pone.0096755.

Khuri, F. R., J. Nemunaitis, I. Ganly, J. Arseneau, I. F. Tannock, L. Romel, M. Gore et al. 2000. a controlled trial of intratumoral ONYX-015, a selectively-replicating adenovirus, in combination with cisplatin and 5-fluorouracil in patients with recurrent head and neck cancer. *Nat Med* 6 (8):879–85. doi: 10.1038/78638.

Kiang, A., Z. C. Hartman, R. S. Everett, D. Serra, H. Jiang, M. M. Frank, and A. Amalfitano. 2006. Multiple innate inflammatory responses induced after systemic adenovirus vector delivery depend on a functional complement system. *Mol Ther* 14 (4):588–98. doi: 10.1016/j.ymthe.2006.03.024.

Kim, K. H., I. P. Dmitriev, S. Saddekni, E. A. Kashentseva, R. D. Harris, R. Aurigemma, S. Bae et al. 2013. A phase I clinical trial of Ad5/3-Delta24, a novel serotype-chimeric, infectivity-enhanced, conditionally-replicative adenovirus (CRAd), in patients with recurrent ovarian cancer. *Gynecol Oncol* 130 (3):518–24. doi: 10.1016/j.ygyno. 2013.06.003.

Kim, K. H., M. J. Ryan, J. E. Estep, B. M. Miniard, T. L. Rudge, J. O. Peggins, T. L. Broadt et al. 2011. A new generation of serotype chimeric infectivity-enhanced conditionally replicative adenovirals: The safety profile of ad5/3-Delta24 in advance of a phase I clinical trial in ovarian cancer patients. *Hum Gene Ther* 22 (7):821–8. doi: 10.1089/hum.2010.180.

Koji, K. and K. Mimura. 2013. Immunogenic tumor cell death induced by chemoradiotherapy in a clinical setting. *Oncoimmunology* 2:1. doi: 10.4161/onco.22197.

Kokoris, M. S. and M. E. Black. 2002. Characterization of herpes simplex virus type 1 thymidine kinase mutants engineered for improved ganciclovir or acyclovir activity. *Protein Sci* 11 (9):2267–72. doi: 10.1110/ps.2460102.

Kondoh, M., N. Ohga, K. Akiyama, Y. Hida, N. Maishi, A. M. Towfik, N. Inoue, M. Shindoh, and K. Hida. 2013. Hypoxia-induced reactive oxygen species cause chromosomal abnormalities in endothelial cells in the tumor microenvironment. *PLoS One* 8 (11):e80349. doi: 10.1371/journal.pone.0080349.

Kono, H. and K. L. Rock. 2008. How dying cells alert the immune system to danger. *Nat Rev Immunol* 8 (4):279–89. doi: 10.1038/nri2215.

Koski, A., L. Kangasniemi, S. Escutenaire, S. Pesonen, V. Cerullo, I. Diaconu, P. Nokisalmi et al. 2010. Treatment of cancer patients with a serotype 5/3 chimeric oncolytic adenovirus expressing GMCSF. *Mol Ther* 18 (10):1874–84. doi: 10.1038/mt.2010.161.

Koski, A., E. Karli, A. Kipar, S. Escutenaire, A. Kanerva, and A. Hemminki. 2013. Mutation of the fiber shaft heparan sulphate binding site of a 5/3 chimeric adenovirus reduces liver tropism. *PLoS ONE* 8 (4):e60032. doi: 10.1371/journal.pone.0060032.

Kostova, Y., K. Mantwill, P. S. Holm, and M. Anton. 2015. An armed, YB-1-dependent oncolytic adenovirus as a candidate for a combinatorial anti-glioma approach of virotherapy, suicide gene therapy and chemotherapeutic treatment. *Cancer Gene Ther* 22 (1):30–43. doi: 10.1038/cgt.2014.67.

Kurihara, T., D. E. Brough, I. Kovesdi, and D. W. Kufe. 2000. Selectivity of a replication-competent adenovirus for human breast carcinoma cells expressing the MUC1 antigen. *J Clin Invest* 106 (6):763–71. doi: 10.1172/JCI9180.

Lal, R., D. Harris, S. Postel-Vinay, and J. de Bono. 2009. Reovirus: Rationale and clinical trial update. *Curr Opin Mol Ther* 11 (5):532–9.

Lamont, J. P., J. Nemunaitis, J. A. Kuhn, S. A. Landers, and T. M. McCarty. 2000. A prospective phase II trial of ONYX-015 adenovirus and chemotherapy in recurrent squamous cell carcinoma of the head and neck (the Baylor experience). *Ann Surg Oncol* 7 (8):588–92.

Ledford, H. 2015. Cancer-fighting viruses win approval. *Nature* 526 (7575):622–3. doi: 10.1038/526622a.

Leja, J., H. Dzojic, E. Gustafson, K. Oberg, V. Giandomenico, and M. Essand. 2007. A novel chromogranin-A promoter-driven oncolytic adenovirus for midgut carcinoid therapy. *Clin Cancer Res* 13 (8):2455–62. doi: 10.1158/1078-0432.CCR-06-2532.

Levatidi L.K. and S. Nicolau. 1922. Sur le culture du virus vaccinal dans les neoplasmes epithelieux. *CR Soc Biol* 86:928.

Li, L. L., H. R. Wang, Z. Y. Zhou, J. Luo, X. Q. Xiao, X. L. Wang, J. T. Li, Y. B. Zhou, and Y. Zeng. 2016. One-prime multi-boost strategy immunization with recombinant DNA, adenovirus, and MVA vector vaccines expressing HPV16 L1 induces potent, sustained, and specific immune response in mice. *Antiviral Res* 128:20–27. doi: 10.1016/j.antiviral.2016.01.014.

Liikanen, I., L. Ahtiainen, M. L. Hirvinen, S. Bramante, V. Cerullo, P. Nokisalmi, O. Hemminki et al. 2013. Oncolytic adenovirus with temozolomide induces autophagy and antitumor immune responses in cancer patients. *Mol Ther* 21 (6):1212–23. doi: 10.1038/mt.2013.51.

Liljenfeldt, L., L. C. Dieterich, A. Dimberg, S. M. Mangsbo, and A. S. Loskog. 2014. CD40L gene therapy tilts the myeloid cell profile and promotes infiltration of activated T lymphocytes. *Cancer Gene Ther* 21 (3):95–102. doi: 10.1038/cgt.2014.2.

Lu, S. 2009. Heterologous prime-boost vaccination. *Curr Opin Immunol* 21 (3):346–51. doi: 10.1016/j.coi.2009.05.016.

Maeda, E., M. Akahane, S. Kiryu, N. Kato, T. Yoshikawa, N. Hayashi, S. Aoki et al. 2009. Spectrum of Epstein-Barr virus-related diseases: A pictorial review. *Jpn J Radiol* 27 (1):4–19. doi: 10.1007/s11604-008-0291-2.

Mantwill, K., U. Naumann, J. Seznec, V. Girbinger, H. Lage, P. Surowiak, D. Beier, M. Mittelbronn, J. Schlegel, and P. S. Holm. 2013. YB-1 dependent oncolytic adenovirus efficiently inhibits tumor growth of glioma cancer stem like cells. *J Transl Med* 11:216. doi: 10.1186/1479-5876-11-216.

Martins, I., M. Michaud, A. Q. Sukkurwala, S. Adjemian, Y. Ma, S. Shen, O. Kepp et al. 2012. Premortem autophagy determines the immunogenicity of chemotherapy-induced cancer cell death. *Autophagy* 8 (3):413–5. doi: 10.4161/auto.19009.

Michaud, M., I. Martins, A. Q. Sukkurwala, S. Adjemian, Y. Ma, P. Pellegatti, S. Shen et al. 2011. Autophagy-dependent anticancer immune responses induced by chemotherapeutic agents in mice. *Science* 334 (6062):1573–7. doi: 10.1126/science.1208347.

Morrison, J., S. S. Briggs, N. Green, K. Fisher, V. Subr, K. Ulbrich, S. Kehoe, and L. W. Seymour. 2008. Virotherapy of ovarian cancer with polymer-cloaked adenovirus retargeted to the epidermal growth factor receptor. *Mol Ther* 16 (2):244–51. doi: 10.1038/sj.mt.6300363.

Morrison, J., S. S. Briggs, N. K. Green, C. Thoma, K. D. Fisher, S. Kehoe, and L. W. Seymour. 2009. Cetuximab retargeting of adenovirus via the epidermal growth factor receptor for treatment of intraperitoneal ovarian cancer. *Hum Gene Ther* 20 (3):239–51. doi: 10.1089/hum.2008.167.

Nemerow, G. R., D. A. Cheresh, and T. J. Wickham. 1994. Adenovirus entry into host cells: A role for alpha(v) integrins. *Trends Cell Biol* 4 (2):52–5.

Nishio, N., I. Diaconu, H. Liu, V. Cerullo, I. Caruana, V. Hoyos, L. Bouchier-Hayes, B. Savoldo, and G. Dotti. 2014. Armed oncolytic virus enhances immune functions of chimeric antigen receptor-modified T cells in solid tumors. *Cancer Res* 74 (18):5195–205. doi: 10.1158/0008-5472.CAN-14-0697.

Nokisalmi, P., S. Pesonen, S. Escutenaire, M. Sarkioja, M. Raki, V. Cerullo, L. Laasonen et al. 2010. Oncolytic adenovirus ICOVIR-7 in patients with advanced and refractory solid tumors. *Clin Cancer Res* 16 (11):3035–43. doi: 10.1158/1078-0432.CCR-09-3167.

Pack, G. T. 1950. Note on the experimental use of rabies vaccine for melanomatosis. *AMA Arch Derm Syphilol* 62 (5):694–5.

Pan, Q. W., S. Y. Zhong, B. S. Liu, J. Liu, R. Cai, Y. G. Wang, X. Y. Liu, and C. Qian. 2007. Enhanced sensitivity of hepatocellular carcinoma cells to chemotherapy with a Smac-armed oncolytic adenovirus. *Acta Pharmacol Sin* 28 (12):1996–2004. doi: 10.1111/j.1745-7254.2007.00672.x.

Rhee, E. G., J. N. Blattman, S. P. Kasturi, R. P. Kelley, D. R. Kaufman, D. M. Lynch, A. La Porte et al. 2011. Multiple innate immune pathways contribute to the immunogenicity of recombinant adenovirus vaccine vectors. *J Virol* 85 (1):315–23. doi: 10.1128/JVI.01597-10.

Rodriguez-Garcia, A., E. Svensson, R. Gil-Hoyos, C. A. Fajardo, L. A. Rojas, M. Arias-Badia, A. S. Loskog, and R. Alemany. 2015. Insertion of exogenous epitopes in the E3-19 K of oncolytic adenoviruses to enhance TAP-independent presentation and immunogenicity. *Gene Ther* 22 (7):596–601. doi: 10.1038/gt.2015.41.

Rodriguez-Rocha, H., J. G. Gomez-Gutierrez, A. Garcia-Garcia, X. M. Rao, L. Chen, K. M. McMasters, and H. S. Zhou. 2011. Adenoviruses induce autophagy to promote virus replication and oncolysis. *Virology* 416 (1–2):9–15. doi: 10.1016/j.virol.2011.04.017.

Rognoni, E., M. Widmaier, C. Haczek, K. Mantwill, R. Holzmuller, B. Gansbacher, A. Kolk et al. 2009. Adenovirus-based virotherapy enabled by cellular YB-1 expression *in vitro* and *in vivo*. *Cancer Gene Ther* 16 (10):753–63. doi: 10.1038/cgt.2009.20.

Rojas, J. J., M. Gimenez-Alejandre, R. Gil-Hoyos, M. Cascallo, and R. Alemany. 2012. Improved systemic antitumor therapy with oncolytic adenoviruses by replacing the fiber shaft HSG-binding domain with RGD. *Gene Ther* 19 (4):453–7. doi: 10.1038/gt.2011.106.

Rubin, H. 2011. The early history of tumor virology: Rous, RIF, and RAV. *Proc Natl Acad Sci U S A* 108 (35):14389–96. doi: 10.1073/pnas.1108655108.

Ruel, J., H. M. Ko, G. Roda, N. Patil, D. Zhang, B. Jharap, N. Harpaz, and J. F. Colombel. 2016. Anal neoplasia in inflammatory bowel disease is associated with HPV and perianal disease. *Clin Transl Gastroenterol* 7:e148. doi: 10.1038/ctg.2016.8.

Stojdl, D. F., B. Lichty, S. Knowles, R. Marius, H. Atkins, N. Sonenberg, and J. C. Bell. 2000. Exploiting tumor-specific defects in the interferon pathway with a previously unknown oncolytic virus. *Nat Med* 6 (7):821–5.

Su, C., L. Peng, J. Sham, X. Wang, Q. Zhang, D. Chua, C. Liu et al. 2006. Immune gene-viral therapy with triplex efficacy mediated by oncolytic adenovirus carrying an interferon-gamma gene yields efficient antitumor activity in immunodeficient and immunocompetent mice. *Mol Ther* 13 (5):918–27. doi: 10.1016/j.ymthe.2005.12.011.

Tahtinen, S., S. Gronberg-Vaha-Koskela, D. Lumen, M. Merisalo-Soikkeli, M. Siurala, A. J. Airaksinen, M. Vaha-Koskela, and A. Hemminki. 2015. Adenovirus improves the efficacy of adoptive T-cell therapy by recruiting immune cells to and promoting their activity at the tumor. *Cancer Immunol Res* 3 (8):915–25. doi: 10.1158/2326-6066. CIR-14-0220-T.

Tang, M., C. Zu, A. He, W. Wang, B. Chen, and X. Zheng. 2015. Synergistic antitumor effect of adenovirus armed with *Drosophila melanogaster* deoxyribonucleoside kinase and nucleoside analogs for human breast carcinoma *in vitro* and *in vivo*. *Drug Des Devel Ther* 9:3301–12. doi: 10.2147/DDDT.S81717.

Taqi, A. M., M. B. Abdurrahman, A. M. Yakubu, and A. F. Fleming. 1981. Regression of Hodgkin's disease after measles. *Lancet* 1 (8229):1112.

Tomko, R. P., R. Xu, and L. Philipson. 1997. HCAR and MCAR: The human and mouse cellular receptors for subgroup C adenoviruses and group B coxsackieviruses. *Proc Natl Acad Sci U S A* 94 (7):3352–6.

Uusi-Kerttula, H., S. Hulin-Curtis, J. Davies, and A. L. Parker. 2015. Oncolytic adenovirus: Strategies and insights for vector design and immuno-oncolytic applications. *Viruses* 7 (11):6009–42. doi: 10.3390/v7112923.

Victor, F. C. and A. B. Gottlieb. 2002. TNF-alpha and apoptosis: Implications for the pathogenesis and treatment of psoriasis. *J Drugs Dermatol* 1 (3):264–75.

Wang, H., Z. Y. Li, Y. Liu, J. Persson, I. Beyer, T. Moller, D. Koyuncu et al. 2011. Desmoglein 2 is a receptor for adenovirus serotypes 3, 7, 11 and 14. *Nat Med* 17 (1):96–104. doi: 10.1038/nm.2270.

Wang, X., C. Su, H. Cao, K. Li, J. Chen, L. Jiang, Q. Zhang et al. 2008. A novel triple-regulated oncolytic adenovirus carrying p53 gene exerts potent antitumor efficacy on common human solid cancers. *Mol Cancer Ther* 7 (6):1598–603. doi: 10.1158/1535-7163.MCT-07-2429.

Wang, Y., G. Hallden, R. Hill, A. Anand, T. C. Liu, J. Francis, G. Brooks, N. Lemoine, and D. Kirn. 2003. E3 gene manipulations affect oncolytic adenovirus activity in immunocompetent tumor models. *Nat Biotechnol* 21 (11):1328–35. doi: 10.1038/nbt887.

Weiss, R. A. and P. K. Vogt. 2011. 100 years of Rous sarcoma virus. *J Exp Med* 208 (12):2351–5. doi: 10.1084/jem.20112160.

Wickham, T. J., P. Mathias, D. A. Cheresh, and G. R. Nemerow. 1993. Integrins alpha v beta 3 and alpha v beta 5 promote adenovirus internalization but not virus attachment. *Cell* 73 (2):309–19.

Wohlleber, D., H. Kashkar, K. Gartner, M. K. Frings, M. Odenthal, S. Hegenbarth, C. Borner et al. 2012. TNF-induced target cell killing by CTL activated through cross-presentation. *Cell Rep* 2 (3):478–87. doi: 10.1016/j.celrep.2012.08.001.

Wonganan, P. and M. A. Croyle. 2010. PEGylated adenoviruses: From mice to monkeys. *Viruses* 2 (2):468–502. doi: 10.3390/v2020468.

Yang, Y., H. Xu, J. Shen, Y. Yang, S. Wu, J. Xiao, Y. Xu, X. Y. Liu, and L. Chu. 2015. RGD-modified oncolytic adenovirus exhibited potent cytotoxic effect on CAR-negative bladder cancer-initiating cells. *Cell Death Dis* 6:e1760. doi: 10.1038/cddis.2015.128.

Zeng, X. and C. R. Carlin. 2013. Host cell autophagy modulates early stages of adenovirus infections in airway epithelial cells. *J Virol* 87 (4):2307–19. doi: 10.1128/JVI.02014-12.

Zhang, J., F. Wei, H. Wang, H. Li, W. Qiu, P. Ren, X. Chen, and Q. Huang. 2010. A novel oncolytic adenovirus expressing *Escherichia coli* cytosine deaminase exhibits potent antitumor effect on human solid tumors. *Cancer Biother Radiopharm* 25 (4):487–95. doi: 10.1089/cbr.2009.0752.

Zhao, L., J. Gu, A. Dong, Y. Zhang, L. Zhong, L. He, Y. Wang et al. 2005. Potent antitumor activity of oncolytic adenovirus expressing mda-7/IL-24 for colorectal cancer. *Hum Gene Ther* 16 (7):845–58. doi: 10.1089/hum.2005.16.845.

Index

233

Printed and bound by CPI Group (UK) Ltd, Croydon, CR0 4YY

24/10/2024

01778308-0004